Kai Velten

**Mathematical Modeling
and Simulation**

Related Titles

Ullmann's Modeling and
Simulation

2007
ISBN: 978–3–527–31605–2

Seppelt, R.

Computer-Based Environmental
Management

2003
ISBN: 978–3–527–30732–6

Kelly, J. J.

Graduate Mathematical Physics

With MATHEMATICA Supplements

2006
ISBN: 978–3–527–40637–1

Landau, R. H., Páez, M. J., Bordeianu, C. C.

Computational Physics

2007
ISBN: 978–3–527–40626–5

Bayin, S.

Mathematical Methods in Science
and Engineering

2006
ISBN: 978–0–470–04142–0

Raabe, D., Roters, F., Barlat, F., Chen, L.-Q.

Continuum Scale Simulation of
Engineering Materials

2004
ISBN: 978–3–527–30760–9

Kai Velten

Mathematical Modeling
and Simulation

Introduction for Scientists and Engineers

WILEY-VCH

WILEY-VCH Verlag GmbH & Co. KGaA

The Author

Prof. Dr. Kai Velten
RheinMain University of Applied Sciences
Geisenheim, Germany
Kai.Velten@gmail.com

Cover

Simulated soil moisture isosurfaces in an asparagus ridge (details are explained in the text). Computer simulation performed by the author.

■ All books published by Wiley-VCH are carefully produced. Nevertheless, authors, editors, and publisher do not warrant the information contained in these books, including this book, to be free of errors. Readers are advised to keep in mind that statements, data, illustrations, procedural details or other items may inadvertently be inaccurate.

Library of Congress Card No.:
applied for

British Library Cataloguing-in-Publication Data
A catalogue record for this book is available from the British Library.

Bibliographic information published by the Deutsche Nationalbibliothek
Die Deutsche Nationalbibliothek lists this publication in the Deutsche Nationalbibliografie; detailed bibliographic data are available on the Internet at <http://dnb.d-nb.de>.

© 2009 WILEY-VCH Verlag GmbH & Co. KGaA, Weinheim

Composition   Laserwords Private Limited, Chennai, India
Printing   Strauss GmbH, Mörlenbach
Binding   Litges & Dopf GmbH, Heppenheim

Printed in the Federal Republic of Germany
Printed on acid-free paper

ISBN: 978-3-527-40758-8

Contents

Mathematical Modeling and Simulation: Introduction for Scientists and Engineers. Kai Velten
Copyright © 2009 WILEY-VCH Verlag GmbH & Co. KGaA, Weinheim
ISBN: 978-3-527-40758-8

The purpose of computing is insight, not numbers.
R.W. Hamming [242]

Preface

"Everyone is an artist" was a central message of the famous twentieth century artist Joseph Beuys. "Everyone models and simulates" is a central message of this book. Mathematical modeling and simulation is a fundamental method in engineering and science, and it is absolutely valid to say that everybody uses it (even those of us who are not aware of doing so). The question is not whether to use this method or not, but rather how to use it effectively.

Today we are in a situation where powerful desktop PCs are readily available to everyone. These computers can be used for any kind of professional data analysis. Even complex structural mechanical or fluid dynamical simulations which would have required supercomputers just a few years ago can be performed on desktop PCs. Considering the huge potential of modeling and simulation to solve complex problems and to save money, one should thus expect a widespread and professional use of this method. Particularly in the field of engineering, however, complex problems are often still treated largely based on experimental data. The amount of money spent on experimental equipment sometimes seems proportional to the complexity and urgency of the problems that are solved, and simple spreadsheet calculations are used to explore the information content of such expensive data. As this book will show, mathematical models and simulations help to reduce experimental costs not only by a partial replacement of experiments by computations, but also by a better exploration of the information content of experimental data.

This book is based on the author's modeling and simulation experience in the fields of science and engineering and as a consultant. It is intended as a first introduction to the subject, which may be easily read by scientists, engineers and students at the undergraduate level. The only mathematical prerequisites are some calculus and linear algebra – all other concepts and ideas will be developed in the course of the book. The reader will find answers to basic questions such as: What is a mathematical model? What types of models do exist? Which model is appropriate for a particular problem? How does one set up a mathematical model? What is simulation, parameter estimation, validation? The book aims to be a practical guide, enabling the reader to setup simple mathematical models on his own and to interpret his own and other people's results critically. To achieve

Mathematical Modeling and Simulation: Introduction for Scientists and Engineers. Kai Velten
Copyright © 2009 WILEY-VCH Verlag GmbH & Co. KGaA, Weinheim
ISBN: 978-3-527-40758-8

this, many examples from various fields such as biology, ecology, economics, medicine, agricultural, chemical, electrical, mechanical and process engineering are discussed in detail.

The book relies exclusively upon open-source software, which is available to everybody free of charge. The reader is introduced into *CAELinux, Calc, Code-Saturne, Maxima, R*, and *Salome-Meca*, and the entire book software – including 3D CFD and structural mechanics simulation software – can be used based on a (free) CAELinux-Live-DVD that is available in the Internet (works on most machines and operating systems, see Appendix A).

While software is used to solve most of the mathematical problems, it is nevertheless attempted to put the reader mathematically on firm ground as much as possible. Trap-doors and problems that may arise in the modeling process, in the numerical treatment of the models or in their interpretation are indicated, and the reader is referred to the literature whenever necessary.

The book is organized as follows. Chapter 1 explains the principles of mathematical modeling and simulation. It provides definitions and illustrative examples of the important concepts as well as an overview of the main types of mathematical models. After a treatment of phenomenological (data-based) models in Chapter 2, the rest of the book introduces the most important classes of mechanistic (process-oriented) models (ordinary and partial differential equation models in Chapters 3 and 4, respectively).

Although it is possible to write a book like this on your own, it is also true that it is impossible to write a book like this on your own . . . I am indebted to a great number of people. I wish to thank Otto Richter (TU Braunschweig), my first teacher in mathematical modeling; Peter Knabner (U Erlangen), for an instructive excursion into the field of numerical analysis; Helmut Neunzert and Franz-Josef Pfreundt (TU and Fraunhofer-ITWM Kaiserslautern), who taught me to apply mathematical models in the industry; Helmut Kern (FH Wiesbaden), for blazing a trail to Geisenheim; Joël Cugnoni (EPFL Lausanne), for our cooperation and an adapted version of CAELinux (great idea, excellent software); Anja Tschörtner, Cornelia Wanka, Alexander Grossmann, H.-J. Schmitt and Uwe Krieg from Wiley-VCH; and my colleagues and friends Marco Günther, Stefan Rief, Karlheinz Spindler, and Aivars Zemitis for proofreading.

I dedicate this book to Birgid, Benedikt, Julia, and Theresa for the many weekends and evenings they patiently allowed me to work on this book, to the Sisters of the Ursuline Order in Geisenheim and Straubing, and, last but not least, to my parents and to my brothers Axel and Ulf, to Bettina and Brigi and, of course, to Felix, for their support and encouragment through so many years.

Geisenheim, May 2008   *Kai Velten*

1

Principles of Mathematical Modeling

We begin this introduction to mathematical modeling and simulation with an explanation of basic concepts and ideas, which includes definitions of terms such as *system*, *model*, *simulation*, *mathematical model*, reflections on the objectives of mathematical modeling and simulation, on characteristics of "good" mathematical models, and a classification of mathematical models. You may skip this chapter at first reading if you are just interested in a hands-on application of specific methods explained in the later chapters of the book, such as regression or neural network methods (Chapter 2) or differential equations (DEs) (in Chapters 3 and 4). Any professional in this field, however, should of course know about the principles of mathematical modeling and simulation. It was emphasized in the preface that everybody uses mathematical models – "even those of us who are not aware of doing so". You will agree that it is a good idea to have an idea of what one is doing. . .

Our starting point is the complexity of the problems treated in science and engineering. As will be explained in Section 1.1, the difficulty of problems treated in science and engineering typically originates from the complexity of the systems under consideration, and models provide an adequate tool to break up this complexity and make a problem tractable. After giving general definitions of the terms *system*, *model*, and *simulation* in Section 1.2, we move on toward mathematical models in Section 1.3, where it is explained that mathematics is *the* natural modeling language in science and engineering. Mathematical models themselves are defined in Section 1.4, followed by a number of example applications and definitions in Sections 1.5 and 1.6. This includes the important distinction between phenomenological and mechanistic models,which has been used as the main organization principle of this book (see Section 1.6.1 and Chapters 2–4). The chapter ends with a classification of mathematical models and Golomb's famous "Don'ts of mathematical modeling" in Sections 1.7 and 1.8.

1.1
A Complex World Needs Models

Generally speaking, engineers and scientists try to understand, develop, or optimize "systems". Here, "system" refers to the object of interest, which can be a part of nature (such as a plant cell, an atom, a galaxy etc.) or an artificial technological system (see Definition 1.2.3 below). Principally, everybody deals with systems in his or her everyday life in a way similar to the approach of engineers or scientists. For example, consider the problem of a table which is unstable due to an uneven floor. This is a technical system and everybody knows what must be done to solve the problem: we just have to put suitable pieces of cardboard under the table legs. Each of us solves an abundant number of problems relating to simple technological systems of this kind during our lifetime. Beyond this, there is a great number of really difficult technical problems that can only be solved by engineers. Characteristic of these more demanding problems is a high complexity of the technical system. We would simply need no engineers if we did not have to deal with complex technical systems such as computer processors, engines, and so on. Similarly, we would not need scientists if processes such as the photosynthesis of plants could be understood as simply as an unstable table. The reason why we have scientists and engineers, virtually their right to exist, is the complexity of nature and the complexity of technological systems.

> **Note 1.1.1 (The complexity challenge)**   It is the genuine task of scientists and engineers to deal with complex systems, and to be effective in their work, they most notably need specific methods to deal with complexity.

The general strategy used by engineers or scientists to break up the complexity of their systems is the same strategy that we all use in our everyday life when we are dealing with complex systems: simplification. The idea is just this: if something is complex, make it simpler. Consider an everyday life problem related to a complex system: A car that refuses to start. In this situation, everyone knows that a look at the battery and fuel levels will solve the problem in most cases. Everyone will do this automatically, but to understand the problem solving strategy behind this, let us think of an alternative scenario. Assume someone is in this situation for the first time. Assume that "someone" was told how to drive a car, that he has used the car for some time, and now he is for the first time in a situation in which the car does not start. Of course, we also assume that there is no help for miles around! Then, looking under the hood for the first time, our "someone" will realize that the car, which seems simple as long as it works well, is quite a complex system. He will spend a lot of time until he will eventually solve the problem, even if we admit that our "someone" is an engineer. The reason why each of us will solve this problem much faster than this "someone" is of course the simple fact that this situation is not new to us. We have experienced this situation before, and from our previous experience we know what is to be done. Conceptually, one can say that we have a simplified picture of the car in our mind similar to Figure 1.1. In the moment when we realize that our car does not start, we do not think of the car as the complex system that it really is, that is, we do not think of this conglomerate of valves, pistons, and all the kind of stuff that can be found under the hood; rather, we have this simplified picture of the car in our mind. We know that this simplified

 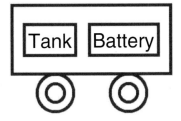

Fig. 1.1 Car as a real system and as a model.

picture is appropriate in this given situation, and it guides us to look at the battery and fuel levels and then to solve the problem within a short time.

This is exactly the strategy used by engineers or scientists when they deal with complex systems. When an engineer, for example, wants to reduce the fuel consumption of an engine, then he will not consider that engine in its entire complexity. Rather, he will use simplified descriptions of that engine, focusing on the machine parts that affect fuel consumption. Similarly, a scientist who wants to understand the process of photosynthesis will use simplified descriptions of a plant focusing on very specific processes within a single plant cell. Anyone who wants to understand complex systems or solve problems related to complex systems needs to apply appropriate simplified descriptions of the system under consideration. This means that anyone who is concerned with complex systems needs models, since simplified descriptions of a system are models of that system by definition.

> **Note 1.1.2 (Role of models)** To break up the complexity of a system under consideration, engineers and scientists use simplified descriptions of that system (i.e. models).

1.2
Systems, Models, Simulations

In 1965, Minsky gave the following general definition of a model [1, 2]:

> **Definition 1.2.1 (Model)** To an observer B, an object A* is a *model* of an object A to the extent that B can use A* to answer questions that interest him about A.

> **Note 1.2.1 (Formal definitions)** Note that Definition 1.2.1 is a *formal definition* in the sense that it operates with terms such as *object* or *observer* that are not defined in a strict axiomatic sense similar to the terms used in the definitions of standard mathematical theory. The same remark applies to several other definitions in this book, including the definition of the term *mathematical model* in Section 1.4. Definitions of this kind are justified for practical reasons, since

they allow us to talk about the formally defined terms in a concise way. An example is Definition 2.5.2 in Section 2.5.5, a concise formal definition of the term *overfitting*, which uses several of the previous formal definitions.

The application of Definition 1.2.1 to the car example is obvious – we just have to identify B with the car driver, A with the car itself, and A* with the simplified tank/battery description of the car in Figure 1.1.

1.2.1
Teleological Nature of Modeling and Simulation

An important aspect of the above definition is the fact that it includes the purpose of a model, namely, that the model helps us to answer questions and to solve problems. This is important because particularly beginners in the field of modeling tend to believe that a good model is one that mimics the part of reality that it pertains to as closely as possible. But as was explained in the previous section, modeling and simulation aims at simplification, rather than at a useless production of complex copies of a complex reality, and hence, the contrary is true:

Note 1.2.2 (The best model) The best model is the simplest model that still serves its purpose, that is, which is still complex enough to help us understand a system and to solve problems. Seen in terms of a simple model, the complexity of a complex system will no longer obstruct our view, and we will virtually be able to look through the complexity of the system at the heart of things.

The entire procedure of modeling and simulation is governed by its purpose of problem solving – otherwise it would be a mere l'art pour l'art. As [3] puts it, "modeling and simulation is always goal-driven, that is, we should know the purpose of our potential model before we sit down to create it". It is hence natural to define fundamental concepts such as the term *model* with a special emphasis on the purpose-oriented or *teleological nature of modeling and simulation*. (Note that teleology is a philosophical discipline dealing with aims and purposes, and the term *teleology* itself originates from the Greek word *telos*, which means end or purpose [4].) Similar teleological definitions of other fundamental terms, such as *system*, *simulation*, and *mathematical model* are given below.

1.2.2
Modeling and Simulation Scheme

Conceptually, the investigation of complex systems using models can be divided into the following steps:

Note 1.2.3 (Modeling and simulation scheme)

Definitions
- Definition of a problem that is to be solved or of a question that is to be answered
- Definition of a system, that is, a part of reality that pertains to this problem or question

Systems Analysis
- Identification of parts of the system that are relevant for the problem or question

Modeling
- Development of a model of the system based on the results of the systems analysis step

Simulation
- Application of the model to the problem or question
- Derivation of a strategy to solve the problem or answer the question

Validation
- Does the strategy derived in the simulation step solve the problem or answer the question for the real system?

The application of this scheme to the examples discussed above is obvious: in the *car example*, the problem is that the car does not start and the car itself is the system. This is the "definitions" step of the above scheme. The "systems analysis" step identifies the battery and fuels levels as the relevant parts of the system as explained above. Then, in the "modeling" step of the scheme, a model consisting of a battery and a tank such as in Figure 1.1 is developed. The application of this model to the given problem in the "simulation" step of the scheme then leads to the strategy "check battery and fuel level". This strategy can then be applied to the real car in the "validation" step. If it works, that is, if the car really starts after refilling its battery or tank, we say that the model is valid or validated. If not, we probably need a mechanic who will then look at other parts of the car, that is, who will apply more complex models of the car until the problem is solved.

In a real modeling and simulation project, the *systems analysis step* of the above scheme can be a very time-consuming step. It will usually involve a thorough evaluation of the literature. In many cases, the literature evaluation will show that similar investigations have been performed in the past, and one should of course try to profit from the experiences made by others that are described in the literature. Beyond this, the system analysis step usually involves a lot of discussions and meetings that bring together people from different disciplines who can answer your questions regarding the system. These discussion will usually show that new data are needed for a better understanding of the system and for the validation of the models in the validation step of the above scheme. Hence, the definition of an experimental program is also another typical part of the systems analysis step.

The *modeling step* will also involve the identification of appropriate software that can solve the equations of the mathematical model. In many cases, it will be possible to use standard software such as the software tools discussed in the next chapters. Beyond this, it may be necessary to write your own software in cases where the mathematical model involves nonstandard equations. An example of this case is the modeling of the press section of paper machines, which involves highly convection-dominated diffusion equations that cannot be treated by standard software with sufficient precision, and which hence need specifically tailored numerical software [5].

In the *validation step*, the model results will be compared with experimental data. These data may come from the literature, or from experiments that have been specifically designed to validate the model. Usually, a model is required to fit the data not only quantitatively, but also qualitatively in the sense that it reproduces the general shape of the data as closely as possible. See Section 3.2.3.4 for an example of a qualitative misfit between a model and data. But, of course, even a model that perfectly fits the data quantitatively and qualitatively may fail the validation step of the above scheme if it cannot be used to solve the problem that is to be solved, which is the most important criterion for a successful validation.

The modeling and simulation scheme (Note 1.2.3) focuses on the essential steps of modeling and simulation, giving a rather simplified picture of what really happens in a concrete modeling and simulation project. For different fields of application, you may find a number of more sophisticated descriptions of the modeling and simulation process in books such as [6–9]. An important thing that you should note is that a real modeling and simulation project will very rarely go straight through the steps of the above scheme; rather, there will be a lot of interaction between the individual steps of the scheme. For example, if the validation step fails, this will bring you back to one of the earlier steps in a *loop-like structure*: you may then improve your model formulation, reanalyze the system, or even redefine your problem formulation (if your original problem formulation turns out to be unrealistic).

> **Note 1.2.4 (Start with simple models!)** To find the best model in the sense of Note 1.2.2, start with the simplest possible model and then generate a sequence of increasingly complex model formulations until the last model in the sequence passes the validation step.

1.2.3
Simulation

So far we have given a definition of the term *model* only. The above modeling and simulation schemes involve other terms, such as *system* and *simulation*, which we may view as being implicitly defined by their role in the above scheme. Can this be made more precise? In the literature, you will find a number of different definitions, for example of the term *simulation*. These differences can be explained by different interests of the authors. For example, in a book with a focus on the so-called *discrete event simulation* which emphasizes the development of a system over time, simulation is defined as "the imitation of the operation of a real-world process or system over time" [6]. In general terms, simulation can be defined as follows:

Definition 1.2.2 (Simulation) *Simulation* is the application of a model with the objective to derive strategies that help solve a problem or answer a question pertaining to a system.

Note that the term *simulation* originates from the Latin word "simulare", which means "to pretend": in a simulation, the model pretends to be the real system. A similar definition has been given by Fritzson [7] who defined simulation as "an experiment performed on a model". Beyond this, the above definition is a *teleological* (purpose-oriented) definition similar to Definition 1.2.1 above, that is, this definition again emphasizes the fact that simulation is always used to achieve some goal. Although Fritzson's definition is more general, the above definition reflects the real use of simulation in science and engineering more closely.

1.2.4
System

Regarding the term *system*, you will again find a number of different definitions in the literature, and again some of the differences between these definitions can be explained by the different interests of their authors. For example, [10] defines a system to be "a collection of entities, for example, people or machines, that act and interact together toward the accomplishment of some logical end". According to [11], a system is "collection of objects and relations between objects". In the context of mathematical models, we believe it makes sense to think of a "system" in very general terms. Any kind of object can serve as a system here if we have a question relating to that object and if this question can be answered using mathematics. Our view of systems is similar to a definition that has been given by [12] (see also the discussion of this definition in [3]): " A system is whatever is distinguished as a system." [3] gave another definition of a "system" very close to our view of systems here: "A system is a potential source of data". This definition emphasizes the fact that a system can be of scientific interest only if there is some communication between the system and the outside world, as it will be discussed below in Section 1.3.1. A definition that includes the teleological principle discussed above has been given by Fritzson [7] as follows:

Definition 1.2.3 (System) A *system* is an object or a collection of objects whose properties we want to study.

1.2.5
Conceptual and Physical Models

The model used in the car example is something that exists in our minds only. We can write it down on a paper in a few sentences and/or sketches, but it does not have any physical reality. Models of this kind are called *conceptual models* [11]. Conceptual models are used by each of us to solve everyday problems such as the car that refuses to start. As K.R. Popper puts it, "all life is problem solving", and conceptual models provide us with an important tool to solve our everyday problems [13]. They are also applied by engineers or scientists to simple problems or questions similar to the car example. If their problem or question is complex enough, however, they rely on experiments, and this leads us to other types of models. To see this, let us use the modeling and simulation scheme (Note 1.2.3) to describe a possible procedure followed by an engineer who wants to reduce the fuel consumption of an engine: In this case, the problem is the reduction of fuel consumption and the system is the engine. Assume that the systems analysis leads the engineer to the conclusion that the fuel injection pump needs to be optimized. Typically, the engineer will then create some experimental setting where he can study the details of the fuel injection process.

Such an experimental setting is then a model in the sense that it will typically be a very simplified version of that engine, that is, it will typically involve only a few parts of the engine that are closely connected with the fuel injection process. In contrast to a conceptual model, however, it is not only an idea in our mind but also a real part of the physical world, and this is why models of this kind are called *physical models* [11]. The engineer will then use the physical model of the fuel injection process to derive strategies – for example, a new construction of the fuel injection pump – to reduce the engine's fuel consumption, which is the simulation step of the above modeling and simulation scheme. Afterwards, in the validation step of the scheme, the potential of these new constructions to reduce fuel consumption will be tested in the engine itself, that is, in the real system. Physical models are applied by scientists in a similar way. For example, let us think of a scientist who wants to understand the photosynthesis process in plants. Similar to an engineer, the scientist will set up a simplified experimental setting – which might be some container with a plant cell culture – in which he can easily observe and measure the important variables, such as CO_2, water, light, and so on. For the same reasons as above, anything like this is a physical model. As before, any conclusion drawn from such a physical model corresponds to the simulation step of the above scheme, and the conclusions need to be validated by data obtained from the real system, that is, data obtained from real plants in this case.

1.3
Mathematics as a Natural Modeling Language

1.3.1
Input–Output Systems

Any system that is investigated in science or engineering must be observable in the sense that it produces some kind of output that can be measured (a system that would not satisfy this minimum requirement would have to be treated by theologians rather than by scientists or engineers). Note that this observability condition can also be satisfied by systems where nothing can be measured directly, such as black holes, which produce measurable gravitational effects in their surroundings. Most systems investigated in engineering or science do also accept some kind of input data, which can then be studied in relation to the output of the system (Figure 1.2a). For example, a scientist who wants to understand photosynthesis will probably construct experiments where the carbohydrate production of a plant is measured at various levels of light, CO_2, water supply, and so on. In this case, the plant cell is the system; the light, CO_2, and water levels are the input quantities; and the measured carbohydrate production is the output quantity. Or, an engineer who wants to optimize a fuel injection pump will probably change the construction of that pump in various ways and then measure the fuel consumption resulting from these modified constructions. In this case, the fuel injection pump is the system, the construction parameters changed by the engineer are the input parameters and the resulting fuel consumption is the output quantity.

> **Note 1.3.1 (Input–output systems)**   Scientists or engineers investigate "input–output systems", which transform given input parameters into output parameters.

Note that there are of course situations where scientists are looking at the system itself and not at its input–output relations, for example when a botanist just wants

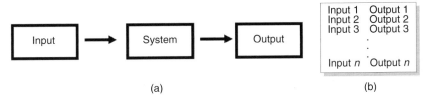

(a)   (b)

Fig. 1.2 (a) Communication of a system with the outside world. (b) General form of an experimental data set.

to describe and classify the anatomy of a newly discovered plant. Typically, however, such purely descriptive studies raise questions about the way in which the system works, and this is when input–output relations come into play. Engineers, on the other hand, are always concerned with input–output relations since they are concerned with technology. The Encyclopedia Britannica defines technology as "the application of scientific knowledge to the practical aims of human life". These "practical aims" will usually be expressible in terms of a system output, and the tuning of system input toward optimized system output is precisely what engineers typically do, and what is in fact the genuine task of engineering.

1.3.2
General Form of Experimental Data

The experimental procedure described above is used very generally in engineering and in the (empirical) sciences to understand, develop, or optimize systems. It is useful to think of it as a means to explore *black boxes*. At the beginning of an experimental study, the system under investigation is similar to such a "black box" in the sense that there is some uncertainty about the processes that happen inside the system when the input is transformed into the output. In an extreme case, the experimenter may know only that "something" happens inside the system which transforms input into output, that is, the system may be really a black box. Typically, however, the experimenter will have some hypotheses about the internal processes, which he wants to prove or disprove in the course of his study. That is, experimenters typically are concerned with systems as gray boxes which are located somewhere between black and white boxes (more details in Section 1.5).

Depending on the hypothesis that the experimenter wants to investigate, he confronts the system with appropriate input quantities, hoping that the outputs produced by the system will help prove or disprove his hypothesis. This is similar to a question-and-answer game: the experimenter poses questions to the system, which is the input, and the system answers to these questions in terms of measurable output quantities. The result is a data set of the general form shown in Figure 1.2b. In rare cases, particularly if one is concerned with very simple systems, the internal processes of the system may already be evident from the data set itself. Typically, however, this experimental question-and-answer game is similar to the questioning of an oracle: we know there is some information about the system in the data set, but it depends on the application of appropriate ideas and methods if one wants to uncover the information content of the data and, so to speak, shed some light into the black box.

1.3.3
Distinguished Role of Numerical Data

Now what is an appropriate method for the analysis of experimental datasets? To answer this question, it is important to note that in most cases experimental data are numbers and can be quantified. The input and output data of Figure 1.2b will typically consist of columns of numbers. Hence, it is natural to think of a system in mathematical terms. In fact, a system can be naturally seen as a mathematical function, which maps given input quantities x into output quantities $y = f(x)$ (Figure 1.2a). This means that if one wants to understand the internal mechanics of a system "black box", that is, if one wants to understand the processes inside the real system that transform input into output, a natural thing to do is to translate all these processes into mathematical operations. If this is done, one arrives at a simplified representation of the real system in mathematical terms. Now remember that a simplified description of a real system (along with a problem we want to solve) is a model by definition (Definition 1.2.1). The representation of a real system in mathematical terms is thus a mathematical model of that system.

> **Note 1.3.2 (Naturalness of mathematical models)** Input–output systems usually generate numerical (or quantifiable) data that can be described naturally in mathematical terms.

This simple idea, that is, the mapping of the internal mechanics of real systems into mathematical operations, has proved to be extremely fruitful to the understanding, optimization, or development of systems in science and engineering. The tremendous success of this idea can only be explained by the naturalness of this approach – mathematical modeling is simply the best and most natural thing one can do if one is concerned with scientific or engineering problems. Looking back at Figure 1.2a, it is evident that mathematical structures emanate from the very heart of science and engineering. Anyone concerned with systems and their input–output relations is also concerned with mathematical problems – regardless of whether he likes it or not and regardless of whether he treats the system appropriately using mathematical models or not. The success of his work, however, depends very much on the appropriate use of mathematical models.

1.4
Definition of Mathematical Models

To understand mathematical models, let us start with a general definition. Many different definitions of mathematical models can be found in the literature. The differences between these definitions can usually be explained by the different scientific interests of their authors. For example, Bellomo and Preziosi [14] define a mathematical model to be a set of equations which can be used to compute the time-space evolution of a physical system. Although this definition suffices for the problems treated by Bellomo and Preziosi, it is obvious that it excludes a great number of mathematical models. For example, many economical or sociological problems cannot be treated in a time-space framework or based on equations only. Thus, a more general definition of mathematical models is needed if one wants to cover all kinds of mathematical models used in science and engineering. Let us start with the following attempt of a definition:

A mathematical model is a set of mathematical statements
$M = \{\Sigma_1, \Sigma_2, \ldots, \Sigma_n\}$.

Certainly, this definition covers all kinds of mathematical models used in science and engineering as required. But there is a problem with this definition. For example, a simple mathematical statement such as $f(x) = e^x$ would be a mathematical model in the sense of this definition. In the sense of Minsky's definition of a model (Definition 1.2.1), however, such a statement is not a model as long as it lacks any connection with some system and with a question we have relating to that system. The above attempt of a definition is incomplete since it pertains to the word "mathematical" of "mathematical model" only, without any reference to purposes or goals. Following the philosophy of the teleological definitions of the terms *model*, *simulation*, and *system* in Section 1.2, let us define instead:

Definition 1.4.1 (Mathematical Model) A mathematical model is a triplet (S, Q, M) where S is a system, Q is a question relating to S, and M is a set of mathematical statements $M = \{\Sigma_1, \Sigma_2, \ldots, \Sigma_n\}$ which can be used to answer Q.

Note that this is again a formal definition in the sense of Note 1.2.1 in Section 1.2. Again, it is justified by the mere fact that it helps us to understand the nature of mathematical models, and that it allows us to talk about mathematical models in a concise way. A similar definition was given by Bender [15]: "A mathematical model is an abstract, simplified, mathematical construct related to a part of reality and created for a particular purpose." Note that Definition 1.4.1 is not restricted to physical systems. It covers psychological models as well that may deal with essentially metaphysical quantities, such as thoughts, intentions, feelings, and so on. Even mathematics itself is covered by the above definition. Suppose, for example, that S is the set of natural numbers and our question Q relating to S is whether there are infinitely many prime numbers or not. Then, a set (S, Q, M) is a mathematical model in the sense of Definition 1.4.1 if M contains the statement "There are infinitely many prime numbers" along with other statements which prove this statement. In this sense, the entire mathematical theory can be viewed as a collection of mathematical models.

The notation (S, Q, M) in Definition 1.4.1 emphasizes the chronological order in which the constituents of a mathematical model usually appear. Typically, a system is given first, then there is a question regarding that system, and only then a mathematical model is developed. Each of the constituents of the triplet (S, Q, M) is an indispensable part of the whole. Regarding M, this is obvious, but S and Q are important as well. Without S, we would not be able to formulate a question Q; without a question Q, there would be virtually "nothing to do" for the mathematical model; and without S and Q, the remaining M would be no more than "l'art pour l'art". The formula $f(x) = e^x$, for example, is such a purely mathematical "l'art pour l'art" statement as long as we do not connect it with a system and a question. It becomes a mathematical model only when we define a system S and a question Q relating to it. For example, viewed as an expression of the exponential growth period of plants (Section 3.10.4), $f(x) = e^x$ is a mathematical model which can be used to answer questions regarding plant growth. One can say it is a genuine property of mathematical models to be more than "l'art pour l'art", and this is exactly the intention behind the notation (S, Q, M) in Definition 2.3.1. Note that the definition of mathematical models by Bellomo and Preziosi [14] discussed above appears as a special case of Definition 1.4.1 if we restrict S to physical systems, M to equations, and only allow questions Q which refer to the space-time evolution of S.

> **Note 1.4.1 (More than "l'art pour l'art")** The system and the question relating to the system are indispensable parts of a mathematical model. It is a genuine property of mathematical models to be more than mathematical "l'art pour l'art".

Let us look at another famous example that shows the importance of Q. Suppose we want to predict the behavior of some mechanical system S. Then the appropriate mathematical model depends on the problem we want to solve, that is, on the question Q. If Q is asking for the behavior of S at moderate velocities, classical (Newtonian) mechanics can be used, that is, $M = \{equations\ of\ Newtonian\ mechanics\}$. If, on the other hand, Q is asking for the behavior of S at velocities close to the speed of light, then we have to set $M = \{equations\ of\ relativistic\ mechanics\}$ instead.

1.5
Examples and Some More Definitions

Generally speaking, one can say we are concerned with mathematical models in the sense of Definition 1.4.1 whenever we perform computations in our everyday life, or whenever we apply the mathematics we have learned in schools and universities. Since everybody computes in his everyday life, everybody uses mathematical models, and this is why it was valid to say that "everyone models and simulates" in the preface of this book. Let us look at a few examples of mathematical models now, which will lead us to the definitions of some further important concepts.

> **Note 1.5.1 (Everyone models and simulates)** Mathematical models in the sense of Definition 1.4.1 appear whenever we perform computations in our everyday life.

Suppose we want to know the *mean age of some group of people*. Then, we apply a mathematical model (S, Q, M) where S is that group of people, Q asks for their mean age, and M is the mean value formula $\bar{x} = \left(\sum_{i=1}^{n} x_i\right)/n$. Or, suppose we want to know the mass X of some substance in the cylindrical tank of Figure 1.3, given

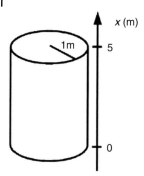

Fig. 1.3 Tank problem.

a constant concentration c of the substance in that tank. Then, a multiplication of the tank volume with c gives the mass X of the substance, that is,

$$X = 5\pi c \tag{1.1}$$

This means we apply a model (S, Q, M) where S is the tank, Q asks for the mass of the substance, and M is Equation 1.1. An example involving more than simple algebraic operations is obtained if we assume that the *concentration c in the tank* of Figure 1.3 depends on the height coordinate, x. In that case, Equation 1.1 turns into

$$X = \pi \cdot \int_0^5 c(x)\, dx \tag{1.2}$$

This involves an integral, that is, we have entered the realms of calculus now.

Note 1.5.2 (Notational convention) Variables such as X and c in Equation 1.1, which are used without further specification are always assumed to be real numbers, and functions such as $c(x)$ in Equation 1.2 are always assumed to be real functions with suitable ranges and domains of definition (such as $c : [0, 5] \to \mathbb{R}_+$ in the above example) unless otherwise stated.

In many mathematical models (S, Q, M) involving calculus, the question Q asks for the optimization of some quantity. Suppose for example we want to *minimize the material consumption* of a cylindrical tin having a volume of 1 l. In this case,

$$M = \{\pi r^2 h = 1, A = 2\pi r^2 + 2\pi rh \to \min\} \tag{1.3}$$

can be used to solve the problem. Denoting by r and h the radius and height of the tin, the first statement in Equation 1.3 expresses the fact that the tin volume is 1 l. The second statement requires the surface area of the tin to be minimal, which is equivalent to a minimization of the metal used to build the tin. The mathematical problem 1.3 can be solved if one inserts the first equation of (1.3) into the second equation of (1.3), which leads to

$$A(r) = 2\pi r^2 + \frac{2}{r} \rightarrow \min \qquad (1.4)$$

This can then be treated using standard calculus ($A'(r) = 0$ etc.), and the optimal tin geometry obtained in this way is

$$r = \sqrt[3]{\frac{1}{2\pi}} \approx 0.54 \; dm \qquad (1.5)$$

$$h = \sqrt[3]{\frac{4}{\pi}} \approx 1.08 \; dm \qquad (1.6)$$

1.5.1
State Variables and System Parameters

Several general observations can be made referring to the examples in the last section. As discussed in Section 1.1 above, the main benefit of the modeling procedure lies in the fact that the complexity of the original system is reduced. This can be nicely seen in the last example. Of course, each of us knows that a cylindrical tin can be described very easily based on its radius r and its height h. This means everyone of us automatically applies the correct mathematical model, and hence, – similar to the car problem discussed in Section 1.1 – everybody automatically believes that the system in the tin problem is a simple thing. But if we do not apply this model to the tin, it becomes a complex system. Imagine a *Martian* or some other extraterrestrial being who never saw a cylinder before. Suppose we would say to this Martian: "Look, here you have some sheets of metal and a sample tin filled with water. Make a tin of the same shape which can hold that amount of water, and use as little metal as possible." Then this Martian will – at least initially – see the original complexity of the problem. If he is smart, which we assume, he will note that infinitely many possible tin geometries are involved here. He will realize that an infinite set of (x, y)-coordinates would be required to describe the sample tin based on its set of coordinates. He will realize that infinitely many measurements, or, equivalently, algebraic operations would be required to obtain the material consumption based on the surface area of the sample tin (assuming that he did not learn about transcendental numbers such as π in his Martian school . . .).

From this original ("Martian") point of view we thus see that the system S of the tin example is quite complex, in fact an infinite-dimensional system. And we see the power of the mathematical modeling procedure which reduces those infinite dimensions to only two, since the mathematical solution of the above problem involves only two parameters: r and h (or, equivalently, r and A). Originally, the system "tin" in the above example is an infinite-dimensional thing not only with respect to its set of coordinates or the other aspects mentioned above, but also with respect to many other aspects which have been neglected in the mathematical model since they are unimportant for the solution of the problem, for example the thickness of the metal sheets, or its material, color, hardness, roughness and so on. All the information which was contained in the original system $S =$ "tin" is reduced to a description of the system as a mere $S_r = \{r, h\}$ in terms of the mathematical model. Here, we have used the notation S_r to indicate that S_r is not the original system which we denote S, but rather the description of S in terms of the mathematical model, which we call the "reduced system". The index "r" indicates that the information content of the original system S is reduced as we go from S to S_r.

Note 1.5.3 (A main benefit) The reduction of the information content of complex systems in terms of *reduced systems* (Definition 1.5.2) is one of the main benefits of mathematical models.

A formal definition of the reduced system S_r can be given in two steps as follows:

Definition 1.5.1 (State variables) Let (S, Q, M) be a mathematical model. Mathematical quantities $s_1, s_2, \ldots, s_n$ which describe the state of the system S in terms of M and which are required to answer Q are called the *state variables* of (S, Q, M).

Definition 1.5.2 (Reduced system and system parameters) Let $s_1, s_2, \ldots,$ s_n be the state variables of a mathematical model (S, Q, M). Let $p_1, p_2, \ldots,$ p_m be mathematical quantities (numbers, variables, functions) which describe properties of the system S in terms of M, and which are needed to compute the state variables. Then $S_r = \{p_1, p_2, \ldots, p_m\}$ is the *reduced system* and $p_1, p_2, \ldots, p_m$ are the *system parameters* of (S, Q, M).

This means that the state variables describe the system properties we are really interested in, while the system parameters describe system properties needed to obtain the state variables mathematically. Although we finally need the state variables to answer Q, the information needed to answer Q is already in the system parameters, that is, in the reduced system S_r. Using S_r, this information is expressed in terms of the state variables by means of mathematical operations, and this is then the final basis to answer Q. For example, in the *tank problem* above we were interested in the mass of the substance; hence, in this example we have one state variable, that is, $n = 1$ and $s_1 = X$. To obtain s_1, we used the concentration c; hence, we have one system parameter in that example, that is, $m = 1$ and $p_1 = c$. The reduced system in this case is $S_r = \{c\}$. By definition, the reduced system contains all information about the system which we need to get the state variable, that is, to answer Q. In the *tin example*, we needed the surface area of the tin to answer Q, that is, in that case we had again one state variable $s_1 = A$. On the other hand, two system parameters $p_1 = r$ and $p_2 = h$ were needed to obtain s_1, that is, in this case the reduced system is $S_r = \{r, h\}$.

$$S_r = \{r\}$$

(a)   (b)

Fig. 1.4 (a) Potted plant. (b) The same potted plant written as a reduced system.

Let us look at another example. In Section 3.10.4 below, a *plant growth model* will be discussed which is intended to predict the time evolution of the overall biomass of a plant. To achieve this, none of the complex details of the system "plant" will be considered except for its growth rate. This means the complex system S = "plant" is reduced to a single parameter in this model: the growth rate r of the plant. In the above notation, this means we have $S_r = \{r\}$ (Figure 1.4). It is not necessary to be a botanist to understand how dramatic this information reduction really is: everything except for the growth rate is neglected, including all kinds of macroscopic and microscopic substructures of the plant, its roots, its stem, its leaves as well as its cell structure, all the details of the processes that happen inside the cells, and so on. From the point of view of such a brutally simplified model, it makes no difference whether it is really concerned with the complex system "plant", or with some shapeless green pulp of biomass that might be obtained after sending the plant through a shredder, or even with entirely other systems, such as a bacteria culture or a balloon that is being inflated.

All that counts from the point of view of this model is that a growth rate can be assigned to the system under consideration. Naturally, botanists do not really like this brutal kind of models, which virtually send their beloved ones through a shredder. Anyone who presents such a model on a botanist's conference should be prepared to hear a number of questions beginning with "Why does your model disregard … ". At this point we already know how to answer this kind of question: we know that according to Definition 1.4.1, a mathematical model is a triplet (S, Q, M) consisting of a system S, a question Q, and a set of mathematical statements M, and that the details of the system S that are represented in M depend on the question Q that is to be answered by the model. In this case, Q was asking for the time development of the plant biomass, and this can be sufficiently answered based on a model that represents the system S = "plant" as $S_r = \{r\}$. Generally one can say that the reduced system of a well-formulated mathematical model will consist of no more than exactly those properties of the original system that are important to answer the question Q that is being investigated.

Note 1.5.4 (Importance of experiments) Typically, the properties (parameters) of the reduced system are those which need experimental characterization. In this way, the modeling procedure guides the experiments, and instead of making the experimenter superfluous (a frequent misunderstanding), it helps to avoid superfluous experiments.

1.5.2
Using Computer Algebra Software

Let us make a few more observations relating to the "1 l tin" example above. The mathematical problem behind this example can be easily solved using software. For example, using the *computer algebra software Maxima*, the problem can be solved as follows:

```
1: A(r):=2 *%pi *r^2 +2/r;
2: define(A1(r),diff(A(r),r));
3: define(A2(r),diff(A1(r),r));
4: solve(A1(r) =0);
5: r:rhs(solve(A1(r)=0)[3]);
6: r,numer;
7: A2(r)>0,pred;
```
(1.7)

   These are the essential commands in the *Maxima* program Tin.mac which you find in the book software. See Appendix A for a description of the book software and Appendix C for a description of how you can run Tin.mac within *Maxima*. In 1.7, the numbers 1:, 2:, and so on are not a part of the code, but just line numbers that we will use for referencing. Line 1 of the code defines the function $A(r)$ from Equation 1.4, which describes the material consumption that is to be minimized (note that %pi is the *Maxima* notation of π). As you know from calculus, you can minimize $A(r)$ by solving $A'(r) = 0$ [16, 17]. The solutions of this equations are the critical points, which can be relative maxima or minima depending on the sign of the second derivative A'' of A. Lines 2 and 3 of the above code define the first and second derivatives of $A(r)$ as the *Maxima* functions A1(r) and A2(r). Line 4 solves $A'(r) = 0$ using *Maxima's* solve command, which gives the result shown in Figure 1.5 if you are using *wxMaxima* (see Appendix C for details on *wxMaxima*).
   As the figure shows, $A'(r) = 0$ gives three critical points. The first two critical points involve the imaginary number i (which is designated as "%i" within *Maxima*), so these are complex numbers which can be excluded here [17]. The third solution in Figure 1.5 is the solution that really solves the tin problem (compare Equation 1.5 above). Line 5 of Equation 1.7 stores this solution in the variable r, using "[3]" to address the third element in the list shown in Figure 1.5. Since

$$(\%o9) \quad [r = \frac{\sqrt{3}\,\%i - 1}{2\,2^{1/3}\%pi^{1/3}}, \quad r = -\frac{\sqrt{3}\,\%i + 1}{2\,2^{1/3}\%pi^{1/3}}, \quad r = \frac{1}{2^{1/3}\%pi^{1/3}}]$$

Fig. 1.5 Result of line 4 of Equation 1.7 in *wxMaxima*.

this element is an equation, rhs is then used to pick the right-hand side of this equation. *Maxima*'s numer command can be used as in line 6 of Equation 1.7 if you want to have the solution in a decimal numerical format. Finally, *Maxima*'s pred command can be used as in line 7 of Equation 1.7 to verify that the value of the second derivative is positive at the critical point that was stored in r (a necessary condition for that critical point to be a minimum [17]). In *Maxima*, line 7 gives "true", which means that the second derivative is indeed positive as required.

1.5.3
The Problem Solving Scheme

In this example – and similarly in many other cases – one can clearly distinguish between the formulation of a mathematical model on the one hand and the solution of the resulting mathematical problem on the other hand, which can be done with appropriate software. A number of examples will show this below. This means that it is not necessary to be a professional mathematician if one wants to work with mathematical models. Of course, it is useful to have mathematical expertise. Mathematical expertise is particularly important if one wants to solve more advanced problems, or if one wants to make sure that the results obtained with mathematical software are really solutions of the original problem and no numerical artifacts. As we will see below, the latter point is of particular importance in the solution of partial differential equations (PDEs). However, people with insufficient mathematical expertise may of course just ask a mathematician. Typically, mathematical modeling projects will have an interdisciplinary character. The important point that we should note here is the fact that the formulation of mathematical models can also be done by nonmathematicians. Above all, the people formulating the models should be experts regarding the system under consideration. This book is intended to provide particularly nonmathematicians with enough knowledge about the mathematical aspects of modeling such that they can deal at least with simple mathematical models on their own.

Note 1.5.5 (Role of software) Typically, the formulation of a mathematical model is clearly separated from the solution of the mathematical problems implied by the model. The latter ("the hard work") can be done by software in many cases. People working with mathematical models hence do not need to be professional mathematicians.

The tin example shows another important advantage of mathematical modeling. After the tin problem was formulated mathematically (Equation 1.4), the powerful and well-established mathematical methods of calculus became applicable. Using the appropriate software (see 1.7), the problem could then be solved with little effort. Without the mathematical model for this problem, on the other hand, an

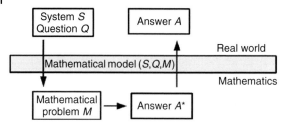

Fig. 1.6 Problem solving scheme.

experimental solution of this problem would have taken much more time. In a similar way, many other problems in science and engineering can be solved effectively using mathematics. From the point of view of science and engineering, mathematics can be seen as a big resource of powerful methods and instruments that can be used to solve problems, and it is the role of mathematical models to make these methods and instruments applicable to originally nonmathematical problems. Figure 1.6 visualizes this process. The starting point is a real-world system S together with a question Q relating to S. A mathematical model (S, Q, M) then opens up the way into the "mathematical universe", where the problem can be solved using powerful mathematical methods. This leads to a problem solution in mathematical terms (A^*), which is then translated into an answer A to the original question Q in the last step.

Note 1.5.6 (Mathematical models as door opener)   Translating originally non-mathematical problems into the language of mathematics, mathematical models virtually serve as a door opener toward the "mathematical universe" where powerful mathematical methods become applicable to originally nonmathematical problems.

As the figure shows, the mathematical model virtually controls the "problem solving traffic" between the real and mathematical worlds, and hence, its natural position is located exactly at the borderline between these worlds. The role of mathematics in Figure 1.6 can be described like a subway train: since it would be a too long and hard way to go from the system S and question Q to the desired answer A in the real world, smart problem solvers go into the "mathematical underground", where powerful mathematical methods provide fast trains toward the problem solution.

1.5.4
Strategies to Set up Simple Models

In many cases, a simple three-step procedure can be used to set up a mathematical model. Consider the following

Problem 1:
Which volumes of fluids A and B should be mixed to obtain 150 l of a fluid C that contains $70\,\mathrm{gl}^{-1}$ of a substance, if A and B contain $50\,\mathrm{gl}^{-1}$ and $80\,\mathrm{gl}^{-1}$, respectively?

For this simple problem, many of us will immediately write down the correct equations:

$$x + y = 150 \tag{1.8}$$
$$50x + 80y = 70 \cdot 150 \tag{1.9}$$

where x [l] and y [l] are the unknown volumes of the fluids A and B. For more complex problems, however, it is good to have a systematic procedure to set up the equations. A well-proven procedure that works for a great number of problems can be described as follows:

Note 1.5.7 (Three steps to setup a model)
- *Step 1:* Determine the number of unknowns, that is, the number of quantities that must be determined in the problem. In many problem formulations, you just have to read the last sentence where the question is asked.
- *Step 2:* Give *precise* definitions of the unknowns, including units. It is a practical experience that this should not be lumped with step 1.
- *Step 3:* Reading the problem formulation sentence by sentence, translate this information into mathematical statements which involve the unknowns defined in step 2.

Let us apply this to *Problem 1* above. In *step 1* and *step 2*, we would ascertain that *Problem 1* asks for two unknowns which can be defined as
- x: volume of fluid A in the mixture [l]
- y: volume of fluid B in the mixture [l]

These steps are important because they tell us about the unknowns that can be used in the equations. As long as the unknowns are unknown to us, it will be hard to write down meaningful equations in step 3. Indeed, it is a frequent beginner's mistake in mathematical modeling to write down equations which involve unknowns that are not sufficiently well defined. People often just pick up symbols that appear in the problem formulation – such as A, B, C in *problem 1* above – and then write down equations like

$$50A + 80B = 70 \tag{1.10}$$

This equation is indeed *almost* correct, but it is hard to check its correctness as long as we lack any precise definitions of the unknowns. The intrinsic problem with equations such as Equation 1.10 lies in the fact that A, B, C are already defined in the problem formulation. There, they refer to the names of the fluids, although they are (implicitly) used to express the volumes of the fluids in Equation 1.10. Thus, let us now write down the same equation using the unknowns x and y defined above:

$$50x + 80y = 70 \tag{1.11}$$

Now the definitions of x and y can be used to check this equation. What we see here is that on the left-hand side of Equation 1.11, the unit is (grams), which results from the multiplication of $50\,\text{gl}^{-1}$ with x [l]. On the right-hand side of Equation 1.11, however, the unit is grams per liter. So we have different units on the different sides of the equation, which proves that this is a wrong equation. At the same time, a comparison of the units may help us to get an idea of what must be done to obtain a correct equation. In this case, it is obvious that a multiplication of the right-hand side of Equation 1.11 with some quantity expressed in liter would solve the unit problem. The only quantity of this kind in the problem formulation is the 150 l volume which is required as the volume of the mixture, and multiplying the 70 in Equation 1.11 with 150 indeed solves the problem in this case.

Note 1.5.8 (Check the units!) Always check that the units on both sides of your equations are the same. Try to "repair" any differences that you may find using appropriate data of your problem.

A major problem in *step 3* is to identify those statements in the problem formulation which correspond to mathematical statements, such as equations, inequalities, and so on. The following note can be taken as a general guideline for this:

Note 1.5.9 (Where are the equations?) The statements of the problem formulation that can be translated into mathematical statements, such as equations, inequalities, and so on, are characterized by the fact that they impose restrictions on the values of the unknowns.

Let us analyze some of the statements in *Problem 1* above in the light of this strategy:

- *Statement 1*: 150 l of fluid C are required.
- *Statement 2*: Fluid A contains 50 gl^{-1} of the substance.
- *Statement 3*: Fluid B contains 80 gl^{-1} of the substance.
- *Statement 4*: Fluid C contains 70 gl^{-1} of the substance.

Obviously, *statement 1* is a restriction on the values of x and y, which translates immediately into the equation:

$$x + y = 150 \tag{1.12}$$

Statement 2 and *statement 3*, on the other hand, impose no restriction on the unknowns. Arbitrary values of x and y are compatible with the fact that fluids A and B contain 50 gl^{-1} and 80 gl^{-1} of the substance, respectively. *Statement 4*, however, does impose a restriction on x and y. For example, given a value of x, a concentration of 70 gl^{-1} in fluid C can be realized only for one particular value of y. Mathematically, *statement 4* can be expressed by Equation 1.9 above. You may be able to write down this equation immediately. If you have problems to do this, you may follow a heuristic (i.e. not 100% mathematical) procedure, where you try to start as close to the statement in the problem formulation as possible. In this case, we could begin with expressing *statement 4* as

$$\{\text{Concentration of substance in fluid C}\} = 70 \qquad (1.13)$$

Then, you would use the definition of a concentration as follows:

$$\frac{\{\text{Mass of substance in fluid C}\}}{\{\text{Volume of the mixture}\}} = 70 \qquad (1.14)$$

The next step would be to ascertain two things:
- The mass of the substance in fluid C comes from fluids A and B.
- The volume of the mixture is 150 l.

This leads to

$$\frac{\{\text{Mass of substance in fluid A}\} + \{\text{Mass of substance in fluid B}\}}{150} = 70 \qquad (1.15)$$

The masses of the substance in A and B can be easily derived using the concentrations given in *Problem 1* above:

$$\frac{50x + 80y}{150} = 70 \qquad (1.16)$$

This is Equation 1.9 again. The heuristic procedure that we have used here to derive this equation is particularly useful if you are concerned with more complex problems where it is difficult to write down an equation like Equation 1.9 just based on intuition (and where it is dangerous to do this since your intuition can be misleading). Hence, we generally recommend the following:

Note 1.5.10 (Heuristic procedure to set up mathematical statements) If you want to translate a statement in a problem formulation into a mathematical statement, such as an equation or inequality, begin by mimicking the statement in the problem formulation as closely as possible. Your initial formulation may involve nonmathematical statements similar to Equation 1.13 above. Try then to replace all nonmathematical statements by expressions involving the unknowns.

Note that what we have described here corresponds to the *systems analysis* and *modeling* steps of the modeling and simulation scheme in Note 1.2.3. Equations 1.8 and 1.9 can be easily solved (by hand and . . .) on the computer using *Maxima*'s `solve` command as it was described in Section 1.5.2 above. In this case, the *Maxima* commands

```
1: solve([
2: x+y =150
3:   ,50*x+80*y=70*150
4:  ]);
```
(1.17)

yield the following result:

```
[[x =50, y =100]]
```
(1.18)

You find the above code in the file `Mix.mac` in the book software (see Appendix A). As you see, the result is written in a nested *list structure* (lists are written in the form "`[a,b,c,... ]`" in *Maxima*): the inner list `[x = 50,y = 100]` gives the values of the unknowns of the solution computed by *Maxima*, while the outer list brackets are necessary to treat situations where the solution is nonunique (see the example in Section 1.5.2 above).

Note that lines 1–4 of Equation 1.17 together form a single `solve` command that is distributed over several lines here to achieve a better readability of the system of equations. Note also that the comma at the beginning of line 3 could also have been written at the end of line 2, which may seem more natural at a first glance. The reason for this notation is that in this way it is easier to generate a larger system of equations, by using copies of line 3 with a "paste and copy" mechanism for example. If you do that and have the commas at the end of each line, your last equation generated in this way will end with a comma which should not be there – so we recommend this kind of notation as a "foolproof" method, which makes your life with *Maxima* and other computer algebra software easier.

1.5.4.1 Mixture Problem

Since *Problem 1* in the last section was rather easy to solve and the various recommendations made there may thus seem unnecessary at least with respect to this particular problem, let us now see how a more complex problem is solved using these ideas:

Problem 2:

Suppose the fluids A, B, C, D contain the substances S_1, S_2, S_3 according to the following table (concentrations in grams per liter):

	A	B	C	D
S_1	2.5	8.2	6.4	12.7
S_2	3.2	15.1	13.2	0.4
S_3	1.1	0.9	2.2	3.1

What is the concentration of S_3 in a mixture of these fluids that contains 75% (percent by volume) of fluids A and B and which contains 4 gl^{-1} and 5 gl^{-1} of the substances S_1 and S_2, respectively?

Referring to *step 1* and *step 2* of the three-step procedure described in Note 1.5.7, it is obvious that we have only one unknown here which can be defined as follows:
- x: concentration of S_3 in the mixture (grams per liter)

Now *step 3* requires us to write down mathematical statements involving x. According to Note 1.5.9, we need to look for statements in the above problem formulation that impose a restriction on the unknown x. Three statements of this kind can be identified:
- *Statement 1*: 75% of the mixture consists of A and B.
- *Statement 2*: The mixture contains 4 gl^{-1} of S_1.
- *Statement 3*: The mixture contains 5 gl^{-1} of S_2.

Each of these statements excludes a great number of possible mixtures and thus imposes a restriction on x. Beginning with *statement 1*, it is obvious that this statement can not be formulated in terms of x. We are here in a situation where a number of auxiliary variables is needed to translate the problem formulation into mathematics.

Note 1.5.11 (Auxiliary variables)   In some cases, the translation of a problem into mathematics may require the introduction of *auxiliary variables*. These variables are "auxiliary" in the sense that they help us to determine the unknowns. Usually, the problem formulation will provide enough information such that the auxiliary variables *and* the unknowns can be determined (i.e. the auxiliary variables will just increase the size of the system of equations).

In this case, we obviously need the following auxiliary variables:
- x_A: percent (by volume) of fluid A in the mixture
- x_B: percent (by volume) of fluid B in the mixture
- x_C: percent (by volume) of fluid C in the mixture
- x_D: percent (by volume) of fluid D in the mixture

Now the above *statement 1* can be easily expressed as

$$x_A + x_B = 0.75 \tag{1.19}$$

Similar to above, *statement 2* and *statement 3* can be formulated as

$$\{\text{Concentration of S1 in the mixture}\} = 4 \tag{1.20}$$

and

$$\{\text{Concentration of S2 in the mixture}\} = 5 \tag{1.21}$$

Based on the information provided in the above table (and again following a similar procedure as in the previous section), these equations translate to

$$2.5x_A + 8.2x_B + 6.4x_C + 12.7x_D = 4 \tag{1.22}$$

and

$$3.2x_A + 15.1x_B + 13.2x_C + 0.4x_D = 5 \tag{1.23}$$

Since x is the concentration of S_3 in the mixture, a similar argumentation shows

$$1.1x_A + 0.9x_B + 2.2x_C + 3.1x_D = x \tag{1.24}$$

So far we have the four equations 1.19, 1.22, 1.23, and 1.24 for the five unknowns x, x_A, x_B, x_C, and x_D, that is, we need one more equation. In this case, the missing equation is given implicitly by the definition of x_A, x_B, x_C, and x_D. These variables express percent values, and hence, we have

$$x_A + x_B + x_C + x_D = 1 \tag{1.25}$$

Altogether, we have now obtained the following system of linear equations:

$$x_A + x_B = 0.75 \tag{1.26}$$
$$2.5x_A + 8.2x_B + 6.4x_C + 12.7x_D = 4 \tag{1.27}$$
$$3.2x_A + 15.1x_B + 13.2x_C + 0.4x_D = 5 \tag{1.28}$$
$$1.1x_A + 0.9x_B + 2.2x_C + 3.1x_D = x \tag{1.29}$$
$$x_A + x_B + x_C + x_D = 1 \tag{1.30}$$

Again, this system of equations can be solved similar to above using *Maxima*. In the *Maxima* program Mix1.mac in the book software (see Appendix A), the problem is solved using the following code

```
1: out:solve([
2: xA+xB=0.75
3: ,2.5*xA+8.2*xB+6.4*xC+12.7*xD=4
4: ,3.2*xA+15.1*xB+13.2*xC+0.4*xD=5
5: ,1.1*xA+0.9*xB+2.2*xC+3.1*xD=x
6: ,xA+xB+xC+xD=1
7: ]);
8: out,numer;
```

$$(1.31)$$

which yields the following results in *Maxima*:

```
         141437        1365        1783       77       14485
(%o6) [[x = ------, xD =----, xC = -----, xB =----, xA =------]]
         98620        19724        9862      4931      19724
(%o7) [[x = 1.434161427702292, xD = 0.06920502940580003, xC = 0.1807949705942,
                 xB = 0.01561549381464206, xA = 0.7343845061853579]]
```

As can be seen, the equation system 1.26–1.30 corresponds to lines 2–6 of the above code and these lines of code are embedded into *Maxima*'s solve command similar to the code in 1.17 that was discussed in the previous section. The only new thing is that the result of the solve command is stored in a variable named out in line 1 of Equation 1.31. This variable out is then used in line 8 of the code to produce a decimal result using *Maxima*'s numer command. This is why the *Maxima* output above comprises of two parts: The output labeled as "(%o6)" is the immediate output of the solve command, and as you can see above the solution is expressed in terms of fractions there. Although this is the most precise way to express the solution, one may prefer decimal numbers in practice. To achieve this, the numer command in line 8 of code (1.31) produces the second part of the above output which is labeled as "(%o7)". So we can finally say that the solution of *problem* 2 above is $x \approx 1.43 \text{ gl}^{-1}$, which is the approximate concentration of S_3 in the mixture.

1.5.4.2 Tank Labeling Problem

When fluids are stored in horizontal, cylindrical tanks similar to the one shown in Figure 1.7b, one typically wants to have labels on the front side of the tank as shown in Figure 1.7a. In practice, this problem is often solved "experimentally", that is, by filling the tank with well-defined fluid volumes, and then setting the labels at the position of the fluid surface that can be seen from outside. This procedure may of course be inapplicable in situations where the fluid surface cannot be seen from outside. More important, however, is the cost argument: this

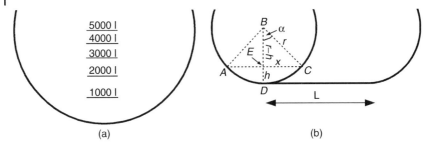

Fig. 1.7 (a) Tank front side with volume labels.
(b) Unknowns and auxiliary variables of the tank labeling
problem.

experimental procedure is expensive in terms of time (working time of the people
who are performing the experiment) and material (e.g. the water that is wasted
during the experiment). It is much cheaper here to apply the mathematical model
that will be developed below. Unfortunately, the situation in this example – where
the problem *could* be solved cheap and efficiently using mathematical models and
open-source software, but where expensive experimental procedures or, in some
cases, expensive commercial software solutions are used – is still rather the rule
than the exception in many fields.

Let us start with the development of an appropriate mathematical model. Let h
(decimeters) be the height of a label at the front side of the tank as indicated in
Figure 1.7b, and let $V(h)$ (cubic decimeters) be the filling volume of the tank that
corresponds to h. If we want to determine the label height for some filling volume
V_f, then the following equation must be solved for h:

$$V(h) = V_f \tag{1.32}$$

Referring to Figure 1.7b, $V(h)$ can be expressed as

$$V(h) = ACD \cdot L \tag{1.33}$$

where ACD (square decimeters) corresponds to the surface at the front side of
the tank that is enclosed by the line segments AC, CD and DA. ACD can be
expressed as

$$ACD = ABCD - ABC \tag{1.34}$$

where the circular segment $ABCD$ is

$$ABCD = \frac{2\alpha}{2\pi}\pi r^2 = \alpha r^2 \tag{1.35}$$

In the last equation, α is expressed in radians (which makes sense here since the
problem is solved based on *Maxima* below, which uses radians in its trigonometric

functions). The surface of the triangle ABC is

$$ABC = x(r - h) \tag{1.36}$$

where

$$x = \sqrt{r^2 - (r - h)^2} \tag{1.37}$$

due to the theorem of Pythagoras. Using the last five equations and

$$\alpha = \cos^{-1}\left(\frac{r - h}{r}\right) \tag{1.38}$$

in Equation 1.32 yields

$$L \cdot \cos^{-1}\left(\frac{r - h}{r}\right) r^2 - L\sqrt{r^2 - (r - h)^2}(r - h) = V_f \tag{1.39}$$

Unlike the equations treated in the last sections, this is now a *transcendental equation* that cannot be solved in closed form using *Maxima*'s `solve` command as before. To solve Equation 1.39, numerical methods such as the bisection method or Newton's method must be applied [18]. In *Maxima*, the `find_root` command can be applied as follows:

```
1: for i:1 thru 4 do
2: (
3: out:find_root(
4: L*acos((r-h)/r)*r^2-L*sqrt(r^2-(r-h)^2)*(r-h)=i*1000
5: ,h,0,r
6: ),
7: print("Label for V=",i*1000,"l:",out,"dm")
8: );
```

$$\tag{1.40}$$

This is the essential part of `Label.mac`, a *Maxima* code which is a part of the book software (see Appendix A), and which solves the tank labeling problem assuming a 10 000 l tank of length $L = 2$ m based on Equation 1.39. Equation 1.39 appears in line 4 of the code, with its right-hand side replaced by `i*1000` which successively generates 1000, 2000, 3000, and 4000 as the right-hand side of the equation due to the `for` command that is applied in line 1, so the problem is solved for 1000, 2000, 3000, and 4000 l of filling volume in a single run of the code (note that the 5000, 6000, and so on labels can be easily derived from this if required). What the `for...thru...do` command in line 1 precisely does is this: it first sets $i = 1$ and then executes the entire code between the brackets in lines 2 and 8, which solves the problem for $V_f = 1000$ l; then, it sets $i = 2$ and executes the entire code between the brackets in lines 2 and 8 again, which solves the problem for $V_f = 2000$, and so on until the same has been done for $i = 4$ (the upper limit given by "thru" in line 1).

Note that the arguments of the `find_root` command are in lines 4 and 5, between the brackets in lines 3 and 6. Its first argument is the equation that is to be solved (line 4), which is then followed by three more arguments in line 5: the variable to be solved for (h in this case), and upper and lower limits for the interval in which the numerical algorithm is expected to look for a solution of the equation (0 and r in this case). Usually, reasonable values for these limits can be derived from the application – in this case, it is obvious that $h > 0$, and it is likewise obvious that we will have $h = r$ for 50001 filling volume since a 10 000–l tank is assumed, which means that we will have $h < r$ for filling volumes below 50001. The `print` command prints the result to the computer screen. Note how text, numbers and variables (such as the variable out that contains the result of the `find_root` command, see line 3) can be mixed in this command. Since the `print` is a part of the for... thru... do environment, it is invoked four times and produces the following result:

```
Label for V= 1000 l: 3.948086422946864 dm
Label for V= 2000 l: 6.410499677168014 dm
Label for V= 3000 l: 8.582542383270068 dm
Label for V= 4000 l: 10.62571600771833 dm
```

1.5.5
Linear Programming

All mathematical models considered so far were formulated in terms of equations only. Remember that according to Definition 1.4.1, a mathematical model may involve any kind of mathematical statements. For example, it may involve inequalities. One of the simplest class of problems involving inequalities are linear programming problems that are frequently used e.g. in operations research. Consider the following problem taken from the linear programming article of *Wikipedia. org*:

Linear programming example
Suppose a farmer has a piece of farm land, say A square kilometers large, to be planted with either wheat or barley or some combination of the two. Furthermore, suppose the farmer has a limited permissible amount F of fertilizer and P of insecticide which can be used, each of which is required in different amounts per unit area for wheat (F_1, P_1) and barley (F_2, P_2). Let S_1 be the selling price of wheat, and S_2 the price of barley. How many square kilometers should be planted with wheat versus barley to maximize the revenue?

Denoting the area planted with wheat and barley with x_1 and x_2 respectively, the problem can be formulated as follows:

$$x_1, x_2 \geq 0 \tag{1.41}$$

$$x_1 + x_2 \leq A \tag{1.42}$$

$$F_1 x_1 + F_2 x_2 \leq F \tag{1.43}$$

$$P_1 x_1 + P_2 x_2 \leq P \tag{1.44}$$

$$S_1 x_1 + S_2 x_2 \to \max \tag{1.45}$$

Here, Equation 1.41 expresses the fact that the farmer cannot plant a negative area, Equation 1.42 the fact that no more than the given A square kilometers of farm land can be used, Equations 1.43 and 1.44 express the fertilizer and insecticide limits, respectively, and Equation 1.45 is the required revenue maximization. Taking Equations 1.41–1.45 as M, the system S as the farm land and the question Q, "How many square kilometers should be planted with wheat versus barley to maximize the revenue?", a mathematical model (S, Q, M) is obtained. For any set of parameter values for A, F, P, ..., the problem can again be easily solved using *Maxima*. This is done in the *Maxima* program Farm.mac which you find in the book software (see Appendix A). Let us look at the essential commands of this code:

```
1: load(simplex);
2: U:[x1>=0
3: ,x2>=0
4: ,x1+x2<=A
5: ,F1*x1+F2*x2 <=F
6: ,P1*x1+P2*x2<=P];
7: Z:S1*x1+S2*x2;
8: maximize_lp(Z,U);
```
(1.46)

Line 1 of this code loads a package required by *Maxima* to solve linear programming problems. Lines 2–6 define the inequalities, corresponding to Equations 1.41–1.44 above. Note that lines 2–6 together make up a single command that stores the list of inequalities in the variable U. Line 7 defines the function Z that is to be maximized, and the problem is then solved in line 8 using *Maxima's* maximize_lp command. Based on the parameter settings in Farm.mac, *Maxima* produces the following result:

```
[100, [x2 = 50, x1 = 0]]
```

This means that a maximum revenue of 100 is obtained if the farmer plants barley only (50 square kilometers).

1.5.6
Modeling a Black Box System

In Section 1.3 it was mentioned that the systems investigated by scientists or engineers typically are "input–output systems", which means they transform the given input parameters into output parameters. Note that the previous examples were indeed referring to such "input–output systems". In the tin example, the radius and height of the tin are input parameters and the surface area of the tin is

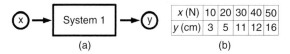

x (N)	10	20	30	40	50
y (cm)	3	5	11	12	16

(a)   (b)

Fig. 1.8 (a) System 1 with input x (N) and output y (cm).
(b) System 1 data (file spring.csv in the book software).

the output parameter. In the plant growth example, the growth rate of the plant and its initial biomass is the input and the resulting time–biomass curve is the output (details in Chapter 3). In the tank example, the geometrical data of the tank and the concentration distribution are input parameters while the mass of the substance is the output. In the linear programming examples, the areas planted with wheat or barley are the input quantities and the resulting revenue is the output. Similarly, all systems in the examples that will follow can be interpreted as input–output systems.

The exploration of an example input–output system in some more detail will now lead us to further important concepts and definitions. Assume a "system 1" as in Figure 1.8 which produces an output length y (centimeters) for every given input force x [N]. Furthermore, assume that we do not know about the processes inside the system that transform x into y, that is, let this system be a "black box" to us as described above. Consider the following problem:

Q: Find an input x that generates an output y = 20 cm.

This defines the question Q of the mathematical model (S, Q, M) that we are going to define. S is the "system 1" in Figure 1.8a, and we are now looking for an appropriate set of mathematical statements M that can help us to answer Q.

All that the investigator of system 1 can do is to produce some data using the system, hoping that these data will reveal something about the processes occurring inside the "black box". Assume that the data in the file spring.csv (which you find in the PhenMod/LinReg directory of the book software, see Appendix A) have been obtained from this system, see Figure 1.8b. To see what happens, the investigator will probably produce a plot of the data as in Figure 1.9a. Note that the plots in Figure 1.9 were generated using the scatter plot option of *OpenOffice.org Calc* (see Appendix A on how you can obtain this software). Figure 1.9a suggests that there is an approximately linear dependence between the x- and y-data. Mathematically, this means that the function $y = f(x)$ behind the data is a straight line:

$$f(x) = ax + b \tag{1.47}$$

Now the investigator can apply a statistical method called *linear regression* (which will be explained in detail in Section 2.2) to determine the coefficients a and b of this equation from the data, which leads to the "regression line"

$$f(x) = 0.33x - 0.5 \tag{1.48}$$

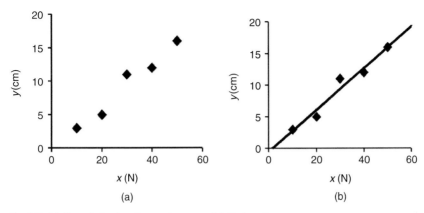

Fig. 1.9 (a) Plot of the data in `spring.csv`. (b) System 1 data with regression line. Both plots generated using *Calc*, see Section 2.1.1.1.

Figure 1.9b shows that there is a good coincidence or, in statistical terminology, a good "fit" between this regression line and the data. Equation 1.48 can now be used as the M of a mathematical model of system 1. The question Q stated above ("Which system input x generates a desired output y = 20 cm? ") can then be easily answered by setting $y = f(x) = 20$ in Equation 1.48, that is,

$$20 = 0.33x - 0.5 \qquad (1.49)$$

which gives $x \approx 62.1$ N. Of course, this is just an *approximate result* for several reasons. First of all, Figure 1.9 shows that there are some deviations between the regression line and the data. These deviations may be due to measurement errors, but they may also reflect some really existing effects. If the deviations are due to measurement errors, then the precise location of the regression line and hence, the prediction of x for y = 20 cm is affected by these errors. If, on the other hand, the deviations reflect some really existing effects, then Equation 1.48 is no more than an approximate model of the processes that transform x into y in system 1, and hence, the prediction of x for y = 20 cm will be only approximate. Beyond this, predictions based on data such as the data in Figure 1.8b are always approximate for principal reasons. The y-range of these data ends at 16 cm, and system 1 may behave entirely different for y-values beyond 16 cm which we would not be able to see in such a data set. Therefore, the experimental validation of predictions derived from mathematical models is always an indispensable part of the modeling procedure (see Section 1.2). See also Chapter 2 for a deeper discussion of the quality of predictions obtained from black box models.

The example shows the *importance of statistical methods* in mathematical modeling. First of all, statistics itself is a collection of mathematical models that can be used to describe data or to draw inferences from data [19]. Beyond this, statistical methods provide a necessary link between nonstatistical mathematical models and the real world. In mathematical modeling, one is always concerned with experimental data, not only to validate model predictions, but also to develop hypotheses about the system, which help to set up appropriate equations. In the example, the data led us to the hypothesis that there is a linear relation between x and y. We have used a plot of the data (Figure 1.9) and the regression method to find the coefficients in Equation 1.48. These are methods of descriptive statistics, which can be used to summarize or describe data. Beyond this, inferential statistics provides methods that allow conclusions to be drawn from data in a way that accounts for randomness and uncertainty. Some important methods of descriptive and inductive statistics will be introduced below (Section 2.1).

> **Note 1.5.12 Statistical methods**   provide the link between mathematical models and the real world.

The reader might say that the estimate of x above could also have been obtained without any reference to models or computations, by a simple tuning of the input using the real, physical system 1. We agree that there is no reason why models should be used in situations where this can be done with little effort. In fact, we do not want to propose any kind of a fundamentalist "mathematical modeling and simulation" paradigm here. A *pragmatic approach* should be used, that is, any problem in science and engineering should be treated using appropriate methods, may this be mathematical models or a tuning of input parameters using the real system. It is just a fact that in many cases the latter cannot be done in a simple way. The generation of data such as in Figure 1.8 may be expensive, and thus, an experimental tuning of x toward the desired y may be inapplicable. Or, the investigator may be facing a very complex interaction of several input and output parameters, which is rather the rule than the exception as explained in Section 1.1. In such cases, the representation of a system in mathematical terms can be the only efficient way to solve the problem.

1.6
Even More Definitions

1.6.1
Phenomenological and Mechanistic Models

The mathematical model used above to describe system 1 is called a *phenomenological model* since it was constructed based on experimental data only, treating the system as a black box, that is, without using any information about the internal processes occurring inside system 1 when x is transformed into y. On the other hand, models that are constructed using information about the system S are called *mechanistic models*, since such models are virtually based on a look into the internal mechanics of S. Let us define this as follows [11]:

Definition 1.6.1 (Phenomenological and mechanistic models) A mathematical model (S, Q, M) is called
- *phenomenological*, if it was constructed based on experimental data only, using no a priori information about S,
- *mechanistic*, if some of the statements in M are based on a priori information about S.

Phenomenological models are also called *empirical models, statistical models, data-driven models* or *black box models* for obvious reasons. Mechanistic models for which all necessary information about S are available are also called *white box models*. Most mechanistic models are located somewhere between the extreme black and white box cases, that is, they are based on some information about S while some other important information is unavailable. Such models are sometimes called *gray box models* or *semi-empirical models* [20].

To better understand the differences between phenomenological and mechanistic models, let us now construct an alternative mechanistic model for system 1 (Figure 1.8). Above, we have treated system 1 as a black box, that is, we have used no information about the way in which system 1 transforms some given input x into the output y (Figure 1.8). Let us now assume that the internal mechanics of system 1 looks as shown in Figure 1.10, that is, assume that system 1 is a mechanical spring, x is a force acting on that spring, and y is the resulting elongation. This is now an *a priori information* about system 1 in the sense of Definition 1.6.1 above, and it can be used to construct a mechanistic mathematical model based on elementary physical knowledge. As is well known, mechanical springs can be described by Hooke's law, which in this case reads

$$x = k \cdot y \tag{1.50}$$

where k is the spring constant (newtons per centimeter), a measure of the elasticity of the spring. The parameter k is either known (e.g. from the manufacturer of the spring), or estimated based on data such as those in Figure 1.8. Now the following mechanistic mathematical model (S, Q, M) is obtained:
- S: System 1
- Q: Which system input x generates a desired output of $y = 20\,\mathrm{cm}$?
- M: Equation 1.50

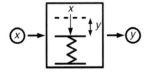

Fig. 1.10 Internal mechanics of system 1.

Based on this model, question Q can be answered as before by setting $y = 20\,\text{cm}$ in the model equation (1.50), which leads to

$$x = k \cdot 20 \tag{1.51}$$

that is, we can answer the question Q depending on the value of the spring constant, k. For example, assuming a value of $k \approx 3.11\,\text{N cm}^{-1}$ for the spring constant, we would get the same estimate $x \approx 62.1\,\text{N}$ as above. The mechanistic model of system 1 has several important advantages compared to the phenomenological model, and these advantages are *characteristic advantages of the mechanistic approach*. First of all, mechanistic models generally allow better predictions of system behavior. The phenomenological model equation (1.48) was derived from the data in Figure 1.8. These data involve forces x between 10 and 50 N. As mentioned below in our discussion of regression methods, this means that one can expect Equation 1.48 to be valid only close to this range of data between 10 and 50 N. The mechanistic model equation (1.50), on the other hand, is based on the well-established physical theory of a spring. Hence, we have good reason to expect its validity even outside the range of our own experimental testing.

Mechanistic models do also allow *better predictions* of modified systems. Assume for example that system 1 in Figure 1.10 is replaced by a system 2 that consists of two springs. Furthermore, assume that each of these system 2 springs has the same spring constant k as the system 1 spring. Then, in the phenomenological approach, the model developed for system 1 would be of no use, since we would not know about the similarity of these two systems (remember that the phenomenological approach assumes that no details are known about the internal mechanics of the system under consideration). This means that a new phenomenological model would have to be developed for system 2. A new data set similar to Figure 1.8 would be required, appropriate experiments would have to be performed, and afterwards, a new regression line similar to Figure 1.9 would have to be derived from the data. In the mechanistic approach, on the other hand, Hooke's law would immediately tell us that in the case of two springs the appropriate modification of Equation 1.50 is

$$x = 2k \cdot y \tag{1.52}$$

Another advantage of mechanistic models is the fact that they usually involve *physically interpretable parameters*, that is, parameters which represent real properties of the system. To wit: the numerical coefficients of the phenomenological model equation 1.47 are just numbers which cannot be related to the system. The parameter k of the mechanistic model equation 1.50, on the other hand, can be related to system properties, and this is of particular importance when we want to optimize system performance. For example, if we want smaller forces x to be required for a given elongation y, then in the phenomenological approach we would have to test a number of systems 2, 3, 4, . . . , until we would eventually arrive at some system with the desired properties. That is, we would have to apply a trial-and-error method. The mechanistic model, on the other hand, tells us exactly what we have to do: we have to replace the system 1 spring with a spring having a smaller spring constant k, and this will reduce the force x required for a given elongation y. In this simple example, it may be hard to imagine that someone would really use the phenomenological approach instead of Hooke's law. But the example captures an essential difference between phenomenological and mechanistic models, and it tells us that we should use mechanistic models if possible.

So, if mechanistic models could be set up easily in every imaginable situation, we would not have to talk about phenomenological models here. However, in many situations, it is not possible or feasible to use mechanistic models. As an essential prerequisite, *mechanistic models need a priori knowledge of the system*. If nothing is known about the system, then we are in the "black box" situation and have to apply phenomenological models. Suppose, for example, we want to understand why some roses wilt earlier than others (this example will be explained in more detail in Section 2.3). Suppose we assume that this is related to the concentrations of certain carbohydrates that can be measured. Then we cannot set up a mechanistic model as long as we do not know all the relevant processes that connect those carbohydrate concentrations with the observed freshness of the rose. Unless these processes are known, all we can do is to produce some data (carbohydrate concentration versus some appropriate measure of rose freshness) and analyze these data using phenomenological models.

This kind of situation where little is known about the system under investigation is rather the rule than the exception, particularly at early stages of a scientific investigation, or at the early stages of a product development in engineering. We may also be in a situation where we principally know enough details about the system under investigation, but where the system is so complex that it would take too much time and resources to setup a mechanistic model. An example is the optimization of the wear resistance of composite materials: Suppose that a composite material is made of the materials $M_1, M_2, \ldots, M_n$, and we want to know how the relative proportions of these materials should be chosen in order to maximize the composite materials resistance to wear. Then, the wear resistance of the composite material can depend in an extremely complex way on its composition. The author has investigated a situation of this kind where mechanistic modeling attempts failed due to the complexity of the overall system, and where a black box-type phenomenological neural network approach (see Section 2.5) was used instead [21]. An important *advantage of phenomenological models* is that they can be used in black box situations of this kind, and that they typically require much less time and resources. Pragmatic considerations should decide which type of model is used in practice. A mechanistic model will certainly be a bad choice if we need three weeks to make it work, and if it does not give substantially better answers to our question Q compared to a phenomenological model which can be set up within a day.

Note 1.6.1 (Phenomenological vs. mechanistic) *Phenomenological models* are universally applicable, easy to set up, but limited in scope. *Mechanistic models* typically involve physically interpretable parameters, allow deeper insights into system performance and better predictions, but they require a priori information on the system and often need more time and resources.

1.6.2
Stationary and Instationary models

It was already mentioned above that the question Q is an important factor that determines the appropriate mathematical model (S, Q, M). As an example, we have considered the alternative treatment of mechanical problems with the equations of classical or relativistic mechanics depending on the question Q that is investigated. In the system 1 example, we have used Q: "Which system input x generates a desired output of $y = 20\,\text{cm}$? ". Let us now modify this Q in order to find other important classes of mathematical models. Consider the following question:

Q: If a constant force x acts on the spring beginning with $t = 0$, what is the resulting elongation $y(t)$ of the spring at times $t > 0$?

This question cannot be answered based on the models developed above. The phenomenological model (Equation 1.48) as well as the mechanistic model (Equation 1.50) both refer to the so-called stationary state of system 1. This means that the elongation y expressed by these equations represents the time-independent (= stationary) state of the spring which is achieved after the spring has been elongated into the state of equilibrium where the force x exactly matches the force of the spring. On the other hand, the above question asks for the instationary (i.e. time-dependent) development of the elongation $y(t)$, beginning with time $t = 0$ when the force x is applied to the spring. To compute this $y(t)$, an instationary mathematical model (S, Q, M) is needed where the mathematical statements in M involve the time t. Models of this kind can be defined based on ordinary differential equations (details in Chapter 3). To make this important distinction between stationary and instationary models precise, let us define

Definition 1.6.2 (Stationary/instationary models)   A mathematical model (S, Q, M) is called
- *instationary*, if at least one of its system parameters or state variables depends on time and
- *stationary* otherwise.

1.6.3
Distributed and Lumped models

Suppose now that the spring in system 1 broke into pieces under normal operational conditions, and that it is now attempted to construct a more robust spring. In such a situation, it is natural to ask the following question:

Q: Which part of the spring should be reinforced?

Naturally, those parts of the spring which bear the highest mechanical stresses should be reinforced. To identify these regions, we need to know the distribution of stresses inside the spring under load. Let $\sigma(x, y, z)$ denote the mechanical stress distribution inside the spring depending on the spatial coordinates x, y, *and* z. Then we need a mathematical model with $\sigma(x, y, z)$ as a state variable. Such a mathematical model can be formulated based on PDEs as will be explained in Chapter 4. The important difference between this model and the previous models of system 1 lies in the fact that in this case the state variable depends on the spatial coordinates. To predict the equilibrium elongation of the spring using Equations 1.47 or 1.50, it was sufficient to describe the spring based on the spring constant k only. These equations, however, cannot be used to derive any spatially distributed information regarding the spring. In this kind of models, all spatial information is lumped together into the parameter k. In the case above, this was justified by the fact that the equilibrium position of a spring can be predicted with sufficient precision using k. On the other hand, if one is asking for the internal stress distribution in the spring, a spatially distributed description of the stresses inside the spring is needed. This motivates the following:

Definition 1.6.3 (Distributed/lumped models)   A mathematical model (S, Q, M) is called
- *distributed*, if at least one of its system parameters or state variables depends on a space variable,
- *lumped* otherwise.

1.7
Classification of Mathematical Models

Based on the examples in the last section, the reader can now distinguish between some basic classes of mathematical models. We will now widen our perspective toward a look at the entire "space of mathematical models", that is, this section will give you an idea of various types of mathematical models that are used in practice.

Note 1.7.1 The practical use of a classification   of mathematical models lies in the fact that you understand "where you are" in the space of mathematical models, and which types of models might be applicable to your problem beyond the models that you have already used.

1.7.1
From Black to White Box Models

The "space of mathematical models" evolves naturally from Definition 1.4.1, where we have defined a mathematical model to be a triple (S, Q, M) consisting of a system S, a question Q, and a set of mathematical statements M. Based on this definition, it is natural to classify mathematical models in an *SQM space*. Figure 1.11a shows one possible approach to visualize this SQM space of mathematical models, based on a classification of mathematical models between black and white box models. Psychological and social systems constitute the "black box" end of the spectrum. Only very vague phenomenological models can be developed for these systems due to their complexity and due to the fact that too many subprocesses are involved which are not sufficiently understood. On the other hand, mechanical systems, electrical circuits etc. are at the white box end of the spectrum since they can be very well understood in terms of mechanistic models (a famous example is Newton's model of planetary motion).

Note that the three dimensions of a mathematical model (S, Q, M) can be seen in the figure: the systems (S) are classified on top of the bar, immediately below the bar there is a list of objectives that mathematical models in each of the segments may have (which is Q), and at the bottom end there are corresponding mathematical structures (M) ranging from algebraic equations (AEs) to differential equations (DEs). Equation 1.47 (Section 1.5.6) is an example of a mathematical model in the form of an AE. As suggested by Figure 1.11, black box regression models of this kind are widely used for the modeling for example, of psychological, social, or economic systems (see Chapter 2 for more on regression models). On the other hand, the wine fermentation model discussed in Section 3.10.2 exemplifies the modeling of a biological/chemical system using ODEs (see Chapters 3 and 4 for more examples of DE models).

The "Q"-criteria in Figure 1.11a illustrate that mathematical models can be used to solve increasingly challenging problems as the model gradually turns from a

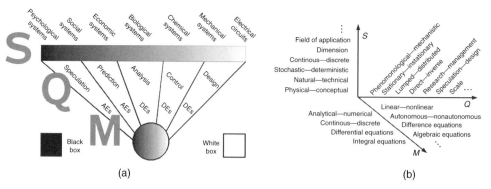

Fig. 1.11 (a) Classification of mathematical models between black and white box models (adapted from [3]). (b) Classification of mathematical models in the SQM space.

black box to a white box model. At the black box end of the spectrum, models can be used to make more or less reliable predictions based on data. For example, you may think here of attempts that have been made to predict share prices using the neural network methods described in Chapter 2 [22]. The model of a biological predator–prey system discussed in Section 3.10.1 is already "white enough" such that it can be used for an analysis of the dynamical system behavior in terms of phase plot diagrams such as Figure 3.17. Beyond this, models of chemical systems can be so precise that they can be used for a control of a process such as the wine fermentation process discussed in Section 3.10.2.

At the white box end of the spectrum, mathematical models can be applied to design, test, and optimize systems and processes on the computer before they are actually physically realized. This is used e.g. in *virtual engineering*, which includes techniques such as interactive design using CFD (see [23] and Section 4.10.3) or *virtual prototyping* [7, 24, 25]. As an example, you may think of the computation of the temperature distribution within a three-dimensional device using finite-element software, as it will be discussed in Section 4.9 below. Based on the method described there, *what-if studies* can be performed, that is, it can be investigated what happens with the temperature distribution if you change certain characteristics of the device virtually on the computer, and this can then be used to optimize the construction of the device so as to achieve certain desired characteristics of the temperature distribution.

1.7.2
SQM Space Classification: S Axis

Since mathematical models are characterized by their respective individual S, Q and M "values", one can also think of each model as being located somewhere in the "SQM space" of Figure 1.11b. On each of the S-, Q- and M-axes of the figure, mathematical models are classified with respect to a number of criteria which were compiled based on various classification attempts in the literature [3, 11, 20, 26–30]. Let us explain these criteria, beginning with the S axis of Figure 1.11b:

Physical – conceptual. Physical systems are part of the real world, for example, a fish or a car. Conceptual systems are made up of thoughts and ideas, for example, a set of mathematical axioms. This book focuses entirely on physical systems.

Natural – technical. Naturally, a natural system is a part of nature, such as a fish or a flower, while a technical system is a car, a machine, and so on. An example of a natural system is the predator–prey system treated in Section 3.10.1, the stormer viscometer treated in Section 2.4 exemplifies a technical system.

Stochastic – deterministic. Stochastic systems involve random effects, such as rolling dice, share prices and so on. Deterministic systems involve no or very little random effects, for example, mechanical systems, such as the planetary system, a pendulum, and so on. In a deterministic system, a particular state A of the system is always followed by one and the same state B, while A may be followed by B,

C or other states in an unpredictable way if the system is stochastic [31]. Below, stochastic models will be considered mainly in Chapter 2 and deterministic models mainly in Chapters 3 and 4.

Continuous – discrete. Continuous systems involve quantities that change continuously with time, such as sugar and ethanol concentrations in a wine fermenter (Section 3.10.2). Discrete systems, on the other hand, involve quantities that change at discrete times only, such as the number of individuals in animal populations (Section 3.10.1). Note that on the M axis of Figure 1.11, continuous systems can be represented by discrete mathematical statements and vice versa (e.g. a continuous mathematical formulation is used in Section 3.10.1 to describe the discrete predator–prey system).

Dimension. Depending on their spatial symmetries, physical systems can be described using 1, 2, or 3 space variables. As will be discussed in Section 4.3.3, the number of space variables used to describe a physical system is called its *dimension* (frequently denoted 1D, 2D, or 3D). Examples: a 1D temperature distribution is computed in Section 4.6 and a 3D temperature distribution in Section 4.9.

Field of application. We can distinguish between chemical systems, physical systems, biological systems, and so on. Systems from these and more fields of application will be considered below.

1.7.3
SQM Space Classification: Q Axis

On the Q- axis of Figure 1.11b, we have the following categories:

Phenomenological – mechanistic. This has been discussed in detail in Section 1.6. Phenomenological models are treated in Chapter 2 and mechanistic models in Chapters 3 and 4.

Stationary – instationary. Again, this has been discussed in Section 1.6. As discussed there, it depends on the question which we are asking (i.e. on the "Q" of a mathematical model (S, Q, M)) whether a stationary (time-independent) or instationary (time-dependent) model is appropriate. See also *Problem* 1 (instationary) and *Problem* 2 (stationary) in Section 4.1.3.

Lumped – distributed. Again, see Section 1.6. As was discussed there, it depends on the question which we are asking (i.e. on the "Q" of a mathematical model (S, Q, M)) whether a lumped (space-independent) or distributed (space-dependent) model is appropriate. The wine fermentation model (Section 3.10.2) is an example of a lumped model since it does not use spatial coordinates. On the other hand, the computation of a 3D temperature distribution in Section 4.9 is based on a distributed model.

Direct – inverse. Consider an input–output system as in Figure 1.2a. If Q assumes given input and system parameters and asks for the output, the model solves a so-called *direct problem* [3]. Most of the models below refer to direct problems. If, on the other hand, Q asks for the input or for parameters of S, the model solves a so-called inverse problem [32]. If Q asks for parameters of S, the resulting problem is also called a *parameter identification problem*. Examples are the regression and neural network models discussed in Chapter 2, and the fitting of ODEs to data discussed in Section 3.9. If Q asks for input parameters, the resulting problem is also called a *control problem*, since in this case the problem is to *control* the input in a way that generates some desired output ([33] and 4.11.3).

Research – management. Research models are used if Q aims at the understanding of S; management models, on the other hand, are used if the focus is on the solution of practical problems related to S. As pointed out in [20], research models tend to be more complex and less manageable from a practical point of view. Depending on Q, the same mathematical equations can be a part of a research or of a management model. For example, the predator–prey model described in Section 3.10.1 is a research model if the investigator just wants to understand the oscillations of the predator and prey populations, and it is a management model if is used to control the predator and prey populations (but as discussed in Section 3.10.1, this model is so simple that it cannot be seriously used as a management model).

Speculation – design. See the above discussion of Figure 1.11a.

Scale. Depending on Q, the model will describe the system on an appropriate scale. For example, depending on Q it can be appropriate to virtually follow a fluid particle on its way through the complex channels of a porous medium, or just to compute the pressure drop across a porous medium based on its permeability. Obviously, these cases correspond to a description of a porous medium on two scales (microscopic/macroscopic). Details of this example will follow in Section 4.10.2.

1.7.4
SQM Space Classification: M Axis

Finally, let us look at the categories on the M-axis of Figure 1.11b:

Linear – nonlinear. In linear models, the unknowns (or their derivatives) are combined using linear mathematical operations only, such as addition/subtraction or multiplication with parameters. Nonlinear models, on the other hand, may involve the multiplication of unknowns, the application of transcendental functions, and so on. Nonlinear models typically have more (and more interesting) solutions but are harder to solve. Examples are linear or nonlinear regression models (Sections 2.2 and 2.4, respectively) and linear or nonlinear ODEs (Section 3.5).

Analytical – numerical. In analytic models, the system behavior can be expressed in terms of mathematical formulas involving the system parameters. Based on these models, qualitative effects of parameters and the entire system behavior can be studied theoretically, without using concrete values for the parameters. Numerical models, on the other hand, can be used to obtain the system behavior for specific parameter values. See Section 3.6 for a general discussion of analytical models (which are also called *closed form models*) versus numerical models.

Autonomous – nonautonomous. This is a mathematical classification of insta-tionary models (see above). If an equation does not depend explicitly on time, it is called *autonomous*, otherwise nonautonomous; see the examples in Section 3.5.

Continuous – discrete. In continuous models, the independent variables may assume arbitrary (typically real) values within some interval. For example, many of the ODE models discussed in Chapter 3 use time (within some time interval) as the independent variable. In discrete models, on the other hand, the independent variables may assume some discrete values only. An example is the *discrete event simulation* technique discussed in Section 2.7.2, or the Nicholson–Bailey host–parasite interaction model discussed in Section 4.11.1, where the time variable just counts the number of breeding seasons instead of expressing the (continuous) physical time.

Difference equations. In difference equations, the quantity of interest is obtained as a sequence of discrete values. Usually, this is expressed in terms of recurrence relations in which each term of the sequence depends on previous terms. Differ-ence equations are frequently used to describe discrete systems. See the examples in Section 4.11.1.

Differential equations. Differential equations are equations involving derivatives of an unknown function. They are a main tool to set up continuous mechanistic models, see the examples in Chapters 3 and 4.

Integral equations. Integral equations are equations involving an integral of an unknown function.

Algebraic equations. AEs are equations involving the usual algebraic operations such as addition, subtraction, division, and so on. Examples are Equations (1.1) or (1.4) in Section 1.5, or the regression equations discussed in Chapter 2.

Note that some of the above categorizations of mathematical models overlap. For example, both phenomenological and mechanistic models can be lumped or distributed, stationary or instationary, and so on. Thus, it may have confused the reader if a single chapter would have been devoted to each of these categorizations. Instead, it was decided to select the categorization between phenomenological mod-els (Chapter 2) and mechanistic models (Chapters 3 and 4) as the main perspective and as a principle to organize the book. The other categorizations are treated within this perspective, that is, they will be referred to in the context of appropriate examples. Note that referring to Figure 1.11b we can say that the categorization of mathematical models between phenomenological and mechanistic models divides the *SQM* space of mathematical models into two different "half-spaces" along the Q-axis. We will repeatedly come back to the above classification of mathematical models in the course of this book, using it like a compass (or, in more up-to-date terminology: like a GPS system) so that the reader will always know about his actual position in the overall space of mathematical models.

1.8
Everything Looks Like a Nail?

To some extent, the modeling and simulation scheme discussed above is just an idealistic theory of how mathematical modeling *should* work, and this must of course be distinguished from the way in which people are dealing with mathematical models in practice. Being aware of this fact, Golomb [34] compiled the following:

Note 1.8.1 (Don'ts of Mathematical Modeling)
1. Don't believe that the model is the reality.
2. Don't extrapolate beyond the region of fit.
3. Don't distort reality to fit the model.
4. Don't retain a discredited model.
5. Don't fall in love with your model.

Don't No. 1 reminds us of the limitations of our models, that is, we should always be aware of the simplifying assumptions made in a model when discussing its implications for the real system. You may know the *cave allegory* of the Greek philosopher Plato, which provides a nice picture of the relationship between a model and the reality, Figure 1.12 [35]. In this allegory, prisoners are chained deep inside a cave in a way that restricts their view to one particular wall of the cave. Behind the prisoners, there is a big fire and some people who are using the light of that fire to project three-dimensional objects such as puppets, animals, and plants onto the cave wall. Plato assumes that the prisoners are chained in the cave since their childhood and thus have never seen anything else apart from the shadows on that cave wall. Thus, they believe that these shadows are the reality, although the

Fig. 1.12 Plato's cave allegory: Don't believe that the model is the reality! (Figure: B. Blüm, idea: *http://commons.wirimedia.org.*)

shadows are of course no more than simplified, two-dimensional models of the real, three-dimensional objects behind them. Very similarly, we must be aware of the fact that we are always "chained" in some way as long as we think about reality in terms of a scientific model, which restricts our view on the real system more or less depending on its inherent assumptions.

Don't No. 2 says that models should be used for prediction only in those regions of the parameter space where they are sufficiently supported by experimental data (see Section 2.2.2 and Note 2.2.3 for more details), while *Don'ts Nos 3–5* basically require us to abandon models that fail to pass the validation step of the modeling and simulation scheme (Note 1.2.3). In [11], the message of *Don't Nos 3–5* is expressed as follows:

> When you have a hammer, you look for a nail.
> When you have a *good* hammer, everything looks like a nail.

You understand the message: People always tend to solve problems similar to the way in which they successfully solved problems in the past. Yesterday, our problem might have been to drive a nail into a piece of wood, and we might have solved this problem adequately using a hammer. Today, however, we may have to drive a *screw* into a piece of wood, and it is of course not quite such a good idea to use the hammer again. Similarly, mathematical models are like tools that help us to solve problems, and we will always tend to reuse the models that helped us to solve our yesterday's problems. This is like a law of nature in mathematical modeling, similar to Newton's law of inertia; let us call it the *"law of inertia of mathematical modeling"*. Forces need to be applied to physical bodies to change their state of motion, and in a similar way forces need to be applied in a mathematical modeler's mind before he will eventually agree to replace established models by more adequate approaches. Even great scientists such as A. Einstein were affected by this kind of inertia. Einstein did not like the idea that the physical universe is probabilistic rather than deterministic (a consequence of the "Copenhagen interpretation" of quantum mechanics), and he expressed this aversion in his famous quote "God does not play dice with the universe" [36]. But do not take this as an excuse for any violation of Golomb's *Don't's*. It just shows that *everybody*, including yourself, should use models with care.

2

Phenomenological Models

Remember the distinction between phenomenological and mechanistic models in Definition 1.6.1: Phenomenological models are constructed based on experimental data only, using no a priori information about S. Mechanistic models, on the other hand, use a priori information about the "internal mechanics" of S, that is, about processes occurring in S. They are treated in Chapters 3 and 4. Thus, in this chapter, our starting point will be a dataset, and we will learn about methods to analyze this dataset. As we will see, most of the methods treated in this chapter can be efficiently implemented using freely available open source software: *Calc* for elementary statistical computations or as an elementary database, and a software package called *R* for professional statistical computations. Real datasets will be used throughout the chapter whenever possible.

As mentioned, the starting point of phenomenological modeling is a dataset, and hence a first natural thing to do is to analyse the dataset itself, for example, in terms of elementary statistical methods. Section 2.1 provides some of the most important statistical methods of elementary data analysis in the form of a "crash course", that is, with no attempt to be exhaustive, focusing on what is needed for this book, and emphasizing practical procedures rather than theory. Section 2.1 will also be used to introduce the reader to the use of *Calc* and *R*.

Sections 2.2–2.4 treat regression models. Basically, regression models provide a mathematical description of input–output systems. The importance of input–output systems has already been pointed out, cf. Figure 1.2 and Sections 1.3 and 1.5. Using regression models, the output of a system can be computed for a given input, which can be used for prediction or interpolation of given data. Linear regression (one input), multiple linear regression (several inputs), and nonlinear regression (nonlinear equations, one or several inputs) will be treated. Everyone concerned with data analysis should know these methods which are really easy to use based on *R*. Even in cases where mechanistic models are being developed, the application of regression methods frequently makes sense since they usually require much less time and resources and allow some quick and rough conclusions to be drawn from a dataset.

Section 2.5 treats neural network models, providing a look beyond classical regression models. Although an in-depth treatment of the various types of neural network models is beyond the scope of this book, Section 2.5 is intended to

Mathematical Modeling and Simulation: Introduction for Scientists and Engineers. Kai Velten
Copyright © 2009 WILEY-VCH Verlag GmbH & Co. KGaA, Weinheim
ISBN: 978-3-527-40758-8

introduce the reader to feedforward neural networks, which can be viewed as a generalized nonlinear regression method. One reason why feedforward neural networks are treated here is their "cost-effectiveness", that is, they are a very useful tool and at the same time easily implemented based on *R*. As we will see, the main advantage of feedforward neural networks compared to classical nonlinear regression is that they can be used without any a priori knowledge of the particular mathematical form of the nonlinearity.

After introducing some basic methods for an appropriate design of experiments in Section 2.6, the chapter ends in Section 2.7 with an overview of other phenomenological approaches that cannot be treated in detail within the scope of this book.

2.1
Elementary Statistics

In Section 1.3 above it was emphasized that a minimum requirement to be satisfied by a system that is investigated in science and engineering is *observability* in the sense that the system produces measurable output. As explained there, most systems do also accept some kind of input, and most investigations in science and engineering (except for more theoretically oriented work) thus begin with a compilation of an input–output dataset having the general form shown in Figure 1.2b. In this sense, it is valid to say that most modeling and simulation work starts with a dataset. Elementary statistical methods offer phenomenological modeling approaches that can be used for a first analysis of datasets.

2.1.1
Descriptive Statistics

The first thing that is usually done with a given dataset is *descriptive statistics*, that is, the application of methods that summarize and describe the data [19]. In many cases, datasets will be given in some spreadsheet format such as *Calc* (which is a part of the open source *OpenOffice* package, see Appendix A) or *Excel* (a part of the commercial *Microsoft Office*). Since the focus of this book is on open source software, we will exclusively refer to *Calc* in the following. This means no restriction for people who want to use *Excel* instead of *Calc*, since *Calc* and *Excel* work almost the same way from a standard user's perspective. Beyond this, *Calc* imports and exports *Excel* data without problems (problems may occur if you are using sophisticated features of *Excel* which a standard user will never see).

Spreadsheet programs such as *Calc* usually offer a number of options for a statistical analysis of the data. Although none of these programs can really compete with a professional and comprehensive statistical software such as *R*, it is frequently efficient to use the statistical facilities of spreadsheet programs. As explained above, you will obtain most of your datasets in a spreadsheet format, and a quick analysis of the data in this original format will be faster in many cases compared to an

Table 2.1 Spring data (see `spring.ods` in the book software).

x	10	20	30	40	50
y	3	5	11	12	16

analysis within *R*, which may involve several more steps beginning with an import of these data into *R* and so on.

As an example, let us start with the spring dataset in Table 2.1. This dataset has already been considered in Section 1.5.6 above, and it can be found in the file `spring.ods` in the "PhenMod/Stat" directory of the book software (see Appendix A). Note that the file extension "ods" ("*open document sheet*") is the standard extension of *Calc* spreadsheet files.

2.1.1.1 Using *Calc*

The simplest thing that one can do with a dataset such as `spring.ods` is to compute *measures of position*, which characterize the approximate location of the data in various ways. The most well known and most frequently used measure of position is the *arithmetic mean*, which is defined as follows:

$$\bar{x} = \frac{\sum_{i=1}^{n} x_i}{n} \tag{2.1}$$

Here, $x_1, x_2, \ldots, x_n$ is some given set of real numbers and $n \in \mathbb{N}$. For example, the arithmetic means of the x and y data in `spring.ods` are $\bar{x} = 30$ and $\bar{y} = 9.4$, respectively, which basically says that the x and y data spread around these values. Let us now see how this computation can be done in *Calc*. Although we cannot provide a general introduction into *Calc* here, we will try to provide enough information such that everything should be understandable even for first-time users (for more information you may refer to the documentation provided under *www.openoffice.org*).

Once you open `spring.ods` in *Calc*, you see a spreadsheet consisting of cells. Individual cells are labeled e.g. as A2 where A refers to the column and 2 refers to the line in which you find that cell. The numbers of the x column of `spring.ods` are in the group of cells A2,A3, ... ,A6, which is also denoted as A2:A6 in *Calc* notation. Now you can use a *Calc* function called AVERAGE to compute the arithmetic mean for this group of cells. To do this, enter the *formula* =AVERAGE(A2:A6) into an empty cell of `spring.ods` (note that every *Calc* formula begins with an "="). After performing this procedure, you will see a "30" in the cell where the formula was entered. To see and eventually edit the formula behind that number, you may use a double click on the cell containing the "30".

An alternative procedure would have been to select and use the AVERAGE function within *Calc*'s *Function Wizard* which you start using the menu option Insert/Function. The *Function Wizard* is particularly useful for inexperienced users since it asks the user for all necessary information and then automatically generates the formula. It also provides a list of the available formulas which are classified into categories such as "Financial", "Mathematical", and "Statistical", along with an explanation of what each particular formula is doing. Within the "Statistical" category, you will find a great number of other functions that can be used to compute alternative measures of position, such as the *median* (*Calc* function MEDIAN()), or the *geometric mean* (*Calc* function GEOMEAN()), see [19, 37] for details.

After computing measures of position, the next step usually is to look at *measures of variation* (or measures of *statistical dispersion*), which basically measure how widely spread the values in a dataset are. The most popular measure of variation is the *sample standard deviation*

$$s = \sqrt{\frac{\sum_{i=1}^{n}(x_i - \bar{x})^2}{n-1}} \tag{2.2}$$

which can be computed similar to above using the *Calc* function STDEV(). STDEV() yields $s \approx 15.8$ and $s \approx 5.3$ when applied to the x and y data of spring.ods, respectively. As Equation 2.2 shows, the sample standard deviation measures the variability of the data in terms of the deviations from the mean, $x_1 - \bar{x}, \ldots, x_n - \bar{x}$. Basically, the sample standard deviation expresses an average of these (squared) deviations. To understand the meaning of this expression a little more, let us look at what is meant by a "sample" here.

Statistical investigations typically focus on a well-defined collection of objects which constitute what is called a *population* [37]. For example, if an investigator wants to characterize the impact of a nutrient on a particular plant species, then his investigation will involve a population consisting of all plants of this species. In many cases, it will be impossible and inefficient to investigate the entire population due to limited time and resources (e.g. the plant species under investigation may cover most of the earth's surface). Statistical investigations will thus typically be restricted to a subset of a population which is called a *sample*. A number of strategies such as *random sampling* (each member of the population has an equal chance of being selected) or *stratified sampling* (which uses a division of the population into subgroups sharing the same characteristics such as gender or age) are used to make sure that the sample represents the entire population as good as possible [19, 37].

The sample standard deviation s refers to a sample $x_1, \ldots, x_n$. Later (in Section 2.1.2) we will understand that such a sample can be thought of as being generated by a random variable X. The variability of a random variable can also be characterized by a standard deviation σ, which expresses a property of the entire population (Section 2.1.2.6). Under certain assumptions discussed in [37], s can be shown to be a reasonable ("unbiased") estimate of σ, and in the same way the arithmetic mean $\bar{x}$ is a reasonable estimate of another property of the population, the expected value μ (see Section 2.1.2 again).

Other frequently used measures of variation include e.g. [37]

- the (sample) **range**, that is, the difference between the maximum and minimum values in the sample, which can be computed using *Calc*'s MAX() and MIN() functions;
- the (sample) **average deviation**, that is, the mean of the absolute deviations $|x_i - \bar{x}|$, which can be computed using *Calc*'s AVEDEV() function; and
- various dimensionless measures such as the (sample) **coefficient of variation** $c_v = s/\bar{x}$, which can be computed using *Calc*'s MEAN() and STDEV() functions as described above.

There are also a number of measures that can be used to characterize the interaction of statistical variables, such as Pearson's *sample correlation coefficient*

$$r = \frac{\sum_{i=1}^{n} x_i y_i - n \cdot \bar{x} \cdot \bar{y}}{\sqrt{\left(\sum_{i=1}^{n} x_i^2 - n \cdot \bar{x}^2\right) \cdot \left(\sum_{i=1}^{n} y_i^2 - n \cdot \bar{y}^2\right)}} \qquad (2.3)$$

which assumes a given sample $(x_1, y_1), \ldots, (x_n, y_n)$, and which expresses the strength of an assumed *linear* correlation of x and y on a scale between -1 and 1 (-1 or 1: the data match a descending or ascending straight line, 0: no interaction, anything in between: data scatter around descending or ascending straight lines depending on the sign of r). Again, r can be interpreted as an approximation of the correlation of random variables, see [37] for more details on that. *Calc*'s CORREL() function can be used to compute r, which yields a value of $r \approx 0.98$ when applied to spring.ods, reflecting the fact that the data in spring.ods almost match an ascending straight line (Figure 1.9).

As you know, "a picture is worth a thousand words". Applied to descriptive statistics, this can be phrased like this: a picture is worth a thousand numerical measures of position, variation, and so on, that is, you should use pictures and graphs to visualize your data whenever possible. *Calc* offers a great number of *graphical plotting options* that can be accessed via the menu "Insert/Chart". This option has, for example, been used to generate Figure 1.9 in Section 1.5.6.

2.1.1.2 Using the *R Commander*

Compared to *Calc*, *R* is a much more professional and much more comprehensive tool to perform a statistical analysis on the computer (see Appendix B). As mentioned above, *Calc*'s main advantage over *R* is that it can often be used very quickly since the analysis can be performed in a spreadsheet format, that is, in the original format of the data in many cases. However, if your intention is a thorough analysis of a dataset that involves statistical models or graphical capabilities beyond *Calc*'s scope, you will have to use *R*. For a beginner, an easy way of getting acquainted with *R* is the *R Commander*, a graphical user interface

(GUI) for *R*. Appendix B explains the way in which you start the *R Commander* within *CAELinux*.

Before starting the analysis, the data should be saved in the "csv" data format, for example, by using *Calc*'s "Save as" menu option. The resulting ".csv" file can then be imported via the *R Commander*'s "Data/Import data/From text file" option (be careful to choose "," as the field delimiter when saving or importing csv files). After this, the same descriptive analysis as in the last section can be performed by using the *R Commander* menu options. For example, the arithmetic mean and the standard deviation can be obtained using the "Statistics/Summaries/Numerical summaries" menu option, graphs can be created via the "Graphs" menu option and so on. ("Graphs/Scatterplot" creates scatterplots similar to Figure 1.9 in Section 1.5.6).

If the standard formatting of the graphs produced by the *R Commander* does not meet your requirements, you can generate a great variety of possible formats based on *R* programs (see Appendix B for details on using *R* programs). Look through the various *R* programs in the book software (Appendix A) to see how this can be done, or consult the literature that is recommended in Appendix B. As an example, consider the program `HeatClos.r` that you find in the MechPDE directory of the book software. This program generates a plot using *R*'s `plot` command, and the plot involves, for example, a nonstandard font size and a nonstandard line width. Looking into the `plot` command in `HeatClos.r`, you will see that the font size can be adjusted using a `par` command which is issued immediately before the `plot` command, and the line width can be set using the `lwd` option of the `plot` command.

The *R* commander provides a *script window* that can be used to facilitate your first steps in *R* programming. Everything you do in the *R Commander* is translated into appropriate *R* code within the script window. For example, after producing a scatterplot using the *R Commander*'s "Graphs/Scatterplot" menu option, you will find a `scatterplot(...)` command in the script window that corresponds exactly to all the choices that you have made in the "Graphs/Scatterplot" window. If you then copy and paste the content of the script window into a text file and save that file with extension ".r", the resulting program can be executed as described in Appendix B, and it will generate exactly the result that you have produced before using the *R Commander*. This *R* program can then be edited and optimized, for example, by using formatting commands as described above.

2.1.2
Random Processes and Probability

Suppose you are interested in some quantity which we denote by X, and which may be temperature, the concentration of some substance, and so on. You will usually need to have precise measurements of that quantity, so let us assume that you have a new measurement device and want to know about the measurement errors produced by that device. Then, a standard procedure is to repeatedly measure that quantity in a situation where the correct result is known (e.g. by using standardized solutions if X is the concentration of some substance). Assuming that the true value of the quantity of interest is 20, the data produced in this way may look like this:

```
20.13443 19.83828 20.01702 19.99835 19.94526 20.01415 19.96707
```

What we see here is that the measurement values oscillate in a random way around the true value. Most measurement devices produce random errors of this kind, which is no problem as long as the amplitude of these oscillations is small enough. Now a natural question regarding the above data is this: what is the probability with which the deviations of the measurement value from the true value will be less than some specified value such as 0.1? In this section, methods will be developed that can be used to answer this kind of questions.

2.1.2.1 Random Variables

In statistical terms, we would say that the above data have been generated by the *random variable X*, where [19]

Definition 2.1.1 (Random variable) A random variable is a variable that has a single numerical value, determined by chance, for each outcome of a procedure.

Everyone of us is concerned with an abundant number of random processes and random variables in this sense, not only as a scientist or engineer. Perhaps the most classical example is the random variable

X_1 : result of a dice

but you may also think of

X_2: waiting time at a bus stop if you arrive there without knowing the time table

and many other examples.

2.1.2.2 Probability

Let us ask for the probability with which a random variable attains certain values. This is an easy thing if one is concerned with simple systems such as a dice. Everyone of us knows that the probability of getting a "3" in a dice play is 1/6 or 16.7%. In statistics, this is usually written as

$$P(X_1 = 3) = \frac{1}{6} \tag{2.4}$$

where P is a "probability function" that yields the probability of the "event" $X_1 = 3$ as a number between 0 and 1. To make this precise, let us define [19]

Definition 2.1.2 (Events and sample space)
• An *event* is any collection of results or outcomes of a procedure.

- A *simple event* is an outcome or an event that cannot be further broken down into simpler components.
- The *sample space* for a procedure consists of all possible *simple events*.

In the dice example, the sample space would be

$$S = \{1, 2, 3, 4, 5, 6\} \tag{2.5}$$

and all subsets $A \subset S$ such as $A_1 = \{1, 2\}$ ("dice result is below 3") or $A_2 = \{1, 3, 5\}$ ("dice result is an odd number") would be events in the sense of the above definition. Examples of simple events would be $A_3 = \{2\}$, $A_4 = \{5\}$, and so on. In the bus-waiting-time example, the sample space would be

$$S = \{x \in \mathbb{R} | 0 \leq x < 15\} \tag{2.6}$$

if we assume that the buses arrive in 15-min intervals, and a possible event would be $[0, 2[$ ("the waiting time is below 2 min").

The probability function P is usually defined based on axioms [37]. A less formal definition, which is sufficient for our purposes, can be given as follows [37]

Definition 2.1.3 (Probability)
Given a sample space S, the *probability function* P assigns to each event $A \subset S$ a number $P(A) \in [0, 1]$, called the *probability* of the event A, which will give a precise measure of the chance that A will occur.

Above it was said that the probability to get "3" as a dice result is 1/6. This is based on the following formula (a consequence of the probability axioms [19, 37]):

Proposition 2.1.1 (Classical approach to probability)
Assume that a given procedure has n different simple events and that each of those simple events has an equal chance of occurring. If event A can occur in s of these n ways, then

$$P(A) = \frac{s}{n} \tag{2.7}$$

This formula works well for the dice and many other similar *discrete random variables* that involve a finite number of equally likely possible results (note that discrete random variables may also involve countable infinitely many possible results, see [19]). It does not work, however, for *continuous random variables* with an infinite number of possible results similar to the random variable X_2 discussed above that describes the bus waiting time. Note that the sample space

$S = \{x \in \mathbb{R} | 0 \leq x < 15\}$ of this example indeed involves an infinite number of continuously distributed possible results between 0 and 15 min. In this case, the following formula can be used [19]:

Proposition 2.1.2 (Relative frequency approximation) Assume that a given procedure is repeated n times, and let $f_n(A)$ denote the relative frequency with which an event A occurs. Then,

$$P(A) = \lim_{n \to \infty} f_n(A) \tag{2.8}$$

This means that if we, for example, want to approximate the probability of bus waiting times between 0 and 2 min (i.e. the probability of $A = [0, 2[)$, the following approximation can be used

$$P(A) \approx f_n(A) \tag{2.9}$$

and the quality of this approximation will increase as n is increased.

2.1.2.3 Densities and Distributions

There is another important approach that can be used to compute probabilities, which is based on an observation that can be made if a random process is repeated a great number of times. Let X be a continuous random variable with sample space $S \subset \mathbb{R}$, and let us consider two disjunct events $[a_1, b_1], [a_2, b_2] \subset S$, that is, $[a_1, b_1] \cap [a_2, b_2] = \emptyset$. Then, suppose that X has been observed $n \in \mathbb{N}$ times, and that the same number of observations has been made within $[a_1, b_1]$ and $[a_2, b_2]$. Now assume that $b_2 - a_2 > b_1 - a_1$. Then, we can say that observations near the interval $[a_1, b_1]$ are more likely compared to observations near the interval $[a_2, b_2]$, since the same number of observations was made in each of the two intervals although $[a_1, b_1]$ is smaller. If m is the number of observations made in each of the intervals, then this difference can be made precise as follows:

$$\frac{m/n}{b_1 - a_1} > \frac{m/n}{b_2 - a_2} \tag{2.10}$$

In this equation, m/n approximates the probability of either of the two events $[a_1, b_1]$ and $[a_2, b_2]$ in the sense of Proposition 2.1.2. These probabilities are the same for both events, but a difference is obtained if they are divided by the respective sizes of the two intervals, which leads to a quantity known as *probability density*. Basically, you can expect more observations within intervals of a given size in regions of the sample space having a high probability density.

Figure 2.1 shows what happens with the probability density if a random experiment (which is based on a "normally distributed" random variable in this case, see below) is repeated a great number of times. The figure is a result of the code RNumbers.r which you find in the book software (Appendix A), and which can be used to simulate a random experiment (more details on this code will follow

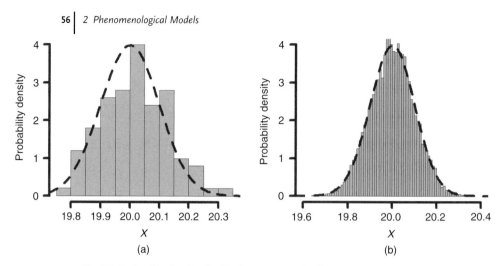

Fig. 2.1 Probability density distributions computed using RNumbers.r with (a) $n = 100$ and (b) $n = 10\,000$.

further below). Figure 2.1a and b shows the distribution of the probability density as a histogram for two cases where the random experiment is repeated (i) 100 and (ii) 10 000 times. Basically, what can be seen here is that as n is increased, the probability density approaches the "bell-shaped" function that is indicated by the dashed line in the figure, and that can be expressed as:

$$f(x) = \frac{1}{\sigma \sqrt{2\pi}} e^{-\frac{1}{2} \left(\frac{x - \mu}{\sigma} \right)^2} \tag{2.11}$$

with $\mu = 20$ and $\sigma = 0.1$ in this case. This function is an example of a *probability density function*. Probability density functions characterize the random behavior of a random variable, that is, the behavior of a random variable can be predicted once we know its probability density function. For example, given a probability densitiy function f of a random variable X with sample space $S \subset \mathbb{R}$, the probability of an event $[a, b] \subset S$ can be computed by the following integral [37]:

$$P(a \leq X \leq b) = \int_a^b f(t)\, dt \tag{2.12}$$

Alternatively, the function

$$F(x) = P(X \leq x) = \int_{-\infty}^x f(t)\, dt \tag{2.13}$$

is also often used to characterize the behavior of a random variable. It is called the *probability distribution* of the random variable. Basically, probability density functions or probability distributions provide a compact way to describe the behavior of a random variable. The probability density function in Equation 2.11, for example, describes the behavior of the random variable based on only two parameters, μ and σ (more details on this distribution and its parameters will follow below).

Probability density functions always satisfy $f(t) \geq 0$ and, if $S = \mathbb{R}$,

$$\int_{-\infty}^{\infty} f(t)\, dt = 1 \tag{2.14}$$

that is, the area under the probability density function is always 1 (note that otherwise Equation 2.12 would make no sense).

2.1.2.4 The Uniform Distribution
The probability density function of the bus waiting time (see above) is

$$f(x) = \begin{cases} \dfrac{1}{15} & \text{if } x \in [0, 15[\\ 0 & \text{otherwise} \end{cases} \tag{2.15}$$

that is, in this case the probability density is constant. As discussed above, this expresses the fact that the same number of observations can be expected within any interval of a given length within $[0, 15]$. This makes sense in the bus-waiting-time example since each particular waiting time between 0 and 15 min is equally likely. Using Equations 2.12 and 2.15, we can, for example, compute the probability of waiting times between 3 and 5 min as follows:

$$P(3 \leq X_2 \leq 5) = \int_3^5 f(t)\, dt = \frac{1}{15} \cdot (5 - 3) = \frac{2}{15} \tag{2.16}$$

Equation 2.15 is the probability density function of the *uniform distribution*, which is written generally as

$$f(x) = \begin{cases} \dfrac{1}{b - a} & \text{if } x \in [a, b[\\ 0 & \text{otherwise} \end{cases} \tag{2.17}$$

It is easy to show that the area under this probability density function is 1 as required by Equation 2.14.

2.1.2.5 The Normal Distribution
Random variables that can be described by the probability density function in Equation 2.11 are said to have the *normal distribution*, which is also known as the *Gaussian distribution* since it was discovered by C.F. Gauss. In a sense, one can say that the normal distribution is called *normal* since it is normal for random processes to be normally distributed. . . A great number of random processes in science and engineering can be described using this distribution. This can be theoretically justified based on the *central limit theorem*, which states that the distribution of a sum of a large number of independent and identically distributed random variables can be approximated by the normal distribution (see [37] for details).

Let us use the notation $X \sim N(\mu, \sigma)$ for a random variable X that is normally distributed with parameters μ and σ (see Equation 2.11; more on these parameters will follow in the next section). Remember that a measurement device was discussed at the beginning of this section, and that we were asking the following question: with what probability will the deviation of the measurement value from the true value be smaller than some specified value such as 0.1? Assuming that the true value of the measured quantity is 20, and assuming normally distributed measurement errors (which is typically true due to the central limit theorem), this question can now be answered using Equations 2.11 and 2.12 as follows:

$$P(19.9 \leq X \leq 20.1) = \int_{19.9}^{20.1} \frac{1}{\sigma\sqrt{2\pi}} e^{-\frac{1}{2}\left(\frac{x-\mu}{\sigma}\right)^2} dt \qquad (2.18)$$

Unfortunately, this integral cannot be solved in closed form, which means that numerical methods must be applied to get the result (see Section 3.6.2 for a general discussion of closed form versus numerical solutions). The simplest way to compute probabilities of this kind numerically is to use spreadsheet programs such as *Calc*. *Calc* offers a function NORMDIST that can be used to compute values either of the probability density function or of the distribution function of the normal distribution. If $F(x)$ is the distribution function of the normal distribution (compare Equation 2.13), the above probability can be expressed as

$$P(19.9 \leq X \leq 20.1) = F(20.1) - F(19.9) \qquad (2.19)$$

which can be obtained using *Calc* as follows:

$$P(19.9 \leq X \leq 20.1) = \text{NORMDIST}(20.1; \mu; \sigma; 1) - \text{NORMDIST}(19.9; \mu; \sigma; 1)$$
$$(2.20)$$

For example, $\mu = 20$ and $\sigma = 0.1$ yield $P(19.9 \leq X \leq 20.1) \approx 68, 3\%$.

We can now explain the background of the code RNumbers.r that was used above to motivate probability density functions. This code simulates a normally distributed random variable based on R's rnorm command. The essential part of this code is the line

```
out=rnorm(n,mu,sigma)
```

where rnorm is invoked with the parameters n (number of random numbers to be generated), mu and sigma (parameters μ and σ of Equation 2.11). R's hist and curve commands are used in RNumbers.r to generate the histogram and the dashed curve in Figure 2.1, respectively (see the code for details).

2.1.2.6 Expected Value and Standard Deviation

Now it is time to understand the meaning of the parameters of the normal distribution, μ and σ. Let us go back to the dice example, and let X_1 be the random variable expressing the result of the dice as before. Suppose two experiments are performed:

Experiment 1
The dice is played five times. The result is: 5, 6, 4, 6, 5.

Experiment 2
The dice is played 10 000 times.

Analyzing Experiment 1 using the methods described in Section 2.1.1, you will find that the average value is $\bar{x} = 5.2$ and the standard deviation is $s \approx 0.84$. Without knowing the exact numbers produced by Experiment 2, it is clear that the average value and the standard deviation in Experiment 2 will be different from $\bar{x} = 5.2$ and $s \approx 0.84$. The relatively high numbers in Experiment 1, and hence, the relatively high average value of $\bar{x} = 5.2$ has been obtained by chance only, and it is clear that such an average value cannot be obtained in Experiment 2. If you get a series of relatively high numbers such as the numbers produced in Experiment 1 as a part of the observations in Experiment 2, then it is highly likely that this will be balanced by a corresponding series of relatively small numbers (assuming a fair dice, of course). As the sample size increases, the average value of a sample stabilizes toward a value that is known as the *expected value* of a random variable X which is usually denoted as $E(X)$ or μ, and which can be computed as

$$\mu = E(X) = \frac{1}{6} \cdot 1 + \frac{1}{6} \cdot 2 + \frac{1}{6} \cdot 3 + \frac{1}{6} \cdot 4 + \frac{1}{6} \cdot 5 + \frac{1}{6} \cdot 6 = 3.5 \qquad (2.21)$$

in the case of the dice example. This means that we can expect $\bar{x} \approx 3.5$ in Experiment 2. The last formula can be generalized to

$$\mu = E(X) = \sum_{i=1}^{n} p_i x_i \qquad (2.22)$$

if X is a discrete random variable with possible values $x_1, \ldots, x_n$ having probabilities $p_1, \ldots, p_n$. Analogously, the (sample) standard deviation of a sample stabilizes toward a value that is known as the *standard deviation* of a random variable X which is usually denoted as σ, and which can be computed as

$$\sigma = \sqrt{\sum_{i=1}^{n} p_i (x_i - \mu)^2} \qquad (2.23)$$

for a discrete random variable. This formula yields $\sigma \approx 2.92$ in the dice example, that is, we can expect $s \approx 2.92$ in Experiment 2. For a continuous random variable with probability density function f, the expected value and the standard deviation can be expressed as follows [37]:

$$\mu = \int_{-\infty}^{\infty} t \cdot f(t) \, dt \tag{2.24}$$

$$\sigma = \sqrt{\int_{-\infty}^{\infty} (t - \mu)^2 \cdot f(t) \, dt} \tag{2.25}$$

As suggested by the notation of the parameters μ and σ of the normal distribution, it can be shown that these parameters indeed express the expected value and the standard deviation of a random variable that is distributed according to Equation 2.11.

2.1.2.7 More on Distributions

Beyond the uniform and normal distributions discussed above, there is a great number of distribution functions that cannot be discussed in detail here, such as *Student's t-distribution* (which is used e.g. to estimate means of normally distributed variables) or the *gamma distribution* (which is used e.g. to describe service times in queuing theory [38]). Note also that all distribution functions considered so far were referring to continuous random variables. Of course, the same concept can also be used for discrete random variables, which leads to *discrete distributions*. An important example is the *binomial distribution*, which refers to *binomial* random processes which have only two possible results. Let us denote these two results by 0 and 1. Then, if p is the (fixed) probability of getting a 1 and the experiment is repeated n times, the distribution function can be written as

$$F(x) = P(X \leq x) = \sum_{j=0}^{floor(x)} \binom{n}{j} p^j (1 - p)^{n-j} \tag{2.26}$$

where *floor(x)* returns the highest integer less than or equal to x. See [19, 37] for more details on the binomial distribution and on other discrete distributions.

2.1.3
Inferential Statistics

While the methods of descriptive statistics are used to describe data, the methods of *inferential statistics*, on the other hand, are used to draw inferences from data (it is as simple as that ...). This is a big topic. Within the scope of this book, we will have to confine ourselves to a treatment of some basic ideas of statistical testing that are required in the following chapters. Beyond this, inferential statistics is e.g. concerned with the estimation of population parameters such as the estimation of the expected value from data, see [19, 37] for more on that.

2.1.3.1 Is Crop A's Yield Really Higher?

Suppose the following yields of crops A and B have been measured (in g):

Crop A: 715, 683, 664, 659, 660, 762, 720, 715
Crop B: 684, 655, 657, 531, 638, 601, 611, 651

These data are in the file `crop.csv` in the book software (Appendix A). The average yield is 697.25 for crop A and 628.5 for crop B, and one may therefore be tempted to say that crop A yields more than crop B. But we need to be careful: can we be sure that crop A's yield is really higher, or is it possible that the difference in the average yields is just a random effect that may be the other way round in our next experiment? And if the data indeed give us a good reason to believe that crop A's yield is higher, can the certainty of such an assertion be quantified? Questions of this kind can be answered by the method of statistical hypothesis testing.

2.1.3.2 Structure of a Hypothesis Test

Statistical hypothesis tests that are performed using software (we do not discuss the traditional methods here, see [19, 37] for that) usually are conducted along the following steps:

- Select the hypothesis to be tested: the *null hypothesis*, often abbreviated as H_0.
- Depending on the test that is performed, you may also have to select an *alternative hypothesis*, which is assumed to hold true if the null hypothesis is rejected as a result of the test. Let H_1 be this alternative hypothesis, or let H_1 be the negation of H_0 if no alternative hypothesis has been specified.
- Select the *significance level* α, which is the probability to erroneously reject a true H_0 as a result of the test. Make α small if the consequences of rejecting a true H_0 are severe. Typical choices are $\alpha = 0.1$, $\alpha = 0.05$, or $\alpha = 0.01$.
- Collect appropriate data and then use the computer to perform an appropriate test using the data. As a result of the test, you will obtain a *p value* (see below).
- If $p < \alpha$, reject H_0. In this case, H_1 is assumed to hold true, and H_1 as well as the test itself are said to be *statistically significant at the level α*.

Note that in the case of a nonsignificant test ($p \geq \alpha$), nothing can be derived from the test. In particular – and you need to be careful regarding this point – the fact that we do not reject H_0 in this case does *not* mean that it has been proved by the test that H_0 is true. The *p* value can be defined as follows [37]:

> **Definition 2.1.4 (*P* value)**
> The *p* value (or observed significance level) is the smallest level of significance at which H_0 would be rejected when a specified test procedure is used on a given dataset.

In view of the above testing procedure, this definition may seem somewhat tautological, so you should note that the "test procedure" in the definition refers to the mathematical details of the testing procedure that cannot be discussed here, see [19, 37].

A hypothesis test may involve two main types of errors: a *type I error*, where a true null hypothesis is rejected, and a *type II error*, which is the error of failing to reject a null hypothesis in a situation where the alternative hypothesis is true. As mentioned above, α is the probability of a type I error, while the probability of a type II error is usually denoted with β. The inverse probability $1 - \beta$, that is, the probability of rejecting a false null hypothesis is called the *power* of the test [19, 37].

2.1.3.3 The *t* test

Coming back to the problem discussed in Section 2.1.3.1 above, let X_1 and X_2 denote the random variables that have generated the data of crop A and crop B, respectively, and let μ_1 and μ_2 denote the (unknown) expected values of these random variables. Referring to the general test structure explained in the last section, let us define the data of a statistical test as follows:

- $H_0: \mu_1 = \mu_2$
- $H_1: \mu_1 > \mu_2$
- $\alpha = 0.05$

Now a *t test* can be used to get an appropriate *p* value [37]. This test can be performed using the program TTest.r in the book software (Appendix A). If you run this program as described in Appendix B, it will produce a few lines of text in which you will read "*p* value $= 0.00319$". Since this is smaller compared to the significance level α assumed above, H_0 is rejected in favor of H_1, and hence the test shows what is usually phrased as follows: "The yield of crop A is statistically significantly higher (at the 5% level) compared to crop B." Note that this analysis assumes normally distributed random variables, see [19, 37] for more details.

The main command in TTest.r that does the computation is t.test, which is used here as follows:

```
t.test(Dataset$x, Dataset$y
       ,alternative="greater",paired=FALSE)
```

Dataset$x and Dataset$y are the data of crop A and crop B, respectively, which TTest.r reads from crop.csv using the read.table command (see TTest.r for details). alternative can be set to alternative="less" if $H_1 : \mu_1 < \mu_2$ is used, and to alternative="two.sided" in the case of $H_1 : \mu_1 \neq \mu_2$. For obvious reasons, t tests using $H_1 : \mu_1 < \mu_2$ or $H_1 : \mu_1 > \mu_2$ are also called *one-sided t tests*, whereas t tests using $H_1 : \mu_1 \neq \mu_2$ are called *two-sided t tests*. paired must be set to true if each of the x values has a unique relationship with one of the y values, for example, if x is the yield of a fruit tree in year 1 and y is the yield of the same tree in year 2. This is a *paired t test*, whereas the above crop yield example – which involves no unique relationships between the data of crop A and crop B – is called an *independent t test* [37]. R's t.test command can also be used to perform *one-sample t tests* where the expected value of a single sample (e.g. the concentration of an air pollutant) is compared with a single value (e.g. a threshold value for that air pollutant). Note that t tests can also be accessed using the "Statistics/Means" menu option in the *R Commander*.

2.1.3.4 Testing Regression Parameters

A detailed treatment of linear regression will follow below in Section 2.2. At this point, we just want to explain a statistical test that is related with linear regression. As we will see below, linear regression involves the estimation of a straight line $y = ax + b$ or of a hyperplane $y = a_0 + a_1x_2 + \cdots + a_nx_n$ from data. Let us focus on the one-dimensional case, $y = ax + b$ (everything is completely analogous in higher dimensions). In the applications, it is often important to know whether a variable y depends on another variable x. In terms of the model $y = ax + b$, the question is whether $a \neq 0$ (i.e. y depends on x) or $a = 0$ (i.e. y does not depend on x). To answer this question, we can set up a statistical hypothesis test as follows:

- $H_0: a = 0$
- $H_1: a \neq 0$
- $\alpha = 0.05$

For this test, the line labeled with an "x" in the regression output in Figure 2.2a reports a p value of $p = 0.00318$. This is smaller than $\alpha = 0.05$, and hence we can say that "a is statistically significantly different from zero (at the 5% level)", which means that y depends statistically significantly on x. In a similar way, the p value $p = 0.71713$ in the line labeled with "Intercept" in the regression output in Figure 2.2a refers to a test of the null hypothesis $H_0 : b = 0$, that is, this p value can be used to decide whether the intercept of the regression line (i.e. the y value for $x = 0$) is significantly different from zero. Note that this analysis assumes normally distributed random variables, and note also that a and b have been used above to denote the random variables that are generating the slope and the intercept of the regression line if the regression procedure described in Section 2.2 is performed based on sample data.

2.1.3.5 Analysis of Variance

The regression test discussed in the previous section can be used to decide about the dependence of y on x in a situation where x is expressed in terms of numbers, which is often phrased like this: "x is at the *ratio level of measurement*" [19]. If x is expressed in terms of names, labels, or categories (the *nominal level of measurement*), the same question can be answered using the *analysis of variance*, which is often abbreviated as *anova*. As an example, suppose that we want to investigate whether fungicides have an impact on the density of fungal spores on plants. To answer this question, three experiments with fungicides A, B, and C and a control experiment with no treatment are performed. Then, these experiments involve what is called a *factor* x which has the *factor levels* "Fungicide A", "Fungicide B", "Fungicide C", and "No Fungicide". At each of these factor levels, the experiment must be repeated a number of times such that the expected values of the respective fungal spore densities are sufficiently characterized.

The results of such an experiment can be found in the file `fungicide.csv` in the book software (see Appendix A). Note that the "Factor" column of this file corresponds to x, while the "Value" column corresponds to y (it reports the result of the measurement, i.e. the density of the fungal spores on the plants in an appropriate unit that we do not need to discuss here). Let X_1, X_2, X_3, and X_4 denote the random variables that have generated these data, and let μ_1, μ_2, μ_3, and μ_4 denote the expected values of these random variables. Then, we can set up a hypothesis test as follows:

- H_0: $\mu_1 = \mu_2 = \mu_3 = \mu_4$
- H_1: There are $i, j \in \{1, 2, 3, 4\}$ s.t. $\mu_i \neq \mu_j$
- $\alpha = 0.05$

Basically, H_0 says that the factor x does not have any impact on the fungal spore density y, while the alternative hypothesis H_1 is the negation of H_0. An appropriate p value for this test can now be computed using the R program `Anova.r` in the book software, which is based on R's `anova` command. If you run this program as described in Appendix B, it will produce a few lines of text in which you read "Pr($>$F) = 0.000376", which means $p = 0.000376$. Again, the test is significant since we have $p < \alpha$. Hence, H_0 can be rejected and we can say that the factor "fungicide" has a statistically significant impact on the fungal spore density (again, at the 5% level).

Note that this analysis assumes random variables that are normally distributed and which have homogeneous variances (i.e. squared standard deviations), see [19, 37] for more details. The above example is called a *one-way analysis of variance* or *single-factor analysis of variance* since it involves one factor x only. R's `anova` command and the `Anova.r` code can also be applied to situations with several factors $x_1, \ldots, x_n$, which is called a *multiway analysis of variance* or *multifactor analysis of variance*. Note that when you perform a multiway analysis of variance using `Anova.r`, you will have to use a data file which provides one column for each of the factors, and one more column for the measurement value. What we have described so far is also known as the *fixed-effects model* of the analysis of variance. Within the general scope of the analysis of variance, a great number of different modeling approaches can be used, for example, *random effects models* which assume a hierarchy of different populations whose differences are constrained by the hierarchy [39].

2.2
Linear Regression

Generally speaking, regression models involve the analysis of a dependent variable in terms of one or several independent variables. In regression, the dependent variable is expressed in terms of the independent variables using various types of regression equations. Parameters in the regression equations are then tuned in a way that fits these equations to data. The idea of the regression method has already been explained based on the spring data `spring.ods` in Section 1.5. As it was discussed there referring to the black box input–output system in Figure 1.8, one can say that regression is a really prototypical method among the existing phenomenological modeling approaches. It contains all the essential ingredients: an input x, an output y, a black box–type system transforming x into y, and the attempt to find a purely data-based mathematical description of the relation between x and y. The term *regression* itself is due to a particular regression study that was performed by Francis Galton who investigated human height data. He found that, independent of their parents' heights, the height of children tends to *regress* toward the typical mean height [40].

2.2.1
The Linear Regression Problem

Assume we have a dataset $(x_1, y_1), (x_2, y_2), \ldots, (x_m, y_m)$ $(x_i, y_i \in \mathbb{R}, \; i = 1, \ldots, m, \; m \in \mathbb{N})$. Then, the simplest thing one can do is to describe the data using a *regression function* or *model function* of the form

$$\hat{y}(x) = ax + b \qquad (2.27)$$

The coefficients a and b in Equation 2.27 are called the *regression coefficients* or the *parameters* of the regression model. x is usually called the *explanatory variable* (or *predictor variable*, or *independent variable*), while $\hat{y}$ (or y) is called the *response variable* (or *dependent variable*). Note that the hat notation is used to distinguish between measured values of the dependent variable (written without hat as y_i) and values of the dependent variable computed using a regression function (which are written in hat notation, see Equation 2.27). A function $\hat{y}(x)$ as in Equation 2.27 is called a *linear regression function* since this function depends linearly on the regression coefficients, a and b [41].

> **Note 2.2.1 (Linear regression)** A general one-dimensional linear regression function $\hat{y}(x)$ computes the response variable y based on the explanatory variable x and regression coefficients $a_0, a_1, \ldots, a_s$ $(s \in \mathbb{N})$. If the expression $\hat{y}(x)$ depends linearly on $a_0, a_1, \ldots, a_s$, it can be fitted to measurement data using linear regression. Higher-dimensional linear regression functions involving multiple explanatory variables will be treated in Section 2.3, the nonlinear case in Sections 2.4 and 2.5.

Equation 2.27 fits the data well if the differences $y_i - \hat{y}(x_i)$ $(i = 1, \ldots, m)$ are small. To achieve this, let us define

$$RSQ = \sum_{i=1}^{m} \left(y_i - \hat{y}(x_i) \right)^2 \tag{2.28}$$

This expression is called the *residual sum of squares* (*RSQ*). Note that *RSQ* measures the distance between the data and the model: if *RSQ* is small, the differences $y_i - \hat{y}(x_i)$ will be small, and if *RSQ* is large, at least some of the differences $y_i - \hat{y}(x_i)$ will be large. This means that to achieve a small distance between the data and the model, we need to make *RSQ* small. In regression, this is achieved by an appropriate tuning of the parameters of the model. Precisely, the parameters a and b are required to solve the following problem:

$$\min_{a,b \in \mathbb{R}} RSQ \tag{2.29}$$

Note that *RSQ* depends on a and b via $\hat{y}$. The solution of this problem can be obtained by an application of the usual procedure for the minimization of a function of several variables to the function $RSQ(a, b)$ (setting the partial derivatives of this function with respect to a and b to zero etc.) [17], which gives [19]

$$a = \frac{\sum_{i=1}^{m} x_i y_i - m\bar{x}\,\bar{y}}{\sum_{i=1}^{m} x_i^2 - m\bar{x}^2} \tag{2.30}$$

$$b = \bar{y} - a\bar{x} \tag{2.31}$$

The use of *RSQ* as a measure of the distance between the model and the data may seem somewhat arbitrary since several other alternative expressions could be used here (e.g. *RSQ* could be replaced by a sum of the absolute differences $|y_i - \hat{y}(x_i)|$). *RSQ* is used here since it leads to *maximum likelihood estimates* of the model parameters, a and b, if one makes certain assumptions on the statistical distribution of the error terms, $y_i - \hat{y}(x_i)$ (see below). Using these assumptions, the maximum likelihood estimates of the model parameters derived from minimizing *RSQ* make the data "more likely" compared to other choices of the parameter values [41].

2.2.2
Solution Using Software

Now let us see how this analysis can be performed using software. We refer to the data in spring.csv as an example again (see Section 1.5.6). Start the *R Commander* as described in Appendix B an then import the data spring.csv into the *R Commander* using the menu option "Data/Import data/From text file". Then choose the menu option "Statistics/Fit models/Linear regression" and select x and y as the explanatory and response variables of the model, respectively. This gives the result shown in Figure 2.2a.

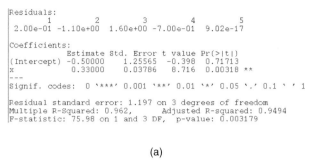

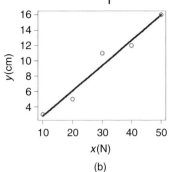

(a)   (b)

Fig. 2.2 (a) Linear regression result obtained using the R Commander and the data in spring.csv. (b) Comparison of the regression line Equation 2.32 with the data spring.csv. Figure produced using LinRegEx1.r.

The R output shown in Figure 2.2a first reports the *residuals* between the data and the model, that is, the differences $y_i - \hat{y}(x_i)$ $(i = 1, \ldots, m)$, which yields 5 values in this case since spring.csv contains 5 lines of data. You can get an idea about the quality of the fit between data and model based on these values, but it is of course better to see this in a plot (see below). After this, R reports on the regression coefficients, in a table comprising two lines, the first one (labeled Intercept) referring to the coefficient b in Equation 2.27 and the second one (labeled x) referring to the coefficient a in the same equation. The labels used in this table are justified by the fact that b describes the intercept of the line given by Equation 2.27, that is, the position where this line crosses the y-axis, while a is the coefficient that multiplies the x in Equation 2.27. In the "Estimate" column of Figure 2.2 you see that $a = -0.5$ and $b = 0.33$ have been obtained as the solution of Problem (2.29). Using Equation 2.27, this means that we have obtained the following regression line:

$$\hat{y}(x) = 0.33x - 0.5 \tag{2.32}$$

Figure 2.2b compares the regression line, Equation 2.32, with the data in spring.csv. This figure has been generated using the R program LinRegEx1.r in the book software. A similar figure can be produced using the R Commander based on the menu option "Graphs/Scatterplot", but you should note that the R Commander offers a limited number of graphical options only. Unlimited graphical options (e.g. to change line thicknesses and colors) can be accessed if you are using R programs such as LinRegEx1.r. You will find a few comments on the content of LinRegEx1.r further below in this section.

Note 2.2.2 (Regression coefficients) Estimates of the regression coefficients in regression equations such as Equation 2.27 can be obtained using formulas such as Equations 2.30 and 2.31, the R-Commander, or R programs such as `LinRegEx1.r`. Note that no formulas are available for general nonlinear regression equations, as discussed in Section 2.4.

As Figure 2.2b shows, the regression line captures the tendency in the data. It is thus reasonable to use the regression line for prediction, extrapolating the tendency in the data using the regression line. Formally, this is done by inserting x values into Equation 2.32. For example, to predict y for $x = 60$, we would compute as follows:

$$\hat{y}(60) = 0.33 \cdot 60 - 0.5 = 19.3 \tag{2.33}$$

Looking at the data in Figure 2.2b, you see that $\hat{y}(60) = 19.3$ indeed is a reasonable extrapolation of the data. Of course, predictions of this kind can be expected to be useful only if the model fits the data sufficiently well, and if the predictions are computed "close to the data". For example, our results would be questionable if we used Equation 2.32 to predict y for $x = 600$, since this x-value would be far away from the data in `spring.csv`. The requirement that predictions should be made close to the data that have been used to construct the model applies very generally to phenomenological models, including the phenomenological approaches discussed in the next sections.

Note 2.2.3 (Prediction)   Regression functions such as Equation 2.27 can be used to predict values of the response variable for given values of the explanatory variable(s). Good predictions can be expected only if the regression function fits the data sufficiently well, and if the given values of the explanatory variable lie sufficiently close to the data.

2.2.3
The Coefficient of Determination

As to the quality of the fit between the model and the data, the simplest approach is to look at appropriate graphical comparisons of the model with the data such as Figure 2.2b. Based on that figure, you do not need to be a regression expert to conclude that there is a good matching between model and data, and that reasonable predictions can be expected using the regression line. A second approach is the coefficient of determination, which is denoted as R^2. Roughly speaking, the coefficient of determination measures the quality of the fit between the model and the data on a scale between 0 and 100%, where 0% refers to very poor fits and 100% refers to a perfect matching between the model and the data. R^2 thus expresses the quality of a regression model in terms of a single number, which is useful e.g. when you want to compare the quality of several regression models, or if you evaluate multiple linear regression models (Section 2.3) which involve higher-dimensional regression functions $\hat{y}(x_1, x_2, \ldots, x_n)$ that cannot be plotted similar to Figure 2.2b (note that a plot of $\hat{y}$ over $x_1, x_2, \ldots, x_n$ would involve an $n + 1$-dimensional space). In the R-output shown in Figure 2.2a, R^2 is the Multiple R Squared value, and hence you see that we have $R^2 = 96.2\%$ in the above example, which reflects the good matching between the model and the data that can be seen in Figure 2.2b. The Adjusted R Squared value in Figure 2.2a will be explained below in Section 2.3.

Formally, the coefficient of determination is defined as [37]

$$R^2 = \frac{\sum_{i=1}^{n} \left(\hat{y}_i - \bar{y}\right)^2}{\sum_{i=1}^{n} \left(y_i - \bar{y}\right)^2} \tag{2.34}$$

where we have used $\hat{y}_i = \hat{y}(x_i)$. For linear regression models, this can be rewritten as

$$R^2 = 1 - \frac{\sum_{i=1}^{n} \left(y_i - \hat{y}_i\right)^2}{\sum_{i=1}^{n} \left(y_i - \bar{y}\right)^2} \tag{2.35}$$

The latter expression is also known as the *pseudo-R^2* and it is frequently used to assess the quality of fit in nonlinear models. Note that R^2 according to Equation 2.35 can attain negative values if the $\hat{y}_i$ values are not derived from a linear regression (whereas Equation 2.34 guarantees $R^2 > 0$), and if these values are "far away" from the measurement values. If you observe negative R^2 values, then your model performs worse than a model that would yield the mean value $\bar{y}$ for every input (i.e. $\hat{y}(x_i) = \bar{y}$, $i = 1, \ldots, m$), since such a mean value model would give $R^2 = 0$ in Equation 2.35.

In a linear model, it can be easily shown that R^2 expresses the ratio between the variance of the predicted values $(\hat{y}_1, \ldots, \hat{y}_1)$ and the variance of the measurement values $(y_1, \ldots, y_n)$. $R^2 = 100\%$ is thus usually expressed like this: "100% of the variance of the measurement data is explained by the model." On the other hand, R^2 values substantially below 100% indicate that there is much more variance in the measurement data compared to the model, and this means that the variance of the data is insufficiently explained by the model, which means that one has to look for additional explanatory variables. For example, if a very poor R^2 is obtained for a linear model $\hat{y} = ax + b$, it may make sense to investigate multiple linear models such as $\hat{y} = ax + bz + c$ which involves an additional explanatory variable z (see Section 2.3 below).

Note 2.2.4 (Coefficient of determination) The coefficient of determination, R^2, measures the quality of fit between a linear regression model and data. On a scale between 0 and 100%, it expresses how much of the variance of the dependent variable measurements is explained by the explanatory variables of the model. If R^2 is small, one can try to add more explanatory variables to the model (see the multiple regression models in Section 2.3) or use nonlinear models (Sections 2.4 and 2.5).

2.2.4
Interpretation of the Regression Coefficients

You may wonder why the values of a and b appear in a column called *Estimates* in the R output of Figure 2.2a, and why standard errors are reported for a and b in the next column. Roughly speaking, this is a consequence of the fact that measurement data typically are affected by measurement errors. For example, a measurement device might exhibit a random variation in its last significant digit, which might lead to values such as 0.856, 0.854, 0.855, and 0.854 when we repeat a particular measurement under the same conditions several times (beyond this, there may be several other systematic and random sources of measurement errors, see [42]). Now suppose that we analyze measurement data $(x_1, y_1), (x_2, y_2), \ldots, (x_m, y_m)$ as above using linear regression, which might lead us to regression coefficients a and b. If we repeat the measurement, the new data will (more or less) deviate from the original data due to measurement errors, leading to (more or less) different values of a and b in a regression analysis. The parameters a and b thus depend on the random errors in the measurement data, and this means that a and b can be viewed as realizations of *random variables* α and β (see Section 2.1.2.1).

Usually, it is assumed that these random variables generate the measurement data $(x_1, y_1), (x_2, y_2), \ldots, (x_m, y_m)$ as follows:

$$y_i = \alpha x_j + \beta + \epsilon_i, \quad i = 1, \ldots, m \tag{2.36}$$

where the ϵ_i expresses the deviation between the model and the data, which includes the measurement error. The error terms ϵ_i are typically assumed to be normally distributed with zero expectation and a constant variance σ^2 independent of i (so-called homoscedastic error terms). Using these assumptions, the standard errors of α and β can be estimated, and these estimates are reported in the column Std.Error of the R output in Figure 2.2a. As you see there, the standard error of α is much smaller than the standard error of β, which means that you may expect larger changes of the estimate of β compared to the estimate of α if you would perform the same analysis again using a different dataset. In other words, the estimate of β is "less sharp" compared to that of α. The numbers reported in the column "t value" of Figure 2.2a refer to the Student's t distribution, and they can be used e.g. to construct confidence intervals of the estimated regression coefficients as explained in [19]. The values in the last column of Figure 2.2a are the p values that have been discussed in Sections 2.1.3.2 and 2.1.3.4.

2.2.5
Understanding LinRegEx1.r

Above we have used the R program LinRegEx1.r to produce Figure 2.2b. See Appendix B for any details on how to use and run the R programs of the book software. The essential commands in this code can be summarized as follows:

```
 1:  eq=y~x
 2:  FileName="Spring.csv"
 3:  Dataset=read.table(FileName,...)
 4:  RegModel=lm(eq,data=Dataset)
 5:  print(summary(RegModel))
 6:  a=10
 7:  b=50
 8:  xprog=seq(a, b, (b-a)/100)
 9:  yprog=predict(RegModel, data.frame(x = xprog))
10:  plot(xprog,yprog,...)
```

$$(2.37)$$

As mentioned before, the numbers "1:","2:", and so on in this code are not a part of the program, but just line numbers that are used for referencing in our discussion. Line 1 defines the regression equation in a special notation that you find explained in detail in R's help pages and in [43]. The command in line 1 stores the regression equation in a variable eq which is then used in line 4 to setup the regression model, so y~x is the part of line 1 that defines the regression equation. Basically, y~x can be viewed as a short notation for Equation 2.27, or for Equation 2.36 (it implies all the statistical assumptions expressed by the last equation, see the above discussion). The part at the left hand of the "~"-sign of such formulas defines the response variable (y in this case), which is then expressed on the right-hand side in terms of the explanatory variable (x in this case). Comparing the formula y~x with Equation 2.27, you see that this formula notation automatically implies the regression coefficients, a and b, which do not appear explicitly in the formula. You should note that the variable names used in these formulas must correspond to the column names that are used in the dataset. In this case, we are using the dataset spring.csv where the column referring to the response variable is denoted as y and the column referring to the explanatory variable is denoted as x. If, for example, we would have used the column names elongation instead of y and force instead of x, then we would have to write the regression equation as elongation~force instead of y~x.

Lines 2 and 3 of program 2.37 read the data from the file spring.csv into the variable Dataset (see LinRegEx1.r for the "long version" of line 3 including all details). Using the regression equation eq from line 1 and Dataset from line 3, the regression is then performed in line 4 using the lm command, and the result is stored in the variable RegModel.

Note 2.2.5 (*R*'s lm function) Linear regression problems (including the multiple linear problems treated in Section 2.3) are solved in R using the lm command.

The variable RegModel is then used in line 5 of program 2.37 to produce the *R* output that is displayed in Figure 2.2a above. Lines 6–9 show how *R*'s predict command can be used to compute predictions based on a statistical model such as RegModel. In lines 6–8, an array xprog is generated using *R*'s seq command that contains 101 equally spaced values between a=10 and b=50 (just try this command to see how it works). The predict command in line 9 then applies the regression equation 2.27 (which it takes from its RegModel argument) to the data in xprog and stores the result in yprog. xprog and yprog are then used in line 10 to plot the regression line using *R*'s plot command as it is shown in Figure 2.2b (again, see LinRegEx1.r for the full details of this command).

2.2.6
Nonlinear Linear Regression

Above it was emphasized that there are more general regression approaches. To see that there is a need for regression equations beyond Equation 2.27, let us consider the dataset gag.csv which you find in the book software. These data are taken from *R*'s MASS library where they are stored under the name GAGurine. As explained in [44, 45] (and in *R*'s help pages), these data give the concentrations of so-called glycosaminoglycans (GAG) in the urine of children aged from 0 to 17 (in units of milligrams per millimole creatinine). GAG data are measured as a screening procedure for a disease called *mucopolysaccharidosis*. As Figure 2.3a shows, the GAG concentration decreases with increasing age. Pediatricians need such data to assess whether a child's GAG concentration is normal. Based on GAG.csv only, it would be relatively time consuming to compare a given GAG concentration with the data. A simpler procedure would be to insert the given GAG concentration into a function that closely fits the data in the sense of regression. So let us try to derive an appropriate regression function similar to above. In GAG.csv, the Age and GAG columns give the ages and GAG concentrations, respectively. Using the regression function Equation 2.27 and proceeding as above, the regression equation can be

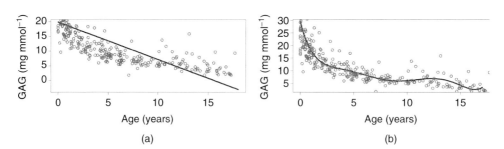

Fig. 2.3 (a) Comparison of the regression line Equation 2.39 with the data GAG.csv. Figure produced using LinRegEx4.r. (b) Comparison of the regression function Equation 2.43 with the data GAG.csv. Figure produced using LinRegEx5.r.

described in R as follows:

$$eq=GAG\sim Age \tag{2.38}$$

Inserting this into `LinRegEx1.r` (and changing the file name etc. appropriately), one arrives at `LinRegEx4.r` which you find in the book software. Analyzing the R output generated by `LinRegEx4.r` as above, we obtain the following equation of the regression line:

$$GAG(x) = -1.27 \cdot Age + 19.89 \tag{2.39}$$

Figure 2.3a compares this linear regression function with the data. As can be seen, the regression function overestimates GAG for ages below about nine years, and it underestimates GAG for higher ages. For ages above 15.7 years, the GAG concentrations predicted by the regression function are negative.

This is due to the fact that the data follow some nonlinear, curved pattern which cannot be described appropriately using a straight line. An alternative is to replace Equation 2.27 by a *polynomial regression function* of the general form

$$\hat{y}(x) = a_0 + a_1 x + a_2 x^2 + \cdots + a_s x^s \tag{2.40}$$

You may wonder why such a regression function is treated here in a section on "linear regression", since the function in Equation 2.40 can of course be a highly nonlinear function of x (depending on s, the degree of the polynomial). But remember, as was explained above in Note 2.2.1, that the term *linear* in "linear regression" does not refer to the regression function's dependence on x, but rather to its dependence on the regression coefficients. Seen as a function of x, Equation 2.40 certainly expresses a function that may be highly nonlinear, but if x is given, $\hat{y}$ is obtained as a linear combination of the regression coefficients $a_0, a_1, \ldots, a_s$. In this sense, all regression functions that can be brought into the general form

$$\hat{y}(x) = a_0 + a_1 f_1(x) + a_2 f_2(x) + \cdots + a_s f_s(x) \tag{2.41}$$

can be treated by linear regression (where $a_0, a_1, \ldots, a_s$ are the regression coefficients as before, and the f_i are arbitrary real functions). Whether linear or nonlinear in x, all these functions can be treated by linear regression, and this explains the title of this subsection ("nonlinear linear regression").

To perform an analysis based on the polynomial regression function 2.40, we can use R similarly as above. Basically, we just have to change the regression equation in the previous R program `LinRegEx4.r`. After some experimentation (or using the more systematic procedure suggested in [45]) you will find that it is a good idea to use a polynomial of degree 6, which is written in the notation required by R analogous to Equation 2.38 as follows:

$$eq=GAG\sim Age+I(Age^2)+I(Age^3)+I(Age^4)+I(Age^5)+I(Age^6) \tag{2.42}$$

In this equation, the R function `I()` is used to inhibit the interpretation of terms like `Age^2` based on the special meaning of the "`^`" operator in R's formula language. Written as "`I(Age^2)`", the "`^`" operator is interpreted as the usual arithmetical exponentiation operator. The R program `LinRegEx5.r` in the book software uses the regression function described in Equation 2.42. Executing this program and analyzing the results as before, the following regression function is obtained (coefficients are rounded for brevity):

$$\begin{aligned}
\mathrm{GAG}(x) = \quad &29.3 - 16.2 \cdot \mathrm{Age} + 6 \cdot \mathrm{Age}^2 - 1.2 \cdot \mathrm{Age}^3 \\
&+ 0.1 \cdot \mathrm{Age}^4 - 5.7\mathrm{e}\text{-}03 \cdot \mathrm{Age}^5 + 1.1\mathrm{e}\text{-}04 \cdot \mathrm{Age}^6
\end{aligned} \tag{2.43}$$

Figure 2.3b compares this regression function with the data. As can be seen, this regression function fits the data much better than the straight line that was used in Figure 2.3a. This is also reflected by the fact that `LinRegEx4.r` (regression using the straight line) reports an R^2 value of 0.497 and 4.55 as the mean (absolute) deviation between the data and the model, while `LinRegEx5.r` (polynomial regression) gives $R^2 = 0.74$ and a mean deviation of 2.8. Therefore the polynomial regression obviously does a much better job in helping pediatricians to evaluate GAG measurement data as described above. Note that beyond polynomial regression, R offers functions for *spline regression* that may yield "smoother" regression functions based on a concatenation of low-order polynomials [45, 46].

Note 2.2.6 (Regression of large datasets) Beyond prediction, regression functions can also be used as a concise way of expressing the information content of large datasets similar to the GAG data example.

2.3
Multiple Linear Regression

In the last section, we have seen how the regression approach can be used to predict a quantity of interest, y, depending on known values of another quantity, x. In many cases, however, y will depend on several independent variables such as $x_1, x_2, \ldots, x_n$ ($n \in \mathbb{N}$). This case can be treated by the *multiple (linear) regression method*. As we will see, the overall procedure is very similar to the approach described in the last section.

2.3.1
The Multiple Linear Regression Problem

Let us begin with an example. Note that we could really take *all* kinds of examples here due to the generality of the regression approach that can be used in all fields of science and engineering. Every dataset could be used that consists of at least three columns: two (or more) columns for the explanatory variables $x_1, x_2, \ldots$, and one column for the response variable y. We will refer here to the file `volz.csv` which you find in the book software. The data in this file have been produced in a PhD thesis which was concerned with the prediction of the *wilting of roses* [47]. More precisely, the intention of this PhD thesis was to find out whether the wilting of roses can be predicted based on the concentrations of certain carbohydrates within a rose. If reliable predictions could be made in this way, then this could serve as a base for the development of a practical tool for the quality control of roses produced on a big scale. Opening `Volz.csv` in *Calc*, you will see 19 columns of data. The first 18 columns (`Conc1–Conc18`) contain concentrations of various carbohydrates measured at some particular time. The last column called `DegWilt` characterizes rose wilting, giving the number of days after the carbohydrate measurements until a certain, fixed degree of wilting is observed (see [47] for details).

Now to treat these data using regression, we need an equation expressing the response variable y (corresponding to `DegWilt`) depending on the explanatory variables $x_1, \ldots, x_{18}$ (corresponding to `Conc1–Conc18`). A straightforward (linear) generalization of Equation 2.27 is

$$\hat{y}(x_1, x_2, \ldots, x_{18}) = a_0 + a_1 x_1 + a_2 x_2 + \cdots + a_{18} x_{18} \tag{2.44}$$

which is the simplest form of a multiple linear regression equation.

> **Note 2.3.1 (Multiple regression)** Multiple regression functions $\hat{y}(\mathbf{x})$ compute a response variable y using explanatory variables $\mathbf{x} = (x_1, \ldots, x_n)$ $(n > 1)$ and regression coefficients $a_0, a_1, \ldots, a_s$. If $\hat{y}(\mathbf{x})$ depends linearly on $a_0, a_1, \ldots, a_s$, it can be fitted to measurement data using multiple linear regression. See Section 2.4 for multiple nonlinear regression.

Similar to Equation 2.41 above, the general form of a multiple linear regression equation involving an arbitrary number of $n \in \mathbb{N}$ explanatory variables is

$$\hat{y}(\mathbf{x}) = a_0 + a_1 f_1(\mathbf{x}) + a_2 f_1(\mathbf{x}) + \cdots + a_s f_s(\mathbf{x}) \tag{2.45}$$

where $\mathbf{x} = (x_1, x_2, \ldots, x_n)^t$ and the f_i are arbitrary real functions. Note that this regression equation is linear since it is linear in the regression coefficients $a_0, \ldots, a_s$, although the f_i may be nonlinear functions (compare the discussion of the GAG data example in Section 2.2.6). For example, a regression function such as

$$\hat{y}(x_1, x_2) = a_0 + a_1 x_1 + a_2 x_2 + a_3 x^2 + a_4 y^2 + a_5 xy \tag{2.46}$$

can be treated using multiple *linear* regression since in this equation $\hat{y}$ is obtained as a linear combination of the regression coeffcients $a_0, \ldots, a_5$, although it depends nonlinearly on the explanatory variables, x and y.

Similar to the discussion of linear regression above, let us assume a general dataset that is given in the form $(x_{i1}, x_{i2}, \ldots, x_{in}, y_i)$ or $(\mathbf{x}_i, y_i)$ where $i = 1, \ldots, m$.

Then, as before, the coefficients $a_0, a_1, \ldots, a_s$ of Equation 2.45 are determined from the requirement that the differences $\hat{y}(x_i) - y_i$ should be small, which is again expressed in terms of the minimization of the RSQ:

$$RSQ = \sum_{i=1}^{m} \left(y_i - \hat{y}(x_i) \right)^2 \tag{2.47}$$

$$\min_{a_0, a_1, \ldots, a_n \in \mathbb{R}} RSQ \tag{2.48}$$

2.3.2
Solution Using Software

To solve this problem using R, the same procedure can be used that was described in Section 2.2 above. You can use the "Statistics/Fit models/Linear regression" menu option of the R *Commander*, selecting `Conc1`,...,`Conc18` as the explanatory variables and `DegWilt` as the response variable. Alternatively, you can use an R program such as `LinRegEx2.r` which you find in the book software. `LinRegEx2.r` works very similarly to `LinRegEx1.r` which was discussed above in Section 2.2 (you will find a few remarks on `LinRegEx2.r` further below). Either of these two ways produces the result shown in Figure 2.4a.

The *interpretation of Figure 2.4a* goes along the same lines as the interpretation of Figure 2.2 above. First of all, you find the estimates of the regression coefficients $a_0, a_1, \ldots, a_{18}$ in the `Estimate` column of Figure 2.4a: $a_0 = 6.478323, a_1 = 0.016486, \ldots, a_{18} = 1.016512$. The regression equation 2.44 thus becomes:

$$\hat{y}(x_1, x_2, \ldots, x_{18}) = 6.478323 + 0.016486x_1 + \cdots + 1.016512x_{18} \tag{2.49}$$

The regression coefficients $a_0, a_1, \ldots, a_{18}$ can be viewed as realizations of corresponding random variables $\alpha_0, \alpha_1, \ldots, \alpha_{18}$, and the "`Std. Error`" and "`t value`" columns of Figure 2.4a report statistical properties of these random variables as discussed in Section 2.2.4 above. Similar to Equation 2.36, this statistical analysis is based on the assumption that the data can be expressed as follows:

$$y_i = \alpha_0 + \alpha_1 x_{i1} + \cdots + \alpha_{18} x_{i18} + \epsilon_i, \quad i = 1, \ldots, m \tag{2.50}$$

Again, the error terms ϵ_i are assumed to be homoscedastic (i.e. normally distributed with zero expectation and a constant variance σ^2).

Figure 2.4b compares the predicted and measured values of `DegWilt` in a type of plot which we call a *predicted-measured plot*. Note the difference between this figure and Figure 2.2 in Section 2.2.2: Figure 2.2 plots the response variable against the explanatory variable, while Figure 2.4b involves the response variable only. A plot of the response variable against the explanatory variables similar to Figure 2.2 cannot be done here since this would involve a 19-dimensional space (18 explanatory variables +1 response variable). Figure 2.4b is an elegant way to get a graphical

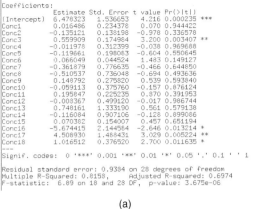

```
Coefficients:
            Estimate Std. Error t value Pr(>|t|)
(Intercept)  6.478323   1.536653   4.216 0.000235 ***
Conc1        0.016486   0.234378   0.070 0.944422
Conc2       -0.135121   0.138198  -0.978 0.336578
Conc3        0.559909   0.174984   3.200 0.003407 **
Conc4       -0.011978   0.312399  -0.038 0.969688
Conc5       -0.119661   0.198083  -0.604 0.550645
Conc6        0.066049   0.044524   1.483 0.149127
Conc7       -0.361879   0.776635  -0.466 0.644850
Conc8       -0.510537   0.736048  -0.694 0.493636
Conc9        0.148792   0.275820   0.539 0.593840
Conc10      -0.059113   0.375760  -0.157 0.876124
Conc11       0.195847   0.225235   0.870 0.391953
Conc12      -0.003367   0.499120  -0.017 0.986744
Conc13       0.748161   1.333190   0.561 0.579138
Conc14      -0.116084   0.907106  -0.128 0.899086
Conc15       0.070382   0.154007   0.457 0.651194
Conc16      -5.674415   2.144584  -2.646 0.013214 *
Conc17       4.508930   1.488431   3.029 0.005224 **
Conc18       1.016512   0.376520   2.700 0.011635 *
---
Signif. codes:  0 '***' 0.001 '**' 0.01 '*' 0.05 '.' 0.1 ' ' 1

Residual standard error: 0.9384 on 28 degrees of freedom
Multiple R-Squared: 0.8158,     Adjusted R-squared: 0.6974
F-statistic:  6.89 on 18 and 28 DF,  p-value: 3.675e-06
```

(a)

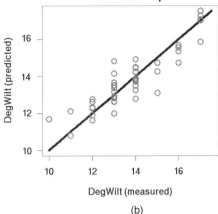

(b)

Fig. 2.4 (a) Result of a multiple regression using R based on the data Volz.csv (response variable: DegWilt, explanatory variables: Conc1,Conc2,...). (b) Comparison of predicted ($\hat{y}$) and measured (y) values of DegWilt. The circles are at the coordinates $(y_i, \hat{y}(\mathbf{x}_i))$ $(i = 1, \ldots, m)$, the line is $\hat{y} = y$. The figure was produced using LinRegEx2.r.

idea about the quality of a regression even in the presence of a great number of explanatory variables. As to the interpretation of Figure 2.4b, note that the line $\hat{y} = y$ displayed in the figure is *not* a regression line. Rather, it can be used to assess the prediction error for each of the predicted values, $\hat{y}(\mathbf{x}_i)$. If $\hat{y}(\mathbf{x}_i)$ coincides with the corresponding measurement value, y_i, we will have $\hat{y}(\mathbf{x}_i) = y_i$ and hence this will generate a circle lying exactly on the line $\hat{y} = y$. On the other hand, any deviations between $\hat{y}(\mathbf{x}_i)$ and y_i will generate corresponding deviations between the circle $(y_i, \hat{y}(\mathbf{x}_i))$ and the line $\hat{y} = y$. Therefore, the data will lie very closely to the line in such a predicted/measured plot if the regression equation matches the data very well, and they will substantially deviate from that line if the regression equation substantially deviates from the data. Figure 2.4b thus tells us that this is an regression model of an average quality: some of the predictions match the data very well, but predictions and measurements may also deviate by several days. See also Figure 2.8 for a comparison of a conventional plot with a predicted measured plot.

Note 2.3.2 (Predicted-measured plot) In a predicted-measured plot such as Figure 2.4b, predicted values ($\hat{y}(\mathbf{x}_i)$) of the response variable are plotted against measured values (y_i) of the response variable. Deviations between data and predictions can thus be seen in terms of deviations from the line $\hat{y} = y$. Predicted-measured plots are particularly useful to evaluate regressions involving more than two explanatory variables, since in that case the response variable cannot be plotted against all explanatory variables.

The average quality of this regression is also reflected by the coefficient of determination shown in Figure 2.4a, $R^2 = 0.8158$. As explained in the previous section, this means that about 20% of the variation of the measurement data are not explained by the explanatory variables of the current regression model (see Note 2.2.4). If we need better predictions, we should thus try to find additional explanatory variables that could then be used in an extended multiple regression model. Note that the data circles in Figure 2.4b follow lines parallel to the y-axis of the plot since the degree of wilting is expressed using integers in Volz.csv.

It can be shown that the R^2 value defined as in Equations 2.34 and 2.35 may increase as additional explanatory variables are incorporated into the model, even if those additional variables do not improve the quality of the model [19]. Therefore, comparisons of R^2 values derived from regression models involving different numbers of explanatory variables can be questionable. A standard way to circumvent this problem is the use of the *adjusted coefficient of determination*, which adjusts the R^2 value with respect to the number of variables and sample size [19]. The adjusted R^2 value appears in the linear regression results produced by R (see Adjusted R-squared in Figure 2.4a).

The R program LinRegEx2.r that was used to produce Figure 2.4b is again based on R's lm command, and it works very similarly to the corresponding program LinRegEx1.r that was discussed in Section 2.2.5 above. As was explained there, the regression equation must be written based on the variable names that are used in the data. In Volz.csv, Conc1−Conc18 are the explanatory variables and DegWilt is the response variable. In analogy to line 1 of program 2.37, the multiple regression equation can thus be written as

$$\text{eq=DegWilt~Conc1+Conc2+Conc3+...+Conc18} \qquad (2.51)$$

If you are using this kind of notation, all explanatory variables must be explicitly written in the code, including Conc4−Conc17 which were left out in Equation 2.51 for brevity. In LinRegEx2.r, Equation 2.51 is written using the abbreviated notation "eq=DegWilt~.". In this notation, the dot serves as a placeholder that stands for all variables in the dataset except for the response variable. This notation can be modified in various ways (see R's help pages). For example, "eq=DegWilt~.-Conc17" results in a multiple regression model that uses all explanatory variables except for Conc17.

2.3.3
Cross-Validation

Although R^2 values and plots such as Figure 2.4b give us some idea regarding the quality of a regression model, they cannot guarantee a good predictive capability of the model. For example, new data may be affected by a new explanatory variable that has been held constant in the regression dataset. Suppose we want to predict the yield of a particular crop based on a regression equation that was obtained using data of crops growing under a constant temperature of $20\,^{\circ}\text{C}$. Although this equation may perform with $R^2 = 1$ on the regression dataset, it will probably be completely useless on another dataset obtained for crops growing under a constant temperature of $10\,°C$. To get at least a first idea as to how a particular regression model performs on unknown data, a procedure called *cross-validation* can be used. Cross-validation approaches mimic "new data" in various ways, for example, based on a partitioning of the dataset into a *training dataset* which is used to obtain the regression equation, and a *test dataset* which is used to assess the regression equation's predictive capability (a so-called *holdout validation* approach, see [48]).

The R program LinRegEx3.r in the book software performs such a cross-validation for the rose wilting data, Volz.csv. This program is very similar to LinRegEx2.r, except for the following lines of code that implement the partitioning of the data into training and test datasets:

```
1:  Dataset=read.table(FileName,...)
2:  TrainInd=sample(1:47,37)
3:  TrainData=Dataset[TrainInd,]
4:  TestData=Dataset[-TrainInd,]                    (2.52)
5:  RegModel=lm(eq,data=TrainData)
6:  DegWiltTrain=predict(RegModel,TrainData)
7:  DegWiltTest=predict(RegModel,TestData)
```

After the data have been stored in the variable Dataset in line 1 of program 2.52, 37 random indices between 1 and 47 (referring to the 47 lines of data in Volz.csv) are chosen in line 2. See [45] and the R help pages for more details on the sample command that is used in line 2. The 37 random indices are stored in the variable TrainInd which is then used in line 3 to assemble the training dataset TrainData based on those lines of Dataset which correspond to the indices in TrainInd. The remaining lines of Dataset are then reassembled into the test dataset TestData in line 4. The regression model RegModel is then computed using the training dataset in line 5 (note the difference to line 4 of program 2.37 where the regression model is computed based on the entire dataset). Then, the predict command is used again to apply the regression equation separately to the training and test datasets in lines 6 and 7 of Equation 2.52.

Figure 2.5 shows an *example result* of LinRegEx3.r. You should note that if you run LinRegEx3.r on your machine, the result will probably be different from the plot shown in Figure 2.5 since the sample command may select different training and test datasets if it is performed on different computers. As explained in [45], the sample command is based on an algorithm generating *pseudorandom numbers*, and the actual state of this algorithm is controlled by a set of integers stored in the R object .Random.seed. As a result of this procedure, LinRegEx3.r may generate different results on different computers depending on the state of the algorithm on each particular computer. Figure 2.5 compares the measured and predicted values of DegWilt similar to Figure 2.4b above. As could be expected, there are larger deviations between the line $\hat{y} = y$ and the data for the test dataset which

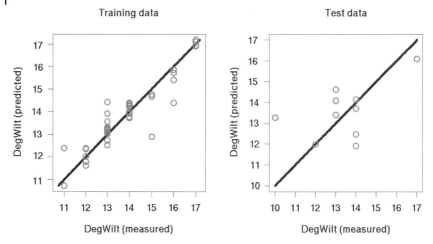

Fig. 2.5 Comparison of predicted ($\hat{y}$) and measured (y) values of DegWilt using a randomly selected training dataset ($n = 37$) and a complementary test dataset ($n = 10$). In each of the plots, the circles are at the coordinates ($y_i, \hat{y}(x_i)$) ($i = 1, \ldots, m$) and the line is $\hat{y} = y$. The figure was produced using LinRegEx3.r.

was not used in the regression procedure. This is also reflected by the R^2 values (training data R^2: 0.86, test data R^2: 0.22) and by the mean deviations (training data: 0.41 days, test data: 1.13 days) computed by LinRegEx3.r. Repeating the cross-validation procedure several times and averaging the results, one gets a fairly good idea of the predictive capability of a regression model, at least referring to data that are similar to the dataset under consideration (which excludes data that have e.g. been obtained using different temperatures etc. as discussed above).

2.4
Nonlinear Regression

2.4.1
The Nonlinear Regression Problem

Until now, we have considered (multiple) linear regression functions of the general form

$$\hat{y}(x) = a_0 + a_1 f_1(x) + a_2 f_1(x) + \cdots + a_s f_s(x) \tag{2.53}$$

where $a_0, a_1, \ldots, a_s$ are the regression coefficients, $x = (x_1, \ldots, x_n)$, and $f_1(x), \ldots,$ $f_s(x)$ are arbitrary real functions (see the discussion of Equation 2.45 in Section 2.3.1 above). As explained above, regression functions of this kind are called *linear* since $\hat{y}(x)$ is obtained as a linear combination of the regression coefficients. In many applications, however, the regression functions will depend in a nonlinear way on the regression coefficients. Using $\mathbf{a} = (a_1, \ldots, a_s)$ this can be expressed in a general form as

$$\hat{y}(\mathbf{x}) = f(\mathbf{x}, \mathbf{a}) \tag{2.54}$$

where f is some general real function (a slightly more general, vectorial form of this equation will be given at the end of this section). Similar to the procedure explained in Sections 2.2.1 and 2.3.1 above, $\hat{y}(\mathbf{x})$ is fitted to measurement data based on a minimization of the RSQ.

2.4.2
Solution Using Software

Let us look at some examples. Figure 2.6 shows *US investment data* (expressing the relative change of investments compared with a reference value) described in [49, 50]. These data are a part of *R*'s Ecdat library, and they are a part of the book software in the file klein.csv. As a result of the economic cycle, the data show an oscillatory, sinusoidal pattern. This means that if we want to describe these data using a regression function, it is natural to apply a general sine function such as

$$\hat{y}(x) = a_0 \cdot \sin(a_1 \cdot (x - a_2)) \tag{2.55}$$

Using

$$f(x, \mathbf{a}) = f(x, a_0, a_1, a_2) = a_0 \cdot \sin(a_1 \cdot (x - a_2)) \tag{2.56}$$

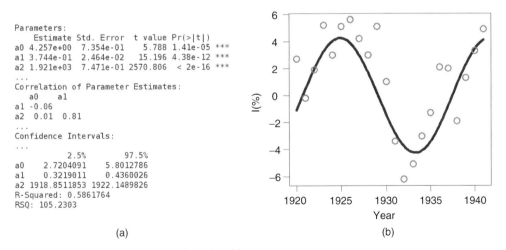

```
Parameters:
     Estimate Std. Error  t value Pr(>|t|)
a0 4.257e+00 7.354e-01    5.788 1.41e-05 ***
a1 3.744e-01 2.464e-02   15.196 4.38e-12 ***
a2 1.921e+03 7.471e-01 2570.806 < 2e-16 ***
...
Correlation of Parameter Estimates:
    a0    a1
a1 -0.06
a2  0.01  0.81
...
Confidence Intervals:
...
            2.5%       97.5%
a0    2.7204091    5.8012786
a1    0.3219011    0.4360026
a2 1918.8511853 1922.1489826
R-Squared: 0.5861764
RSQ: 105.2303
```

(a)   (b)

Fig. 2.6 (a) Nonlinear regression result produced by *R*'s nls function based on Equation 2.55, Klein.csv and NonRegEx1.r. (b) Comparison of the regression function Equation 2.55 (line) with the data Klein.csv (circles). Figure produced using NonRegEx1.r.

it is seen that Equation 2.55 has the general form of Equation 2.54, and thus it is a nonlinear regression function. Note that it cannot be brought into the linear form of Equation 2.53 since a_1 and a_2 appear in the (nonlinear) sine function. Three regression coefficients can be used to fit this function to the data: a_0 determines the amplitude of the function, a_1 determines its period, and a_2 moves the sine along the x-axis.

Using *R*, a nonlinear regression based on Equation 2.55 and the data in Klein.csv can be performed by a simple editing of LinRegEx1.r which was discussed in Section 2.2 above. This leads to the *R* program NonRegEx1.r in the book software. Let us look at the essential commands in NonRegEx1.r that do the nonlinear regression (everything else is very similar to LinRegEx1.r):

```
1:  eq=inv~a0*sin(a1*(year-a2))
2:  parstart=c(a0=5,a1=2*pi/15,a2=1920)
3:  FileName="Klein.csv"
4:  Dataset=read.table(FileName,...)
5:  RegModel=nls(eq,data=Dataset,start=parstart)
```
$$(2.57)$$

Line 1 defines the regression function according to Equation 2.55. Note that $\hat{y}$ and x have been replaced by the appropriate column names of Klein.csv, inv, and year, respectively. The equation eq defined in line 1 is then used in line 5 to compute the regression model based on *R*'s nls function. This is the essential difference to the linear regression models in the previous sections, which could all be treated using *R*'s lm function.

Note 2.4.1 (*R*'s nls function)   In contrast to the lm function which was used for linear regression above, the nls function determines the parameter estimates based on an *iterative numerical procedure*. This means that the computation begins with certain *starting values* of the parameters which are then improved step by step until the problem is solved with sufficient accuracy.

The required accuracy can be controlled via the *R* function nls.control, see *R*'s help pages. Details about the iterative procedure used by nls can be found in [43, 51]. The starting values must be provided by the user, which is done in line 2 of program 2.57. Generally, the iterative procedure called by nls will converge better if the starting values of the parameters are chosen close to the solution of the regression problem. This means that if nls does not converge, you should try other starting values of the parameters until you obtain convergence. It may also happen that you do not get convergence for any set of starting values, which usually means that your regression equation is inappropriate, so try another model in that case.

To *choose the starting values*, you should of course use any kind of available a priori information on the parameters that you can get. For example, if you know certain limits for the parameters based on theoretical considerations it is usually a good idea to choose the starting value exactly between those limit values. On the other hand, you may use parameter values from the literature or try to derive estimated values from the data. In our case, reasonable estimates can be derived from the data in Figure 2.6. Looking at the data you see that the amplitude of the oscillation is about ±5, so it is a good idea to choose $a_0 = 5$. Furthermore, the data suggest a period length of about 15 years, so we set $a_1 = 2\pi/15$ since R's sine function expects its argument in radians. Finally, the period begins at the x coordinate 1920 and hence we set $a_2 = 1920$. Exactly these starting values for the parameters are defined in line 2 of program 2.57, and they are then used in line 5 as an argument of the nls function. Running NonRegEx1.r using these starting values, the result shown in Figure 2.6a and b is obtained.

Figure 2.6b compares the regression function Equation 2.55 that is obtained using the estimates of the coefficients a_0, a_1, and a_2 from Figure 2.6a with the data Klein.csv. As can be seen, the regression function correctly describes the general tendency of the data. A substantial (but inevitable) scattering of the data around the regression function remains, which is also expressed by an R^2 value of only 0.58 (Figure 2.6a). The *confidence intervals* in Figure 2.6a have been generated using R's confint command, see NonRegEx1.r. These confidence intervals refer to the random variables which generate the estimates of a_0, a_1, and a_2, and which we denote as α_0, α_1, and α_2 analogous to Section 2.2.4. For example, for α_0 we have a confidence interval of $[2.7204091, 5.8012786]$ which means that this interval covers the unknown "true" expected value of α_0 with a probability of 95% [19]. In this way, we get an idea of how sharply α_0 and the other parameters can be estimated from the data. As Figure 2.6a shows, we have smaller confidence intervals around α_1 and α_2, which means that these parameters can be estimated with a higher precision from the data compared to α_0.

The nls function also reports *correlations* between the parameter estimates (Figure 2.6b). For example, Figure 2.6a reports a correlation of 0.01 between α_0 and α_1 and a correlation of 0.81 between α_1 and α_2 (where we have used Greek letters α_0, α_1, and α_2 to denote the random variables which generate the estimates of a_0, a_1, and a_2, analogous to Section 2.2.4). Such correlations can be used to improve the experimental design with respect to an improved estimability of the parameters as discussed in [41]. As discussed there, particularly high correlations between two estimated parameters may indicate that the information content in the dataset does not suffice for a discrimination between those two parameters, or they may indicate degeneracies in the model formulation.

2.4.3
Multiple Nonlinear Regression

The procedure explained in the last section covers a great number of examples which involve a single independent variable. Similar to linear regression, however, nonlinear regression may of course also involve several independent variables. As an example, let us consider the calibration of a Stormer viscometer. Appropriate data are a part of R's MASS library, and you will also find these data in the file stormer.csv in the book software. In [45], the principle of a stormer viscometer is explained as follows (see [52] for more details):

Note 2.4.2 (Stormer viscometer) A stormer viscometer measures the viscosity of a fluid by measuring the time taken for an inner cylinder in the mechanism to perform a fixed number of revolutions in response to an actuating weight. The viscometer is calibrated by measuring the time taken with varying weights while the mechanism is suspended in fluids of accurately known viscosity.

The calibration dataset thus comprises three columns of data: the viscosity v [10^{-1} Pa.s], the weight w [g], and the time T [s] which correspond to the three columns Viscosity, Wt, and Time of the file stormer.csv. It is known from theoretical considerations that v, w, and T are related as follows [45]:

$$T = \frac{a_1 v}{w - a_2} \tag{2.58}$$

Once a_1 and a_2 are known, this equation can be used to determine the viscosity v from known values of w and T. To determine a_1 and a_2 from the calibration dataset stormer.csv, we can perform a nonlinear regression using Equation 2.58. Using the identifications $\hat{y} = T$, $x_1 = v$, and $x_2 = w$, Equation 2.58 can be written using the above notation as

$$\hat{y}(x_1, x_2) = \frac{a_1 x_1}{x_2 - a_2} \tag{2.59}$$

Note that this equation cannot be written in the form of Equation 2.53 since a_2 appears in the denominator of the fraction on the right-hand side of Equation 2.59, and hence it is a nonlinear regression equation. Basically, this regression problem can be treated by a simple editing of NonRegEx1.r, replacing line 1 of program 2.57 by

```
eq=Time~a1*Viscosity/(Wt-a2)
```
$\hfill (2.60)$

Note that Equation 2.60 corresponds exactly to Equation 2.58 above as the columns Viscosity, Wt, and Time of the file stormer.csv correspond to v, w, and T. Note also that if formulas such as Equation 2.60 are used in the nls command, all operators such as *, / and so on, will have their usual arithmetical meaning (see [45] and R's help pages). This means that you do not have to use R's inhibit function I() similar to Equation 2.42 to make sure that all operators are used as arithmetical operators. The R program NonRegEx2.r implements the stormer viscometer regression problem using Equation 2.58. It uses $a_1 = 1$ and $a_2 = 0$ as starting values. nls converges without problems for these values, and you should note that these values have been chosen without using any a priori knowledge of the system: $a_1 = 1$ has been chosen based on the simple idea that we

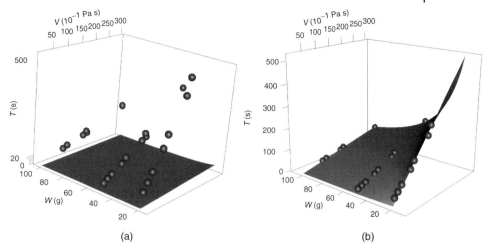

Fig. 2.7 (a) $T(v, w)$ according to Equation 2.58 using the starting values $a_1 = 1$ and $a_2 = 0$ (surface) compared with the data in `stormer.csv` (spheres). (b) Same plot, but using the estimates $a_1 = 29.4013$ and $a_2 = 2.2183$ obtained by nonlinear regression using R's `nls` function. Plots generated by `NonRegEx2.r`.

need positive T values (which gives $a_1 > 0$ if we assume $w > a_2$), and the choice $a_2 = 0$ basically expresses that nothing is known about that parameter. In fact, there is some a priori knowledge on these parameters that can be used to get more realistic starting values (see [45]), but this example nevertheless shows that `nls` may also converge using very rough estimates of the parameters.

Figures 2.7 and 2.8 show the results produced by `NonRegEx2.r`. First of all, Figure 2.7a shows that there is substantial deviation between the regression function Equation 2.58 and the data in `stormer.csv` if the above starting values of a_1 and a_2 are used. Figure 2.7b shows the same picture using the estimates of a_1 and a_2 obtained by R's `nls` function, and you can see by a comparison of these two plots that the nonlinear regression procedure virtually deforms the regression surface defined by Equation 2.58 until it fits the data. As Figure 2.7b shows, the fit between the model and the data is almost perfect, which is also reflected by the R^2 value computed by `NonRegEx2.r` ($R^2 = 0.99$). Figure 2.8 compares the regression result displayed in the conventional plot that really shows the regression function $T(v, w)$ (Figure 2.8a) with the predicted-measured plot that was introduced in Section 2.3 above (Figure 2.8b). The message of both plots in Figure 2.8 is the same: an almost perfect coincidence between the regression function and the data.

Beyond this, Figure 2.8a is of course more informative compared to Figure 2.8b since you can identify the exact location of any data point in the v/w space, for example, the location of data points showing substantial deviations from the regression function. Also, a plot such as Figure 2.8a allows you to assess whether the regression function is sufficiently characterized by data, and in which regions

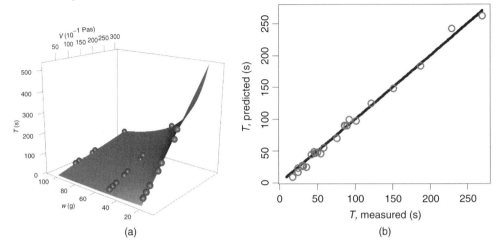

Fig. 2.8 Comparison of the regression equation with the data (a) in a conventional plot as in Figure 2.7 and (b) in a predicted-measured plot. Plots generated by NonRegEx2.r.

of the v/w space additional experimental data are needed. Looking at Figure 2.8a, for example, the nonlinearity of the regression surface obviously is sufficiently well characterized by the three "rows" of data. You should note, however, that conventional plots such as Figure 2.8a are only available for regressions involving up to two independent variables. Regressions involving more than two independent variables are usually visualized using a predicted-measured plot such as Figure 2.8b, or by using a conventional plot such as Figure 2.8a that uses two of the independent variables of the regression and neglects the other independent variables (of course, such conventional plots must be interpreted with care).

2.4.4
Implicit and Vector-Valued Problems

So far we have discussed two examples of nonlinear regressions referring to regression functions of the form

$$\hat{y}(\mathbf{x}) = f(\mathbf{x}, \mathbf{a}) \tag{2.61}$$

The particular form of the regression function in the investment data example was

$$f(x, a_0, a_1, a_2) = a_0 \cdot \sin(a_1 \cdot (x - a_2)) \tag{2.62}$$

In the viscometer example we had

$$f(x_1, x_2, a_1, a_2) = \frac{a_1 x_1}{x_2 - a_2} \tag{2.63}$$

This can be generalized in various ways. For example, the regression function may be given implicitly as the solution of a differential equation, see the example in Section 3.9. $\hat{y}$ may also be a vector-valued function $\hat{\mathbf{y}} = (\hat{y}_1, \ldots, \hat{y}_r)$. An example of this kind will be discussed below in Section 3.10.2. The nonlinear regression function then takes the form

$$\hat{\mathbf{y}}(\mathbf{x}) = \mathbf{f}(\mathbf{x}, \mathbf{a}) \tag{2.64}$$

where $\mathbf{x} = (x_1, \ldots, x_n) \in \mathbb{R}^n$, $\mathbf{a} = (a_1, \ldots, a_s) \in \mathbb{R}^s$, and $\hat{\mathbf{y}}(\mathbf{x})$ and $\mathbf{f}(\mathbf{x}, \mathbf{a})$ are real vector functions $\hat{\mathbf{y}}(\mathbf{x}) = (\hat{y}_1(\mathbf{x}), \ldots, \hat{y}_r(\mathbf{x}))$, $\mathbf{f}(\mathbf{x}, \mathbf{a}) = (f_1(\mathbf{x}, \mathbf{a}), \ldots, f_r(\mathbf{x}, \mathbf{a}))$ $(r, s, n \in \mathbb{N})$.

2.5
Neural Networks

If we perform a nonlinear regression analysis as described above, we need to know the explicit form of the regression function. In our analysis of the investment data `klein.csv`, the form of the regression function (a sine function) was derived from the sinusoidal form of the data in a graphical plot. In some cases, we may know an appropriate form of the regression function based on a theoretical reasoning, as was the case in our above analysis of the stormer viscometer data `stormer.csv`. But there are, of course, situations where the type of regression function cannot be derived from theory, and where graphical plots of the data are unavailable (e.g. because there are more than two independent variables). In such cases, we can try for example, polynomial or spline regressions (see the example in Section 2.2), or so-called (artificial) neural networks (ANN).

> **Note 2.5.1 (Application to regression problems)** Among other applications (see below), neural networks can be used as particularly flexible nonlinear regression functions. They provide a great number of tuning parameters that can be used to approximate any smooth function.

2.5.1
General Idea

To explain the idea, let us reconsider the multiple linear regression function

$$\hat{y} = a_0 + a_1 x_1 + a_2 x_2 + \cdots + a_n x_n \tag{2.65}$$

As was discussed above, this equation is a black box–type model of an input–output system (Figure 1.2), where $x_1, \ldots, x_n$ are the given input quantities and y is the output quantity computed from the inputs. Graphically, this can be interpreted as shown in Figure 2.9a. The figure shows a *network of nodes* where each of the nodes corresponds to one of the quantities $x_1, \ldots, x_n$ and y. The nodes are

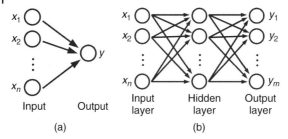

Fig. 2.9 (a) Graphical interpretation of multiple regression.
(b) Artificial neural network with one hidden layer.

grouped together into one *layer* comprising the input nodes $x_1, \ldots, x_n$, and a second layer comprising the output node y. The arrows indicate that the information flow is from the input nodes toward the output node, similar to Figure 1.2 above. Now the multiple linear regression equation (2.65) can be viewed as expressing the way in which the output node processes the information that it gets from the input nodes: each of the input node levels $x_1, \ldots, x_n$ is multiplied with a corresponding constant $a_1, \ldots, a_n$, the results are added up, and the level of the output node y is then obtained as this sum plus a constant (the so-called *bias*) a_0.

So far, this is no more than a graphical interpretation of multiple regression. This interpretation becomes interesting in view of its *analogy with neural networks in biological tissues* such as the human brain. Formally, such neural networks can also be described by figures similar to Figure 2.9a, that is, as a system of interconnected nodes (corresponding to the biological neurons) which exchange information along their connections [53, 54]. Since the information exchange in biological neural networks is of great importance if one e.g. wants to understand the functioning of the human brain, a great deal of research has been devoted to this topic in the past. As a part of this research effort, mathematical models have been developed that describe the information exchange in interconnected networks such as the network shown in Figure 2.9a, but of course involving more complex network topologies than the one shown in Figure 2.9a, and more complex (nonlinear) equations than the simple multiple linear regression equation, Equation 2.65.

It turned out that these mathematical models of interconnected networks of nodes are useful for their own sake, that is, independently of their biological interpretation, e.g. as a flexible regression approach that is apt to approximate any given smooth function. This class of mathematical models of interconnected networks of nodes are called *(artificial) neural network models* or *ANN models*, or simply *neural networks*. They may be applied in their original biological context, or in a great number of entirely different applications such as general *regression analysis* (e.g. prediction of tribological properties of materials [55–60], or the permeability prediction example below), *time series prediction* (stock prediction etc.), *classification and pattern recognition* (face identification, text recognition, etc.), *data processing* (knowledge discovery in databases, e-mail spam filtering, etc.) [53, 54].

> **Note 2.5.2 (Analogy with biology)**   Multiple linear regression can be interpreted as expressing the processing of information in a network of nodes (Figure 2.9a). Neural networks arise from a generalization of this interpretation, involving additional layer(s) of nodes and nonlinear operations (Figure 2.9b). Models of this kind are called *(artificial) neural networks (ANN's)* since they have been used to describe the information processing in biological neural networks such as the human brain.

2.5.2
Feed-Forward Neural Networks

The diversity of neural network applications corresponds to a great number of different mathematical formulations of neural network models, and to a great number of more or less complex network topologies used by these models [53, 54]. We will confine ourselves to the simple network topology shown in Figure 2.9b. This network involves an input and an output layer similar to Figure 2.9a, and in addition to this there is a so-called *hidden layer* between the input and output layers. As indicated by the arrows, the information is assumed to travel from left to right only, which is why this network type is called a *feedforward neural network*. This is one of the most commonly used neural network architectures, and based on the mathematical interpretation that will be given now (using ideas and notation from [45]) it is already sufficiently complex e.g. to approximate arbitrary smooth functions.

Let us assume that there are $n \in \mathbb{N}$ input nodes corresponding to the given input quantities $x_1, \ldots, x_n$, $H \in \mathbb{N}$ hidden nodes and $m \in \mathbb{N}$ output nodes corresponding to the output quantities $y_1, \ldots, y_m$. Looking at the top node in the *hidden layer* of Figure 2.9b, you see that this node receives its input from all nodes of the input layer very similar to Figure 2.9a. Let us assume that this node performs the same operation on its input as was discussed above referring to Figure 2.9a, multiplying each of the inputs with a constant, taking the sum over all the inputs and then adding a constant. This leads to an expression of the form

$$\sum_{k=1}^{n} w_{ik;h1} x_k + b_{h1} \tag{2.66}$$

Here, the so-called *weight* $w_{ik;h1}$ denotes the real coefficient used by the hidden node 1 (index $h1$) to multiply the kth input (index ik), and b_{h1} is the *bias* added by hidden node 1. Apart from notation, this corresponds exactly to the multiple linear regression equation 2.65 discussed above. The network would thus be no more than a complex way to express multiple (linear) regression if all nodes in the network would do no more than the arithmetics described by Equation 2.66. Since this would make no sense, the hidden node 1 will apply a nonlinear real function ϕ_h to Equation 2.66, giving

$$\phi_h \left(\sum_{k=1}^{n} w_{ik;h1} x_k + b_{h1} \right) \tag{2.67}$$

The application of this so-called *activation function* is the basic trick that really activates the network and makes it a powerful instrument far beyond the scope of linear regression. The typical choice for the activation function is the logistic function

$$f(x) = \frac{e^x}{1 + e^x} \tag{2.68}$$

The state of the hidden nodes $l = 1, \ldots, H$ after the processing of the inputs can be summarized as follows:

$$\phi_h \left(\sum_{k=1}^{n} w_{ik;hl} x_k + b_{hl} \right), \quad l = 1, \ldots, H \tag{2.69}$$

These numbers now serve as the input of the *output layer* of the network. Assuming that the output layer processes this input the same way as the hidden layer based on different coefficients and a different nonlinear function ϕ_o, the output values are obtained as follows:

$$y_j = \phi_o \left(b_{oj} + \sum_{l=1}^{H} w_{hl,oj} \cdot \phi_h \left(b_{hl} + \sum_{k=1}^{n} w_{ik;hl} \cdot x_k \right) \right), \quad j = 1, \ldots, m \tag{2.70}$$

Similar as above, the weights $w_{hl,oj}$ denote the real coefficient used by the output node j (index oj) to multiply the input from the hidden node l (index hl), and b_{oj} is the bias added by output node j. Below, we will us R's nnet command to fit this equation to data. This command is restricted to single-hidden-layer neural networks such as the one shown in Figure 2.9b. nnet nevertheless is a powerful command since the following can be shown [45, 61–63]:

Note 2.5.3 (Approximation property) The single-hidden-layer feedforward neural network described in Equation 2.70 can approximate any continuous function $f : \Omega \subset \mathbb{R}^n \to \mathbb{R}^m$ uniformly on compact sets by increasing the size of the hidden layer (if linear output units ϕ_o are used).

The nnet command is able to treat a slightly generalized version of Equation 2.70 which includes so-called *skip-layer connections*:

$$y_j = \phi_o \left(b_{oj} + \sum_{k=1}^{n} w_{ik;oj} \cdot x_k + \sum_{l=1}^{H} w_{hl,oj} \cdot \phi_h \left(b_{hl} + \sum_{k=1}^{n} w_{ik;hl} \cdot x_k \right) \right), \tag{2.71}$$

$$j = 1, \ldots, m$$

Referring to the network topology in Figure 2.9b, skip-layer connections are direct connections from each of the input units to each of the ouput units, that is, connections which skip the hidden layer. Since you can probably imagine how Figure 2.9b will look after adding these skip-layer connections, you will understand why we skip this here... As explained in [45], the important point is that skip layer connections make the neural network more flexible, allowing it to construct the regression surface as a perturbation of a linear hyperplane (again, if ϕ_o is linear). In Equation 2.71, the skip layer connections appear in the terms $w_{ik;oj} \cdot x_k$, which is the result of input node k after processing by the output node j. Again, the weights $w_{ik;oj}$ are real coefficients used by the output nodes to multiply the numbers received by the input nodes along the skip-layer connections.

Similar to above, Equation 2.71 can be fitted to data by a *minimization of RSQ* (similar to the discussion in Section 2.2.1, see also [45] for alternative optimization criteria provided by the nnet command that will be treated in Section 2.5.3 below). Altogether, the number of weights and biases appearing in Equation 2.71 is

$$N_p = H(n+1) + mH + mn + 1 \tag{2.72}$$

So you see that there is indeed a great number of "tuning" parameters that can be used to achieve a good fit between the model and the data, which makes it plausible that a statement such as Note 2.5.3 can be proved.

2.5.3
Solution Using Software

As a first example, let us look at Klein's investment data again (klein.csv, see Section 2.4). Above, a sine function was fitted to the data, leading to a residual sum of squares of $RSQ = 105.23$ (Figure 2.6). Let us see how a neural network performs on these data. An appropriate *R program* is NNEx1.r, which you find in the book software. Basically, NNEx1.r is obtained by just a little editing of NonRegEx1.r that was used in Section 2.4 above: we have to replace the nonlinear regression command nls used in NonRegEx1.r by the nnet command that computes the neural network. Let us look at the part of NNEx1.r that does the neural network computing:

```
1:  eq=inv~year
2:  FileName="Klein.csv"
3:  Data=read.table(FileName,...)
4:  Scaled=data.frame(year=Data$year/1941,inv=Data$inv)
5:  NNModel=nnet(eq,data=Scaled,size=3,decay=1e-4,
5:                  linout=T, skip=T, maxit=1000, Hess=T)
6:  eigen(NNModel$Hessian)$values
```

$$\tag{2.73}$$

In line 1, year and inv are specified as the input and output quantities of the model, respectively. Note that in contrast to our last treatment of these data in

Section 2.4, we do not need to specify the nonlinear functional form of the data which will be detected automatically by the neural network (compare line 1 of program 2.73 with line 1 of program 2.57). After the data have been read in lines 2 and 3, the explanatory variable `year` (which ranges between 1920 and 1941) in `Klein.csv` is rescaled to a range between 0 and 1, which is necessary for the reasons explained in [45]. A little care is necessary to distinguish between scaled and unscaled data particularly in the plotting part of the code (see `NNEx1.r`).

Note 2.5.4 (R's nnet command) In *R*, the `nnet` command can be used to compute a single-hidden layer feedforward neural network (Equation 2.71). Before using this command, the input data should be scaled to a range between 0 and 1.

The neural network model `NNModel` is obtained in line 5 using *R*'s `nnet` command. `nnet` uses the equation `eq` from line 1 and the scaled data `Scaled` from line 4. Beyond this, the `size` argument determines the number of nodes in the hidden layer; the `decay` argument penalizes overfitting, which will be discussed below; `linout=T` defines linear activation functions for the output units (i.e. ϕ_o is linear); `skip=T` allows skip-layer connections; `maxit=1000` restricts the maximum number of iterations of the numerical procedure, and `Hess=T` instructs `nls` to compute the Hessian matrix which can be used to check if a secure local minimum was achieved by the algorithm (see below). Note that `linout` and `skip` are so-called *logical variables* which have the possible values "T" (true) or "F" (false). Line 6 of the code is again related to the Hessian matrix and will be discussed below.

2.5.4
Interpretation of the Results

When you execute `NNEx1.r` in *R*, the `nnet` command will determine the parameters of Equation 2.71 such that the y_j computed by Equation 2.71 lie close to the data `klein.csv`, by default in the sense of a minimal RSQ as explained above. Remember that there are two kinds of parameters in Equation 2.71: the weights $w_{ik;oj}$, $w_{hl,oj}$, and $w_{ik;hl}$ which are used by the nodes of the network to multiply their input values, and the biases b_{oj} and b_{hl} which are added to the weighted sums of the values of the hidden layer or of the input layer. `NNEx1.r` uses a network with 3 nodes in the hidden layer (line 5 of program 2.73), which means that in Equation 2.71 we have $n = 1$, $H = 3$ and $m = 1$. Using Equation 2.72, you see that this gives a total number of $N_p = 11$ parameters that must be determined by `nnet`. Since Equation 2.71 is nonlinear, these parameters are determined by an iterative numerical procedure similar to the one discussed in Section 2.4 above [45, 64]. This procedure needs starting values as discussed above. In some cases, you may know appropriate starting values, which can be supplied to `nnet` in a way similar to the one used above for the `nls` command (see Section 2.4 and *R*'s help pages on

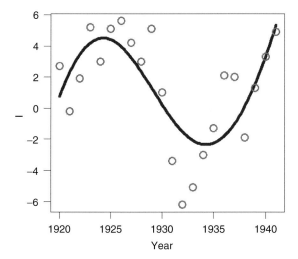

Fig. 2.10 Comparison of the neural network Equation 2.71 based on the parameters in program 2.73 (line) with the data in klein.csv (circles). Figure produced using NNEx1.r.

the nnet command). If you do not specify those starting values yourself, the nnet command uses automatically generated random numbers as starting values.

Note 2.5.5 (Random choice of starting values) Similar to the nls command that can be used for nonlinear regression (see Section 2.4), R's nnet command fits a neural network to data using an iterative procedure. By default, the starting values for the network weights and biases are chosen randomly, which implies that the results of subsequent runs of nnet will typically differ.

Executing NNEx1.r several times, you will see that some of the results will be unsatisfactory (similar to a mere linear regression through the data), while other runs will produce a picture similar to Figure 2.10. This figure is based on the skip-layer neural network equation 2.71 using the following parameters:

$$
\begin{array}{lllll}
\text{b->h1} & \text{i1->h1} \\
124.29 & -125.32 \\
\text{b->h2} & \text{i1->h2} \\
357.86 & -360.13 \\
\text{b->h3} & \text{i1 ->h3} \\
106.45 & -107.35 \\
\text{b->o} & \text{h1->o} & \text{h2->o} & \text{h3->o} & \text{i1->o} \\
41.06 & -195.48 & 136.35 & -156.09 & 48.32
\end{array}
\tag{2.74}
$$

An output similar to 2.74 is a part of the results produced by nnet. The correspondence with the parameters of Equation 2.71 is obvious: for example, "b->h1" refers to the bias added by the hidden layer node 1, which is b_{h1} in the notation of Equation 2.71, and hence 2.74 tells us that $b_{h1} = 124.29$. "i1->h1" refers to the weight used by the hidden layer node 1 to multiply the value of input node 1, which is $w_{i1;h1}$ in the notation of Equation 2.71, and hence 2.74 tells us that $w_{i1;h1} = -125.32$. Note that "i1->o" is the weight of the skip-layer connection (i.e. we have $w_{i1;o1} = 48.32$).

Note that the results in Figure 2.10 are very similar to the results obtained above using a sinusoidal nonlinear regression function (Figure 2.6). The difference is that in this case the sinusoidal pattern in the data was correctly found by the neural network without the need to find an appropriate expression of the regression function before the analysis is performed (e.g. based on a graphical analysis of the data as above). As explained above, this is particularly relevant in situations where it is hard to get an appropriate expression of the regression function, for example, when we are concerned with more than two input quantities where graphical plots involving the response variable and all input quantities are unavailable. The RSQ produced by the network shown in Figure 2.10 ($RSQ = 103.41$) is slightly better than the one obtained for the nonlinear regression function in Figure 2.6 ($RSQ = 105.23$). Comparing these two figures in detail, you will note that the shape of the neural network in Figure 2.10 is not exactly sinusoidal: its values around 1940 exceed its maximum values around 1925. This underlines the fact that neural networks are governed by the data only (if sufficient nodes in the hidden layer are used): the neural network in Figure 2.10 describes an almost sinusoidal shape, but it also detects small deviations from a sinusoidal shape. In this sense, neural networks have the potential to perform better compared to nonlinear regression functions such as the one used in Figure 2.6 which is restricted to an exact sinusoidal shape.

Note 2.5.6 (Automatic detection of nonlinearities) Neural networks describe the nonlinear dependency of the response variable on the explanatory variables without a previous explicit specification of this nonlinear dependency (which is required in nonlinear regression, see Section 2.4).

The nnet command determines the parameters of the network by a minimization of an appropriate fitting criterion [45, 64]. Using the default settings, the RSQ will be used in a way similar to the above discussion in Sections 2.2 and 2.4. The numerical algorithm that works inside nnet thus minimizes e.g. RSQ as a function of the parameters of the neural network, Equation 2.71, that is, as a function of the weights $w_{ik;oj}$, $w_{hl,oj}$, $w_{ik;hl}$ and of the biases b_{oj} and b_{hl}. Formally, this is the minimization of a function of several variables, and you know from calculus that if a particular value of the independent variable is a local minimum of such a function, the Hessian matrix at that point is positive definite, which means that the *eigenvalues of the Hessian matrix* at that point are positive [65]. In line 6 of 2.73, the eigenvalues of the Hessian matrix are computed (referring to the particular weights and biases found by nnet), and the result corresponding to the neural network in Figure 2.10

is this:

```
Eigenvalues of the Hessian:
362646.5 2397.25 16.52111
1.053426 0.01203984 0.003230483
0.00054226 0.0004922841 0.0003875433
0.0001999698 7.053957e-05
```
(2.75)

Since all eigenvalues are positive, we can conclude here that this particular neural network corresponds to a secure local minimum of RSQ.

2.5.5
Generalization and Overfitting

The decay parameter of the nnet command remains to be discussed. In NNetEx1.r, decay=1e-4 was used (line 5 of program 2.73). If you set decay=0 instead, you can obtain results such as the one shown in Figure 2.11a. Comparing this with Figure 2.10, you see that this fits the data much better, which is also reflected by an improved RSQ value ($RSQ = 45.48$ in Figure 2.11a compared to $RSQ = 103.4$ in Figure 2.10). Does this mean that the best results are obtained for decay=0? To answer this question, let us increase the size of the hidden layer. Until now, size=3 was used in all computations, that is, a hidden layer comprising of three nodes. Increasing this parameter, we increase the number of weights and biases that can be tuned toward a better fit of the neural network and the data, and hence we increase the flexibility of the neural network in this way. Figure 2.11b shows a result obtained for a hidden layer with nine nodes (size=9). In terms of

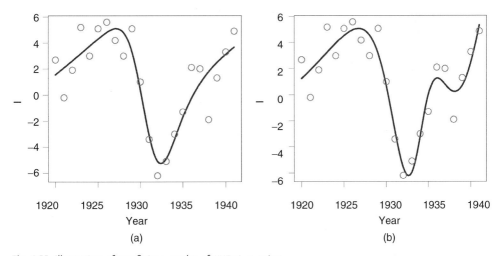

Fig. 2.11 Illustration of overfitting: results of NNEx1.r using (a) decay=0 and size=3 (b) decay=0 and size=9.

the RSQ, this neural network is again better than the previous one ($RSQ = 30.88$ in Figure 2.11b compared to $RSQ = 45.48$ in Figure 2.11a). Obviously, this improvement is achieved by the fact that the neural network in Figure 2.11b follows an extra curve compared to the network in Figure 2.11a, attempting to "catch" as many data points as possible as closely as possible.

This behavior is usually not desired since it restricts the predictive capability of a neural network. Usually, one wants neural networks to have the following

> **Definition 2.5.1 (Generalization property)**   Suppose two mathematical models (S, Q, M) and (S, Q, M^*) have been setup using a *training dataset* D_{train}. Then (S, Q, M) is said to *generalize better* than (S, Q, M^*) on a *test dataset* D_{test} with respect to some error criterion E, if (S, Q, M) produces a smaller value of E on D_{test} compared to (S, Q, M^*).

You may think of (S, Q, M) and (S, Q, M^*) as being regression or neural network models, and of E as being the RSQ as discussed above. Note that the mathematical models compared in Definition 2.5.1 refer to the same system S and to the same question Q since the generalization property pertains to the "mathematical part" of a mathematical model. The definition emphasizes the fact that it is not sufficient to look at a mathematical model's performance on the dataset which was used to construct the model if you want to achieve good predictive capabilities (compare the discussion of cross-validation in Section 2.3.3). To evaluate the predictive capabilities, we must of course look at the performance of the model on datasets that were *not* used to setup the model, and this means that we must ask for the generalization property of a model. Usually, better predictions are obtained from mathematical models which describe the essential tendency of the data (such as the neural networks in Figures 2.10 and 2.11a) instead of following random oscillations in the data similar to Figure 2.11b. The phenomenon of a neural network fitting the data so "well" that it follows random oscillations in the data instead of describing the general tendency of the data is known as *overfitting* [45]. Generally, overfitting is related with an increased "roughness" of the neural network function, since overfitted neural networks follow extra curves in an attempt to catch as many data points as possible as described above. The overfitting phenomenon can be defined as follows [66]:

> **Definition 2.5.2 (Overfitting)**   A mathematical model (S, Q, M) is said to *overfit* a training dataset D_{train} with respect to an error criterion E and a test dataset D_{test}, if another model (S, Q, M^*) with a larger error on D_{train} generalizes better to D_{test}.

For example, the neural network model behind Figure 2.11b will overfit the training data in the sense of the definition if it generates a larger error on unknown data e.g. compared to the neural network model behind Figure 2.10.

There are several strategies that can be used to reduce overfitting [45, 64]. So-called *regularization methods* use modified fitting criteria that penalize the "roughness" of the neural network, which means that these fitting criteria consider for example, both the RSQ and the roughness of the neural network. In terms of such a modified fitting criterion, a network such as the one in Figure 2.11a can be better than the "rougher" network in Figure 2.11b (although the RSQ of the second network is smaller). One of these regularizations methods called *weight decay* makes use of the fact that the roughness of neural networks is usually associated with "large" values of its weight parameters, and this is why this method includes the sum of squares of the network weights in the fitting criterion. This is the role of the decay parameter of the nnet command: decay=0 means there is no penalty for large weights in the fitting criterion. Increasing the value of decay, you increase the penalty for large weights in the fitting criterion. Hence decay=0 means that you may get overfitting for neural networks with sufficiently many nodes in their hidden layer, while positive values of the decay parameter decrease the "roughness" of the neural network and will generally improve its predictive capability. Ripley suggests to use decay values between 10^{-4} and 10^{-2} [67, 68]. To see the effect of this parameter, you may use a hidden layer with nine nodes similar to Figure 2.11b, but with decay=1e-4. Using these settings, you will observe that the result will look similar to Figures 2.10 and 2.11a, which means you get a much smoother (less "rough") neural network compared to the one in Figure 2.11b.

2.5.6
Several Inputs Example

As a second example which involves several input quantities, we consider the data in rock.csv which you find in the book software. These data are part of the *R* package, and they are concerned with petroleum reservoir exploration. To get oil out of the pores of oil-bearing rocks, petroleum engineers need to initiate a flow of the oil through the pores of the rock toward the exploration site. Naturally, such a flow consumes more or less energy depending on the overall flow resistance of the rock, and this is why engineers are interested in a *prediction of flow resistance* depending on the rock material. The file rock.csv contains data that were obtained from 48 rock sample cross-sections, and it relates geometrical parameters of the rock pores with its permeability, which characterizes the ease of flow through a porous material [69]. The geometrical parameters in rock.csv are: area, a measure of the total pore spaces in the sample (expressed in pixels in a 256×256 image); peri, the total perimeter of the pores in the sample (again expressed in pixels); and shape, a measure of the average "roundness" of the pores (computed as the smallest perimeter divided by the square root of the area for each individual pore; approx. 1.1 for an ideal circular pore, smaller for noncircular shapes). Depending on these geometrical parameters, rock.csv reports the rock permeability perm expressed in units of milli Darcy ($= 10^{-3}$ Darcy, see [70]).

In a first attempt to describe these data using a neural network, let us consider *two explanatory variables*, area and peri, neglecting the third geometrical variable, shape. With this restriction we will be able to generate 3D graphical plots of the neural network below. Moreover, we will take log(perm) as the response variable since `perm` covers several orders of magnitude (see `rock.csv`). Note that within R, `log` denotes the natural logarithm. To compute the neural network, we can proceed as above and start e.g. with `NNEx1.r`, editing the model and the name of the data file. This has been done in the R program `NNEx2.r` which you find in the book software. The core of the code in `NNEx2.r` that does the neural network computing is very similar to program 2.73 above. Basically, we just have to replace line 1 of 2.73 with

$$eq=log(perm)\sim area+peri \tag{2.76}$$

Again, a scaling of the explanatory variables must be applied similar to the one in line 4 of program 2.73 (see Note 2.5.4), but we leave out these technicalities here (see `NNEx2.r` for details). If you run `NNex2.r` within R, you will get results similar to the one shown in Figure 2.12 (note that you may obtain slightly different results for the reasons explained in the discussion of `NNEx1.r`). Figure 2.12 compares the neural network with the data in a way similar to the one in Section 2.4 above, using a predicted-measured plot and a conventional 3D plot. The 3D plot shows how the neural network builds up a nonlinear, three-dimensional surface that attains a shape that follows the essential tendency in the data much better than what could be achieved by multiple regression (note that in this case multiple regression amounts to fitting a flat surface to the data). This is also reflected by the residual sums of squares computed by `NNex2.r`: $RSQ = 32.7$ in the multiple linear model, and $RSQ = 15$ for the neural network model.

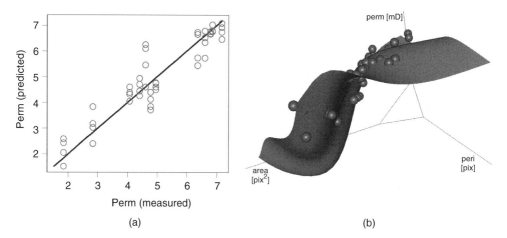

(a)   (b)

Fig. 2.12 Comparison of a neural network predicting `perm` depending on `area` and `peri` with the data in `rock.csv` (a) in a predicted-measured plot and (b) in a conventional plot in the `area-peri-perm` 3d-space. Plots generated by `NNEx2.r`.

Until now we have left out the `shape` variable of `rock.csv` as an explanatory variable. Changing the model within `NNEx2.r` to

$$eq=\log(perm)\sim area+peri+shape \qquad (2.77)$$

a neural network involving the *three explanatory variables* `area`, `peri` and `shape` is obtained, and in this way fits with an even better *RSQ* around 10 can be obtained. Similar to `NNEx1.r`, `NNEx2.r` plots the eigenvalues of the Hessian matrix so that you can check if a secure local minimum of the fitting criterion has been achieved as discussed above (e.g. Figure 2.12 refers to a neural network which has positive eigenvalues of the Hessian only as required). Finally, you should note that to evaluate the predictive capabilities of the neural networks discussed in this section, cross-validation can be used similar to above (Section 2.3.3).

Neural networks are useful not only for prediction, but also to *visualize and better understand a dataset*. For example, it can be difficult to understand the effects of two explanatory variables on a dependent variable (e.g. the effects of `area` and `peri` on `perm` in the above example) based on a 3D scatterplot of the data (the spheres in Figure 2.12b) only. In this case, a three-dimensional nonlinear neural network surface that approximates the data similar to Figure 2.12b can help us to see the general nonlinear form described by the data. Of course, linear or nonlinear regression plots such as Figure 2.8 can be used in a similar way to visualize datasets.

Note 2.5.7 (Visualization of datasets)   Neural networks (and mathematical models in general) can be used to visualize datasets. An example is Figure 2.12b, where the model surface highlights and accentuates the nonlinear effects of two explanatory variables on a dependent variable.

2.6
Design of Experiments

Suppose you want to perform an experiment to see if there are any differences in the durability of two house paintings A and B. In an appropriate experiment, you would e.g. paint five wall areas using paint A and five other wall areas using paint B. Then, you would measure the durability of the paintings in some suitable way, for example, by counting the number of defects per surface area after some time. The data could then be analyzed using the t test (Section 2.1). This example is discussed in [71], and the authors comment on it as follows:

You only have to paint a house once to realize the importance of this experiment.

Since the time and effort caused by an experiment as well as the significance of its results depend very much on an appropriate design of the experiment, this can also be phrased as follows:

Note 2.6.1 (Importance of experimental design) You only have to paint a house once to realize the importance of an appropriate design of experiments.

Indeed, the *design of experiments* – often abbreviated as *DOE* – is an important statistical discipline. It encompasses a great number of phenomenological models which all focus on an increase of the efficiency and significance of experiments. Only a few basic concepts can be treated within the scope of this book, and the emphasis will be on the practical software-based use of these methods. The reader should refer to books such as [41, 72, 73] to learn about more advanced topics in this field.

2.6.1
Completely Randomized Design

Figure 2.13 shows a possible experimental design that could be used in the house painting example. The figure shows 10 square test surfaces on a wall which are labeled according to the paint (A or B) that was applied on each of these test surfaces. This is a "naive" experimental design in the sense that it reflects the first thought which many of us may have when we think about a possible organization of these test surfaces. And, as it is the case with many of our "first thoughts" in many fields, this is a bad experimental design, even the worst one imaginable. The point is that most experiments that are performed in practice are affected by *nuisance factors* which, in many cases, are unknown and out of the control of the experimenter at the time when the experiment is designed. A great number of such possible nuisance factors may affect the wall painting experiment. For example, there may be two different rooms of a house behind the left ("A") and right ("B") halves of the wall shown in Figure 2.13, and the temperature of one of these rooms and, consequently, the temperature of one half of the wall may be substantially higher compared to the temperature of the other room and the other half of the wall. Since temperature may affect the durability of the painting, any conclusions drawn from such an experiment may hence be wrong.

This kind of error can very easily be avoided by what is called a *completely randomized design* or *CRD* design. As the name suggests, a completely randomized design is a design where the positions of the A and B test surfaces are determined

Fig. 2.13 House painting example: naive experimental design, showing a wall (large rectangle) with several test surfaces which are painted using paints A or B.

randomly. Based on the usual technical terminology, this can be phrased as follows: A completely randomized design is a design where the *treatments* or *levels* (corresponding to A and B in this case) of the *factor* under investigation (corresponding to the paint) are assigned randomly to the *experimental units* (corresponding to the test surfaces).

Using software, this can be done very easily. For example, *Calc*'s rand() function can be used as follows (see Section 2.1.1 and Appendix A for details about *Calc*):

Completely randomized design using *Calc*
- Generate a *Calc* spreadsheet with three columns labeled as *Experimental unit, Random number* and *Factor level*
- Write the desired factor levels in the *Factor level* column, for example, "A" in five cells of that column and "B" in another five cells in the case of the wall painting example.
- Enter the command =rand() in the cells of the *Random number* column (of course, it suffices to enter this into the top of that column, which can then be copied to the other cells using the mouse – see *Calc*'s help pages for details).
- Use *Calc*'s "Data/Sort" menu option to sort the data with respect to the *Random number* column.
- Write 1, 2, 3, ... in the *Experimental unit* column of the spreadsheet, corresponding to an enumeration of the experimental units that was determined before this procedure was started.

In the wall painting example, this procedure yields e.g. the result shown in Figure 2.14, which defines a random assignment of the test surfaces (which we can think of as being enumerated from 1 to 10 as we move from the left to the right side of the wall in Figure 2.13) to the factor levels A and B. Figure 2.14 has been generated using the *Calc* file CRD.ods in the book software (see Appendix A). Note that when you generate your own completely randomized designs using this file, you will have to enter the rand() command into the cells of the *Random number* column of that file again.

Assuming a higher temperature of the left part of the wall in Figure 2.13 as discussed above, this higher temperature would affect both the A and B test surfaces based on the completely randomized design in Figure 2.14. Hence, although the results still would be affected by the temperature variation along the wall since the variance of the data would be higher compared to an isothermal experiment, the completely randomized design would at least prevent us from wrong conclusions caused by the fact that the higher temperatures would be attributed to one of the factor levels only as discussed above.

A completely randomized design can also be generated using the design.crd function which is a part of *R*'s agricolae package. For the house painting example,

	A	B	C
1	**Experimental unit**	**Calc random number (sorted)**	**Factor level**
2	1	0,03	B
3	2	0,3	A
4	3	0,33	B
5	4	0,36	A
6	5	0,63	B
7	6	0,64	B
8	7	0,71	A
9	8	0,81	A
10	9	0,9	B
11	10	0,98	A

Fig. 2.14 Completely randomized design for the house painting example, computed using *Calc*'s rand() function. See the file CRD.ods in the book software.

this can be done using the following code (see CRD.r in the book software):

```
1:  library(agricolae)
2:  levels=c("A", "B")
3:  rep=c(5,5)
4:  out=design.crd(levels,rep,number=1)
5:  print(out)
```
(2.78)

After the agricolae package is loaded in line 1, the levels (A and B in the above example) and the number of replications of each level (5 replications for A and B) are defined in the variables levels and rep, which are then used in the design.crd command in line 4 to generate the completely randomized design. The design is then stored in the variable out, which is printed to the screen using *R*'s print command in line 5. The result may look like this (may: depending on the options that you choose for random number generation, see below):

```
   plots levels r        plots levels r
1    1     A    1     6    6     B    3
2    2     B    1     7    7     B    4
3    3     A    2     8    8     A    4
4    4     B    2     9    9     A    5
5    5     A    3    10   10     B    5
```

Here, the "plots" and "levels" columns correspond to the "Experimental unit" and "Factor level" column in Figure 2.14, while the "r" column counts the number of replications separately for each of the factor levels. Note that the result depends on the method that generates the random numbers that are used to randomize the design. A number of such methods can be used within the agricolae package (see the documentation of this package).

2.6.2
Randomized Complete Block Design

Consider the following experiment that is described in [72]: A hardness testing machine presses a rod with a pointed tip into a metal specimen with a known force. The depth of the depression caused by the tip is then used to characterize the hardness of the specimen. Now suppose that four different tips are used in the hardness testing machine and that it is suspected that the hardness readings depend on the particular tip that is used. To test this hypothesis, each of the four tips is used four times to determine the hardness of identical metal test coupons. In a first approach, one could proceed similar to the previous section. The hardness experiment involves one factor (the tip), four levels of the factor (tip 1–tip 4), and four replications of the experiment at each of the factor levels. Based on this information, a completely randomized design could be defined using the methods described above.

However, there is a problem with this approach. Sixteen different metal test coupons would be used in such a completely randomized design. Now it is possible that these metal test coupons differ slightly in their hardness. For example, these metal coupons may come from long metal strips, and temperature variations during the manufacturing of these strips may result in a nonconstant hardness of the strips and hence of the metal coupons. These hardness variations would then potentially affect the comparison of the four tips in a completely randomized experiment.

To remove the effects of possible hardness variations among the metal coupons, a design can be used that uses only four metal test coupons and that tests each of the four tips on each of these four test coupons. This is called a *blocked experimental design*, since it involves four "blocks" (corresponding to the four metal test coupons) where all levels of the factor (tip 1–tip 4) are tested in each of these blocks. Such blocked designs are used in many situations in order to achieve more homogeneous experimental units on which to compare the factor levels. Within the blocks, the order in which the factor levels are tested should be chosen randomly for the same reasons that were discussed in the previous section, and this leads to what is called a *randomized complete block design (RCBD)*.

In *R*, a RCBD for the above example can be computed using the following code (see RCBD.r in the book software):

```
1:  library(agricolae)
2:  levels=c("Tip 1", "Tip 2", "Tip 3", "Tip 4")
3:  out=design.rcbd(levels,4,number=1)
5:  print(out)
```

(2.79)

This code is very similar to the code 2.78 above, except for the fact that the command design.rcbd is used here instead of design.crd. Note that the second argument of design.rcbd gives the number of blocks (4 in this case). The code 2.79 may yield the following result in *R* (may: see the above discussion of

Equation 2.78):

	plots	block	levels		plots	block	levels
1	1	1	Tip 2	9	9	3	Tip 3
2	2	1	Tip 1	10	10	3	Tip 2
3	3	1	Tip 4	11	11	3	Tip 4
4	4	1	Tip 3	12	12	3	Tip 1
5	5	2	Tip 1	13	13	4	Tip 2
6	6	2	Tip 3	14	14	4	Tip 4
7	7	2	Tip 4	15	15	4	Tip 3
8	8	2	Tip 2	16	16	4	Tip 1

This result can be interpreted similar to the corresponding result of `design.crd` that was discussed in the previous section. Again, the "plots" column just counts the experiments, the "block" column identifies one of the four blocks corresponding to the four metal test coupons, and the "levels" column prescribes the factor level to be used in each experiment. Figure 2.15 visualizes this experimental design. Note that each of the tips is used exactly once on each of the metal test coupons as required.

2.6.3
Latin Square and More Advanced Designs

Again, there may be a problem with the experimental design described in the last section. As Figure 2.15 shows, tip 4 is tested three times in the third run of the experiment that is performed on the metal test coupons A–C. Now it may very well be that the result of the hardness measurement depends on the number of hardness measurements that have already been performed on the same test coupon. Previous measurements that have been performed on the same test coupon may have affected the structure and rigidity of the test coupon in some way. To avoid this as much as possible, the experimenter may test the four tips on different locations on the test coupon with a maximum distance between any two

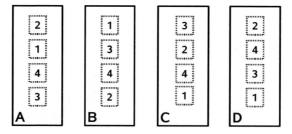

Fig. 2.15 Randomized complete block design computed using `design.rcbd` in R (hardness testing example): metal test coupons A–D, and numbers indicating the randomly chosen sequence of the tips 1–4.

of these locations. But then, the location itself may affect the hardness reading. Referring to the metal test coupon "A" in Figure 2.15, for example, the hardness measurements using tips 2 and 3 may be somewhat influenced by the fact that their measurement positions are more close to the ends of the metal coupon compared to the measurement positions of tips 1 and 4.

If the hardness measurement really depends on the number of measurements that have been previously performed on the same metal test coupon, or if the measurement depends on the measurement location on the metal test coupon, a design such as the one shown in Figure 2.15 – where one of the tips is used more than once in one particular position of the measurement order (such as tip 4) – is obviously an unsuitable design. Based on such a design, differences between the hardness readings produced by the tips may be observed which are caused by the particular position of the measurement order where the tips are used (rather than by the tips itself). Obviously, a better experimental design should randomly distribute the factor levels on each of the metal test coupons in a way such that any factor level appears only once in one particular position on the metal test coupons. A design of this kind is known as a *Latin square design*.

In *R*, a Latin square design for the above example can be computed using the following code (see LSD.r in the book software):

$$
\begin{array}{ll}
\texttt{1: library(agricolae)} & \\
\texttt{2: levels=c("Tip 1", "Tip 2", "Tip 3", "Tip 4")} & \quad (2.80) \\
\texttt{3: out=design.lsd(levels,number=1)} & \\
\texttt{5: print(out)} &
\end{array}
$$

This code may yield the following result (may: see the discussion of code (2.6.1) in Section 2.6.1):

plots	row	col	levels		plots	row	col	levels
1	1	1	1	Tip 1	9	9	3	1	Tip 4
2	2	1	2	Tip 2	10	10	3	2	Tip 1
3	3	1	3	Tip 3	11	11	3	3	Tip 2
4	4	1	4	Tip 4	12	12	3	4	Tip 3
5	5	2	1	Tip 3	13	13	4	1	Tip 2
6	6	2	2	Tip 4	14	14	4	2	Tip 3
7	7	2	3	Tip 1	15	15	4	3	Tip 4
8	8	2	4	Tip 2	16	16	4	4	Tip 1

Figure 2.16 visualizes this result similar to Figure 2.15 above. As can be seen, every tip occurs only once in each single row and column of this experimental design as required.

The design discussed so far can be generalized in various ways. While the Latin square can be used to treat situations with two sources of extraneous variability (in the above example, the metal test coupon and the position of a tip in the test sequence), the *Graeco–Latin square design* can treat a similar situation with three

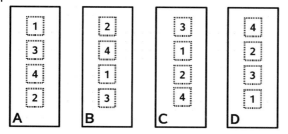

Fig. 2.16 Latin square design computed using design.lsd
in *R* (hardness testing example): metal test coupons A–D,
and numbers indicating the randomly chosen sequences of
the tips 1–4.

sources of variability. In *R*'s `agricolae` package, Graeco–Latin square designs
can be used based on the `design.graeco` command in a similar way as the
corresponding commands that were discussed above. In some cases, a RCBD may
be too demanding in terms of time and resources, that is, we may be unable to
test every factor level in each block of the design. Then, a *randomized balanced
incomplete block design* can be used, which can be obtained using the `design.bib`
function of *R*'s `agricolae` package.

2.6.4
Factorial Designs

All experimental designs considered so far involved one factor only. If an experiment
involves two factors or more, *factorial designs* or *response surface designs* are applied
[73]. Factorial designs are preferentially used in situations where each factor is
varied on two levels only, a lower level which is typically designated as "−",
and an upper level designated as "+". Designs of this type are called *two-level
factorial designs*. Response surface methods, on the other hand, focus on situations
where the factors are varied on more than two levels. They provide procedures
that can be used to decide about an optimal choice of the factor levels, based
on an approximation of the response of the system (y) depending on the factors
($x_1, x_2, \ldots$) e.g. using polynomials. We will confine ourselves here to the basics of
factorial designs (see [72, 73] for more on response surface methods).

A factorial design involving $n \in \mathbb{N}$ factors on $m \in \mathbb{N}$ levels is usually denoted as a
"m^n design". Hybrid factorial designs such as a 2×3^2 design are also used, which
involves one factor that is varied on two levels and two factors that are varied on three
levels in this case. As was already mentioned, the most frequently used factorial designs are 2^n designs, that is, designs involving n factors that are varied on two levels.

As a simple example, consider a chemical reactor which is used to produce some
product, and which is affected by two factors: the amount of a catalyst that is used
and the temperature. Let us denote the catalyst with A and the temperature with B,
and let us assume that an experiment is performed where both factors are varied
at a low and a high level, respectively, which we denote as "−" and "+". Table 2.2

Table 2.2 Example of a 2^2 factorial design.

No.	A	B	Yield
1	−	−	50
2	+	−	54
3	−	+	64
4	+	+	90

shows a possible result of such an experiment (the yield of the product is given in appropriate units which we do not need to discuss here). In the terminology introduced above, this is a 2^2 design as it involves two factors being varied on two levels. Experimental designs that involve all possible combinations of the factor levels, such as the design in Table 2.2, are also called *full factorial designs*.

Full factorial designs such as the design in Table 2.2 can be generated very easily using *R*'s expand.grid command as follows:

```
1:  levels=c("-", "")+
2:  expand.grid(A=levels,B=levels)
```

(2.81)

which yields the design that was used in Table 2.2:

```
  A B           A B
1 - -         3 - +
2 + -         4 + +
```

For the same reasons that were explained in Section 2.6.2 above, the experimenter may decide to use a *randomized block design* in a factorial experiment. In *R*, such a design can be realized e.g. using the design.ab command of *R*'s agricolae package. The package contains an example experiment that involves the "perricholi", "canchan", and "tomasa" potato varieties which are cultivated using three nitrogen levels. A full factorial 3^2 design involving five replications of each experiment is used, and the replications are organized in randomized blocks. To compute such a design, the following code can be used in *R* (see FacBlock.r in the book software):

```
1:  library(agricolae)
2:  variety=c("perricholi", "canchan", "tomasa")
3:  nitrogen=c(40,80,120)
4:  out=design.ab(variety, nitrogen, 5, number=1)
5:  print(out)
```

(2.82)

This code works very similar to the codes discussed in the previous section. Since we use a 3^2 design here which involves five replications, you can easily anticipate that this code will result in a total of $3^2 \cdot 5 = 45$ experimental runs. The first part of the output looks like this:

	plots	block	variety	nitrogen		plots	block	variety	nitrogen
1	1	1	tomasa	80	10	10	2	canchan	80
2	2	1	perricholi	120	11	11	2	canchan	40
3	3	1	canchan	120	12	12	2	perricholi	120
4	4	1	perricholi	80	13	13	2	tomasa	80
5	5	1	canchan	80	14	14	2	perricholi	40
6	6	1	perricholi	40	15	15	2	canchan	120
7	7	1	tomasa	40	16	16	2	perricholi	80
8	8	1	canchan	40	17	17	2	tomasa	40
9	9	1	tomasa	120	18	18	2	tomasa	120 ...

As can be seen, the code produces a complete 3^2 design involving all possible combinations of the potato varieties and nitrogen levels (randomly ordered) in each of the blocks.

Of course, time and resources restrictions may prevent us from performing full factorial experiments involving 45 different experimental runs, and similar to the "randomized balanced incomplete block design" discussed in Section 2.6.3 above, the experimenter will be interested in clever ways to reduce the overall number of experimental runs in a way that does not affect the significance of his results. Factorial designs that do not consider all possible combinations of the factor levels are called *fractional factorial designs*. Simple fractional factorial designs are usually denoted as "m^{n-k} designs", where m (the number of factor levels) and n (the number of factors) have the same meaning as before, and k expresses the fact that the total number of experimental runs has been reduced by a factor of $1/2^k$. Example: While a full factorial 2^5 design would require $2^5 = 32$ runs, a fractional 2^{5-2} design needs only $2^3 = 8$ runs. To generate such fractional factorial designs, one can use R's ffDesMatrix command (a part of the BHH2 contributed package). See [72, 73] for more theoretical background on factorial and fractional factorial designs.

2.6.5
Optimal Sample Size

An important issue in experimental design is the proper selection of sample size, that is, of the number of repetitions of a particular experiment that are necessary to achieve the desired results. This is very well supported by a number of R functions such as power.t.test, power.anova.test, power.prop.test, etc., which are a part of R's standard distribution. We will confine ourselves here to a simple example that demonstrates the way in which these functions can be used. Suppose the yield of crop variety A is suspected to be higher than the yield of crop variety B, and suppose you want to show this using a one-sided t test with $\alpha = 0.1$ and

$\beta = 0.9$ (see Section 2.1.3 for details on the testing procedure). Then, the following R command can be used to compute an optimal sample size:

```
power.t.test(sig.level=0.1,power=0.9,delta=2,sd=1
                     ,alternative="one.sided")
```

This yields an optimal sample size of $n \approx 3.9$ in R, which means that $n = 4$ replications should be used for each of the crop varieties. The `delta` argument is the "true difference in means", that is, $\mu_1 - \mu_2$ (see Section 2.1.3). Of course, you do not know this true difference in means a priori in a practical situation, so it should be set to a difference that you want to be detected by the experiment. For example, a difference in the yields of the varieties A and B may be practically important only if it is larger than $2\,\mathrm{kg\,m^{-2}}$, so in this situation you would set `delta=2` as above. Of course, it is more difficult to get a significant test result if `delta` is small, which is reflected by the fact that n increases as `delta` decreases (for example, $n = 14$ if `delta=1` is used in the above command). Hence `delta` should be chosen as large as possible such that it still satisfies the above requirement. The `sd` argument of `power.t.test` is the standard deviation of the random variables that generate the data, which is assumed to be constant here across the crop varieties. Again, this is not known a priori, and in a practical situation you would set `sd` according to the experience made in similar prior experiments (if available), or you would have to guess an appropriate order of magnitude based on your knowledge of the system and the measurement procedure.

2.7
Other Phenomenological Modeling Approaches

You should note that there is a great number of phenomenological modeling approaches beyond the ones introduced above. Only a few of these topics can be briefly addressed in the following sections: soft computing approaches in Section 2.7.1, discrete event simulation in Section 2.7.2, and signal processing in Section 2.7.3.

2.7.1
Soft Computing

Soft computing is used as a label for relatively new computational techniques such as *artificial neural networks (ANNs)*, *fuzzy logic*, *evolutionary algorithms*, but also for recent developments in fields such as *rough sets* and *probabilistic networks* [74, 75]. As it is explained in [75], a common feature of these techniques is that, unlike conventional algorithms, they are tolerant of imprecision, uncertainty, and partial truth.

Artificial neural networks have already been introduced in Section 2.5. The above discussion was restricted to neural networks used as a tool for nonlinear regression. As mentioned there, there is a great number of applications in various other fields such as time series prediction, classification and pattern recognition, or data processing. An important feature of neural networks is their ability to "learn" from data. We have seen in Section 2.5 that a neural network is able to "learn" the nonlinear shape of a function from a dataset, that is, there is no need to specify this nonlinear shape as a mathematical function as it is required in the classical nonlinear regression approach. This kind of adaptivity, that is, the ability of a model to adapt to a changing problem environment, is a characteristic feature of soft computing approaches in general [75].

As explained above, artificial neural networks were originally inspired by an analogy with biological neural networks (Note 2.5.2). In a similar way, *evolutionary algorithms* encompass a class of stochastic optimization algorithms that were originally inspired by an analogy with the biological ideas of genetic inheritance and the Darwinian law of the "survival of the fittest". In *genetic algorithms* – the most widely used type of evolutionary algorithms – individuals are represented as arrays of binary digits that can take on the values 0 or 1. Basically, these arrays can be thought of as representing the genes of the individuals. After a random initial population has been generated, an iterative process starts where new generations of the population are generated from the previous population by applying a certain number of stochastic operators to the previous population, which basically can be thought of as reflecting the Darwinian law of the "survival of the fittest". Similar to neural networks, this bio-inspired approach turned out to be extremely fruitful in the applications. Evolutionary algorithms have been applied in bio–informatics, phylogenetics, computer science, engineering, economics, chemistry, manufacturing, mathematics, physics, and other fields. See the examples in [76–78], and [75] for a detailed case study involving a financial application (portfolio optimization). Note that evolutionary algorithms are a part of the larger field of evolutionary computation which includes other techniques such as *swarm intelligence*, which describe the collective behavior of decentralized, self-organized systems [79]. Evolutionary computation itself is usually classified as a subfield of *artifical intelligence*, a discipline of computer science.

As regards *software* for soft computing applications, we have already used *R*'s nnet package in Section 2.5 above to do neural network-based nonlinear regression. The same package can also be used to solve classification problems, see [45]. *R*'s contributed package *genalg* (R Based Genetic Algorithm) can be used to implement genetic algorithms. While nnet comes as a standard part of the *R* distribution, genalg can be obtained from *R*'s internet site (*www.r-project.org*). *R* may also serve as a platform for the implementation of fuzzy models (there are a number of fuzzy-based contributed packages, see the list on *www.r-project.org*).

2.7.1.1 Fuzzy Model of a Washing Machine

We end this section on soft computing with an example of a fuzzy model. Based on the ideas developed in [80], fuzzy models use logical variables that can take on any value between 0 and 1. In this sense, these models allow for "uncertainty", as opposed to the usual concept where logical variables can take on the values 0 and 1 only. This is of interest in many technological applications where a system needs to be controlled in a smooth way, that is, not based on conventional "on/off" switches, but rather based on a kind of control that allows a smooth transition between the various states of a system.

An example of a washing machine controlled by a fuzzy model is discussed in [81]. In that example, the amount of detergent that is used by the machine must be determined based on the dirtiness of the load (as measured by the opacity of the washing water using an optical sensor system), and based on the weight of the laundry (as measured by a pressure sensor system). Using practical experiences, a set of control rules can be established which determines the amount of detergent that should be used as the dirtiness varies between the classes of a so-called *fuzzy subset*:

```
Almost_Clean, Dirty, Soiled, Filthy
```

and as the weight of the laundry varies between the classes of another fuzzy subset:

```
Very_Light, Light, Heavy, Very_Heavy
```

Now practical experience may tell us that the amount of detergent should be increased by a certain amount in a situation where the weight of the laundry is in the class Light while the dirtiness increases from Soiled to Filthy. Then, a fuzzy model will define a smooth transition of the amount of detergent that is used as the dirtiness changes its class from Soiled to Filthy. Basically, it will allow a classification of dirtiness partially between the classes Soiled and Filthy, that is, it will allow for uncertainty as explained above, and the amount of detergent that is used will reflect this uncertainty in the sense that it will be an average amount between the amounts that are defined in the rules for the classes Soiled and Filthy. Control strategies of this kind often produce better results compared to classical "on/off" strategies. See [75, 81] for a number of other examples such as vacuum cleaners, antilock brakes, and so on.

Fuzzy models can be described as an approach that leaves the realms of conventional mathematics to some extent. Beyond fuzzy models, there is a great variety of other modeling approaches that are located somewhere in a transition zone between quantitative (i.e. mathematical model based) and *qualitative modeling approaches*. See [82] for examples of qualitative approaches used in the social sciences (such as narrative analysis, action research, critical ethnography, etc.), and [83, 84] for a case study approach that uses both quantitative and qualitative methods.

2.7.2
Discrete Event Simulation

In this book, you have already seen (and will continue to see) that there is a great number of different approaches in mathematical modeling and simulation. With this in mind, you may be surprised to know that books such as the "Handbook of Simulation" by J. Banks (ed.) [6] or "Modern Simulation and Modeling" by Rubinstein and Melamed [85] both are devoted to one of these approaches only: discrete event simulation. Two things can be derived from this: (i) These discrete event simulation people have a very substantial self-esteem and self-confidence, and they know that what they do is an important part of the overall effort in modeling and simulation. (ii) *You* should know what they are doing!

Discrete event simulation can be defined as follows [10]:

> **Definition 2.7.1 (Discrete event simulation)** Discrete event simulation concerns the modeling of a system as it evolves over time by a representation in which the state variables changes instantaneously at a (countable number of) separate points in time. These points in time are the ones at which an *event* occurs, where event is defined as an instantaneous occurrence that may change the state of a system.

To understand the way in which discrete event simulation is used in practice, consider the following example which we cite here (almost unchanged) from [86]: Suppose that a single server (such as a clerk, a machine, or a computer) services randomly arriving customers (people, parts, or jobs). The order of service is first in, first out. The time between successive arrivals has the stationary distribution F, and the service time distribution is G. At time 0, the system is empty.

We may simulate this system as follows. First, generate a random number A_1 using the distribution F, and then a random number B_1 using G. See Section 2.1.2.3 for details on distributions, the discussion of `RNumbers.r` in Sections 2.1.2.3 and 2.1.2.5 for an example of random number generation using R, and [86] for the theoretical background of random number generation. Customer 1 enters the system at time A_1 (which is an *event* in the sense of the definition above) and leaves at time $A_1 + B_1$ (another event). Next, generate two more random numbers A_2 and B_2. Customer 2 arrives at $A_1 + A_2$. If $A_1 + A_2 \geq A_1 + B_1$, he starts service right away and finishes at $A_1 + B_1 + B_2$. And so on. A simulation of this kind can be used to estimate the number of customers that are served up to a given time, their average waiting time, and so on.

In a similar way, discrete event simulations can be used to simulate a great number of systems in various fields. They are used for example, to optimize and increase the efficiency of systems in manufacturing and material handling, logistics and transportation, and healthcare, see [6, 10]. Discrete event simulations are supported on a number of commercial and open source software platforms such as:

- `simcol`, open source, contributed package of R, see
  www.r-project.org and [87]
- *OpenModelica*, open source, see
  www.ida.liu.se/labs/pelab/modelica/OpenModelica.html and [7]
- *jemula*, open source, see jemula.origo.ethz.ch/
- *eM-Plant*, commercial, see *www.ugsplm.com*

- *SIMPROCESS*, commercial, see *www.caci.com*
- *simul8*, commercial, see *www.simul8.com/*

See Section 4.11.2 for an application of *R*'s `simcol` package.

2.7.3
Signal Processing

As it will become clear in the following chapters on differential equations, many mathematical models involve rates of changes of quantities of interest (see Note 3.1.1). Hence, it is a natural task to compute rates of changes from experimental data. As an example, let us reconsider the dataset `spring.csv` that was already analyzed in Section 1.5.6 above. Table 2.3 shows the x and y data of this dataset, and let us assume that we are interested in the rate of change of y with respect to x, that is, in the derivative $y'(x)$. Now the question is how $y'(x)$ can be derived from discrete data of the general form $(x_1, y_1), (x_2, y_2), \ldots, (x_n, y_n)$. In a naive approach, we could use the approximation

$$y'(x_i) \approx \frac{\Delta y_i}{\Delta x_i} = \frac{y_i - y_{i-1}}{x_i - x_{i-1}} \tag{2.83}$$

for $i = 2, \ldots, n$, which is based on the definition of the derivative as

$$y'(x) = \lim_{h \to 0} \frac{y(x + h) - y(x)}{h} \tag{2.84}$$

Using this approximation in `spring.csv` yields the values labeled as "$\Delta y / \Delta x$" in Table 2.3. As can be seen, these values scatter substantially in an interval between 0.1 and 0.6, that is, it seems that there are substantial changes of $y'(x)$ as x is increased. Looking at the plot of the data in Figure 1.9 (Section 1.5.6), however, it seems more likely here that the scattering of the $\Delta y / \Delta x$-data computed from the naive approach is caused by measurement errors only, which make the data swing a little bit around the regression line in Figure 1.9b that represents the "true" relation between x and y. Thus, we see that the naive approach is very sensitive to measurement errors, which makes this approach unusable for what is called the *numerical differentiation* of data, that is, for the determination of approximate derivatives from discrete datasets.

A better approach is already suggested by our analysis of `spring.csv` in Section 1.5.6 above, where we used the regression line in Figure 1.9b as a mathematical

Table 2.3 Naively computed rates of change in the dataset `spring.csv`.

x	10	20	30	40	50
y	3	5	11	12	16
$\Delta y / \Delta x$		0.2	0.6	0.1	0.4

model of the data, which was expressed as

$$\tilde{y}(x) = 0.33x - 0.5 \tag{2.85}$$

As explained there, this equation expresses the general (linear) tendency of the data, but it "damps out" the oscillations in the data that are e.g. induced by measurement errors. Equation 2.85 tells us that the rate of change of y with respect to x is constant:

$$\tilde{y}'(x) = 0.33 \tag{2.86}$$

This contradicts the conclusion that was drawn above from the naive approach, but a constant rate of change of y with respect to x obviously is in much better coincidence with the data that we see in Figure 1.9. The idea of differentiating regression functions is used quite generally as one of the standard procedures for the numerical differentiation of data. In the general case, the data may of course be nonlinear, which means that one will have to use, for example, polynomials as regression functions instead of the regression line that was used above. To obtain an approximation of $y'(x_i)$, one will typically fit a polynomial to a few datapoints around x_i only. This procedure can be viewed as an implementation of certain *low pass filters*, termed variously as *Savitzky–Golay smoothing filters, least squares filters*, or *DISPO* (digital smoothing polynomial) filters [88–90]. See [91] for an example application of this method to an analysis of the resin transfer molding (RTM) process, a process that is used in the manufacturing of fiber-reinforced composite materials.

There are many other approaches that can be used to "damp out" measurement error-induced oscillations of data, such as the *moving average approach* where a particular measurement value y_i basically is replaced by the average of a certain number of neighboring measurement values, see [92]. These methods are a part of the large field of *signal processing*, which is concerned with the analysis, interpretation, and manipulation of signals, preferentially signals in the form of sounds, images, biological signals such as the electrocardiogram (ECG), radar signals, and so on, but its methods are also used for the analysis of general experimental datasets [93]. Anyone who has used a MP3 player knows about the power of modern signal processing methods. Signal processing covers approaches such as the famous *Fourier analysis*, which allows a decomposition of a function in terms of sinusoidal functions, and which is used e.g. to remove unwanted frequencies or artifacts from audio or video recordings, or the recently developed *wavelet transforms*, which can be roughly described as a further development of the general idea of the Fourier transform and which have a similarly broad range of application (they are particularly well known for their use in data compression applications).

The open source software R provides a number of signal processing–related packages. Examples are the decompose package that can be used for the decomposition of a time series dataset into seasonal, trend, and irregular components using moving averages, or the `fft` package that can be used to compute fast discrete Fourier transformations. While these packages are a part of *R*'s base distribution, there is also a number of contributed signal processing packages that can be accessed via *R*'s internet site at *www.r-project.org*, including several packages that can be used to perform wavelet-based signal processing (examples are the `wavelets`, `waveslim`, and `wavetresh` packages).

3
Mechanistic Models I: ODEs

3.1
Distinguished Role of Differential Equations

As was explained, *mechanistic models* use information about the internal "mechanics" of a system (Definition 1.6.1). Referring to Figure 1.2, the main difference between phenomenological models (discussed in Chapter 2) and mechanistic models lies in the fact that phenomenological models treat the system as a black box, while in the mechanistic modeling procedure one virtually takes a look inside the system and uses this information in the model. This chapter and the following Chapter 4 treat *differential equations*, which is probably the most widely used mathematical structure of mechanistic models in science and engineering. Differential equations arise naturally, for example, as mathematical models of physical systems. Roughly speaking, differential equations are simply "equations involving derivatives of an unknown function". Their distinguished role among mechanistic models used in science and engineering can be explained by the fact that both scientists and engineers aim at the understanding or optimization of processes within systems.

The word "process" itself already indicates that a process involves a situation where "something happens", that is, where some quantities of interest change their values. Absolutely static "processes" where virtually "nothing happens" would be hardly of any interest to scientists or engineers. Now if it is true that some quantities of interest relating to a process under consideration change their values, then it is also true that such a process involves rates of changes of these quantities, which means in mathematical terms that it involves derivatives – and this is how "equations containing derivatives of an unknown function" or differential equations come into play. In many of the examples treated below it will turn out that it is natural to use rates of changes to formulate the mathematics behind the process, and hence to write down differential equations, while it would not have been possible to find appropriate equations without derivatives.

Mathematical Modeling and Simulation: Introduction for Scientists and Engineers. Kai Velten
Copyright © 2009 WILEY-VCH Verlag GmbH & Co. KGaA, Weinheim
ISBN: 978-3-527-40758-8

Note 3.1.1 (Distinguished role of differential equations)
1. Mechanistic models consider the processes running inside a system.
2. Typical processes investigated in science and engineering involve rates of changes of quantities of interest.
3. Mathematically, this translates into equations involving derivatives of unknown functions, i.e. differential equations.

Differential equations are classified into *ordinary and partial differential equations*. It is common to use *ODE* and *PDE* as abbreviations for ordinary and partial differential equations, respectively. This section is devoted to ODEs that involve derivatives with respect to only one variable (time in many cases), while PDEs (treated in Chapter 4) involve derivatives with respect to more than one variable (typically, time and/or space variables). In Section 3.2, mechanistic modeling is introduced as some kind of "systems archaeology", along with some first simple ODE examples that are used throughout this chapter. The procedure to set up ODE models is explained in Section 3.4, and Section 3.5 provides a theoretical framework for ODEs. Then, Sections 3.6–3.8 explain how you can solve ODEs either in closed form (i.e. in terms of explicit formulas) or using numerical procedures on the computer. ODE models usually need to be fitted to experimental data, that is, their parameters need to be determined such that the deviation of the solution of the ODE from experimental data is minimized (similar to the regression problems discussed in Chapter 2). Appropriate methods are introduced in Section 3.9, before a number of additional example applications are discussed in Section 3.10.

3.2
Introductory Examples

3.2.1
Archaeology Analogy

If one wants to explain what it really is that makes mechanistic modeling a very special and exciting thing to do, then this can hardly be done better than by the "archaeology analogy" of the French twentieth century philosopher Jacques Derrida [94]:

Note 3.2.1 (Derrida's archaeology analogy) "Imagine an explorer arrives in a little-known region where his interest is aroused by an expanse of ruins, with remains of walls, fragments of columns, and tablets with half-effaced and unreadable inscriptions. He may content himself with inspecting what lies exposed to his view, with questioning the inhabitants (...) who live in the vicinity, about what tradition tells them of the history and meaning of these archaeological remains, and with noting what they tell him – and he proceeds upon his journey. But he may act differently. He may have brought picks, shovels, and spades with him, and he may set the inhabitants to work with these implements. Together with them he may start upon the ruins, clearing away rubbish and, beginning from the visible remains, uncover what is buried. If his work is crowned with success, the discoveries are self-explanatory: the ruined walls are part of the ramparts of a palace or a treasure house; fragments of columns can be filled out into a temple; the numerous inscriptions, which by good luck, may be bilingual, reveal an alphabet and a language, and, when they have been deciphered and translated, yield undreamed-of information about the events of the remote past..."

Admittedly, one may not necessarily consider archaeology as an exciting thing to do, particularly when it is about sitting for hours at inconvenient places, scratching dirt from pot sherds, and so on. However, what Derrida describes is what might be called the exciting part of archaeology: revealing secrets, uncovering the buried, and exploring the unknown. And this is exactly what is done in mechanistic modeling. A mechanistic modeler is what might be called a *system archaeologist*. Looking back at Figure 1.2, he is someone who virtually tries to break up the solid system box in the figure, thereby trying to uncover the hidden internal system mechanics. A phenomenological modeler, in contrast, just walks around the system, collecting and analyzing the data which it produces. As Derrida puts it, he contents himself "with inspecting what lies exposed to view".

The exploration of subsurface structures by archeologists based on ground-penetrating radar provides a nice allegory for the procedure in mechanistic modeling. In this method, the archaeologist walks along a virtual x axis, producing scattered data along that x axis similar to a number of datasets that are investigated below. In the phenomenological approach, one would be content with an explanation of these data in terms of the input signal sent into the soil, for example, using appropriate methods from Chapter 2, and with no attempt toward an understanding of the soil structures generating the data. What the archaeologist does, however, is mechanistic modeling: based on appropriate models of the measurement procedure, he gains information about subsurface structures. Magnetic resonance imaging (MRI) and computed tomography (CT) are perhaps the most fascinating technologies of this kind – everybody knows these fantastically detailed pictures of the inside of the human body.

Note 3.2.2 (Objective of mechanistic modeling) Datasets contain information about the internal mechanics of the data-generating system. Mechanistic modeling means to uncover the hidden internal mechanics of a system similar to an archaeologist, who explores subsurface structures using ground-penetrating radar data.

3.2.2
Body Temperature

Now let us try to become system archaeologists for ourselves, starting with simple data sets and considerations. To begin with, suppose you do not feel so good today and decide to measure your body temperature. Using a modern clinical thermometer, you will have the result within a few seconds, usually indicated by a beep signal of your thermometer. You know that your thermometer needs these few seconds to bridge the gap between room and body temperature. Modern clinical thermometers usually have a display where this process of adjustment can be monitored, or better: *could* be monitored if you could see the display during measurement. Be that as it may, the dataset in Figure 3.1a shows data produced by the author using a clinical thermometer. The figure was produced using the dataset `fever.csv` and the Maxima program `FeverDat.mac` from the book software (see the description of the book software in Appendix A). `FeverDat.mac` does two things: it reads the data from `fever.csv` using *Maxima*'s `read_nested_list` command, and then it plots these data using the `plot2d` command (see `FeverDat.mac` and *Maxima*'s help pages for the exact syntax of these commands).

3.2.2.1 Phenomenological Model

Remembering what we have learned about phenomenological modeling in the previous chapter, it is quite obvious what can be done here. The data points follow a very simple and regular pattern, and hence it is natural to use an explicit function $T(t)$ describing that pattern, which can then be fitted to the data using *nonlinear regression* as described in Section 2.4. Clearly, the data in Figure 3.1a describe an essentially exponential pattern (imagine a $180°$ counterclockwise rotation of the data). Mathematically, this pattern can be described by the function

$$T(t) = T_b - (T_b - T_0) \cdot e^{-r \cdot t} \tag{3.1}$$

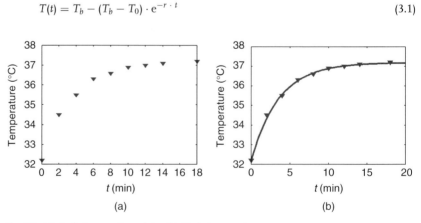

Fig. 3.1 (a) Body temperature data. (b) Body temperature data (triangles) and function $T(t)$ from Equation 3.4.

The parameters of this function have natural interpretations: T_0 is the initial temperature since $T(0) = T_0$, T_b is the body temperature since $\lim_{t \to \infty} T(t) = T_b$, and r controls the rate of temperature adjustment between T_0 and T_b. As Figure 3.1a shows, the values of T_0 and T_b should be slightly above 32 and 37 °C, respectively. Based on `fever.csv`, let us set $T_0 = 32.2$ and $T_b = 37.2$. To estimate r, we can, for example, substitute the datapoint $(t = 10, T = 36.9)$ from `fever.csv` in Equation 3.1

$$36.9 = 37.2 - 5 \cdot e^{-r \cdot 10} \tag{3.2}$$

which leads to

$$r = -\frac{\ln(0.06)}{10} \approx 0.281 \tag{3.3}$$

Note that Equation 3.3 can also be obtained using *Maxima*'s `solve` command, see the code `FeverSolve.mac` in the book software. Similar to the code (1.7) discussed in Section 1.5.2, the `solve` command produces several solutions here. Nine of these solutions are complex numbers, while one of the solutions corresponds to Equation 3.3. Using Equation 3.3, $T(t)$ can now be written as

$$T(t) = 37.2 - 5 \cdot e^{\ln(0.06)/10 \cdot t} \tag{3.4}$$

Plotting this function together with the body temperature data from Figure 3.1a, Figure 3.1b is obtained. Again, this plot was generated using *Maxima*: see `FeverExp.mac` in the book software. As the figure shows, the function $T(t)$ fits the data very well. Remember our discussion of nonlinear regression in Section 2.4 where a quantity called *pseudo-R^2* was introduced in formula (2.35) as a measure of the quality of fit. Here, the *Maxima* program `FeverExp.mac` computes an pseudo-R^2 value of 99.8%, indicating an almost perfect fit of the model to the data. Comparing Equation 2.35 with its implementation in `FeverExp.mac`, you will note that $\sum_{i=1}^{n}(y_i - \hat{y}_i)^2$ is realized in the form `(y-yprog).(y-yprog)`, where the "." denotes the scalar product of vectors, which multiplies vectors with components $y_i - \hat{y}_i$ in this case (see the *Maxima* help pages for more details on *Maxima*'s vector operation syntax). Note that the parameters of the model $T(t)$ have been obtained here using heuristic arguments. Alternatively, they could also be estimated using the nonlinear regression procedure described in Section 2.4.

3.2.2.2 Application

The model in Equation 3.4 can now be used to answer all kinds of questions related to the body temperature data. For example, it could be used to estimate the variation of the total measurement time (i.e. the time until the final measurement value is achieved) with varying starting temperatures of the thermometer. Or, it could be used to accelerate the measurement procedure using estimates of T_b based on the available data, and so on. Remember our definition of mathematical models in Section 1.4 above: a mathematical model is a set of mathematical statements that

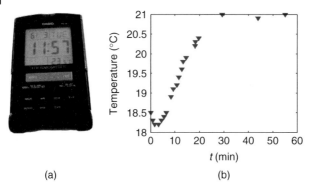

(a)   (b)

Fig. 3.2 (a) Alarm clock with temperature sensor. (b) Room temperature data.

can be used to answer a question which we have related to a system. As was pointed out there and in Note 1.2.2, the best mathematical model is the smallest and simplest set of mathematical statements that can answer the given question. In this sense, we can say that the phenomenological model (3.4) is the best mathematical model of the body temperature data probably with respect to most questions that we might have regarding the body temperature data. In Section 3.4.1, however, we will see that Equation 3.4 can also be derived from a mechanistic modeling approach.

3.2.3
Alarm Clock

Let us consider now a data set very similar to the body temperature data, but with a little complication that will lead us beyond the realms of phenomenological modeling. Suppose you enter a warm room with a temperature sensor in your hand, and you write down the temperature output of that sensor beginning with time $t = 0$ corresponding to the moment when you enter the warm room. At a first glance, this is a situation perfectly similar to the body temperature measurement, and you would probably expect a qualitative pattern of your data similar to Figure 3.1a. Now suppose that your data look as shown in Figure 3.2b; that is, your data are qualitatively different from those in Figure 3.1a, showing an initial decrease in the temperature even after you entered the warm room at time 0. Figure 3.2b has been produced using the *Maxima* code RoomDat.mac and the data room.csv in the book software (similar to FeverDat.mac discussed in Section 3.2.2).

3.2.3.1 Need for a Mechanistic Model
In principle, these data could be treated using a phenomenological model as before. To achieve this, we would just have to find some suitable function $T(t)$, which exhibits the same qualitative behavior as the data shown in Figure 3.2b. For example, a polynomial could be used for $T(t)$ (see the polynomial regression example in Section 2.2.6) or $T(t)$ could be expressed as a combination of a function

similar to Equation 3.1 with a second-order polynom. Afterwards, the parameters of $T(t)$ would have to be adjusted such that $T(t)$ really matches the data, similar to our treatment of the body temperature data. As before, the function $T(t)$ could then be used, for example, to estimate the total measurement time depending on the starting temperature, and so on. However, it is obvious that any estimate obtained in this way would be relatively uncertain as long as we do not understand the initial decrease in the temperature in Figure 3.2b. For example, if we would use the phenomenological model $T(t)$ to estimate the total measurement time for a range of starting temperatures, then we would implicitly assume a similar initial decrease in the temperature for the entire range of starting temperatures under consideration – but can this be assumed? We do not know unless we understand the initial decrease in the temperature.

The initial decrease in the temperature data shown in Figure 3.2b contains information about the system that should be used if we want to answer our questions regarding the system with a maximum of precision. The data virtually want to tell us something about the system, just as ground-penetrating radar data tell the archaeologist something about subsurface structures. To construct a phenomenological model of temperature data, only the data themselves are required, that is, one virtually just looks at the display of the device generating the temperature data. Now we have to change our point of view toward a look at the data-generating device itself, and this means we shift toward mechanistic modeling.

Note 3.2.3 (Information content of "strange effects") Mechanistic models should be used particularly in situations where "strange effects" similar to the alarm clock data can be observed. They provide a means to explain such effects and to explore the information content of such data.

3.2.3.2 Applying the Modeling and Simulation Scheme

Figure 3.2a shows the device that produced the data in Figure 3.2b: an alarm clock with temperature display. The data in Figure 3.2b were produced when the author performed a test of the alarm clock's temperature sensor, measuring the temperature inside a lecture room. Initially, he was somewhat puzzled by the decrease in the measurements after entering the warm lecture room, but of course the explanation is simple. The alarm clock was cheap and its temperature sensor an unhasty and lethargic one – an unbeatable time span of 30 min is required to bridge the gap between 18 and 21 °C in Figure 3.2b. Before the author entered the lecture room at time $t = 0$, he and the alarm clock were outdoors for some time at an ambient temperature around 12 °C. The initial decrease in the temperature measurements, thus, obviously meant that the author had disturbed the sensor inside the alarm clock when it still tried to attain that 12 °C.

Now to set up a mechanistic mathematical model that can describe the pattern of the data in Figure 3.2b, we can follow the steps of the *modeling and simulation scheme* described in Note 1.2.3 (Section 1.2.2). This scheme begins with the *definitions step*, where a question to be answered or a problem to be solved is defined. Regarding Figure 3.2b, a natural question would be

Q_1: How can the initial decrease of the temperature data be explained?

Alternatively, we could start with the problem

Q_2: Predict the final temperature value based on the first few data points.

In the *systems analysis step* of the scheme in Note 1.2.3, we have to identity those parts of the system that are relevant for Q_1 and Q_2. Remember the car example in Section 1.1, where the systems analysis step led us from the system "car" in its entire complexity to a very simplified car model comprising only tank and battery (Figure 1.1). Here, our starting point is the system "alarm clock" in its entire complexity, and we need to find a simplified model of the alarm clock now in the systems analysis step, guided by our questions Q_1 and Q_2. Obviously, any details of the alarm clock not related to the temperature measurement can be skipped in our simplified model – just as any details of the car not related to the problem "The car is not starting" were skipped in the simplified model of Figure 1.1.

Undoubtedly, the temperature sensor is an indispensable ingredient of any simplified model that is expected to answer our questions Q_1 and Q_2. Now remember that we stated in Note 1.2.2 that the simplest model is the best model, and that one, thus, should always start with the simplest imaginable model. We have the simplest possible representation of the temperature sensor in our model if we just consider the temperature T_s displayed by the sensor, treating the sensor's internal construction as a black box. Another essential ingredient of the simplified model is the ambient temperature T_a that is to be measured by the sensor. With these two ingredients, we arrive at the simplified alarm clock model in Figure 3.3, which we call *Model A*. Note that Model A is not yet a mathematical model, but rather what we have called a *conceptual model* in Section 1.2.5. Model A represents an intermediate step that is frequently used in the development of mathematical models. It identifies state variables T_s and T_a of the model to be developed and it provides an approximate sketch of their relationship, with T_s being drawn inside a

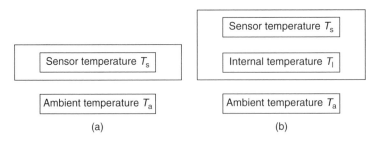

(a)   (b)

Fig. 3.3 Simplified models of the alarm clock: (a) Model A and (b) Model B.

rectangle symbolizing the alarm clock, T_a outside that rectangle. But although we already know the system S and the question Q of the mathematical model to be developed, Model A is still not a complete description of a mathematical model (S, Q, M) since mathematical statements M that could be used to compute the state variables are missing.

3.2.3.3 Setting Up the Equations

In this case, it is better to improve Model A a little bit before going into a formulation of the mathematical statements, M. Based on Model A, the only thing that the sensor (represented by T_s) "sees" is the ambient air temperature. But if this were true, then the initial decrease in the temperature data in Figure 3.2b would be hard to explain. If the sensor "sees" only the ambient air temperature, then its temperature should increase as soon as we enter the warm room at time $t = 0$. The sensor obviously somehow memorizes the temperatures of the near past, and this *temperature memory* must be included into our alarm clock model if we want to reproduce the data of Figure 3.2b. Now there are several possibilities how this temperature memory could be physically realized within the alarm clock. First of all, the temperature memory could be a consequence of the temperature sensor's internal construction. As a first, simple idea one might hypothesize that the temperature sensor always "sees" an old ambient temperature $T_a(t - t_{lag})$ instead of the actual ambient temperature $T_a(t)$. If this were true, the above phenomenological model for temperature adaption, Equation 3.1, could be used as follows. First, let us write down the ambient temperature T_a for this case

$$T_a(t) = \begin{cases} T_{a_1} & t < t_{lag} \\ T_{a_2} & t \geq t_{lag} \end{cases} \tag{3.5}$$

Here, T_{a_1} is the ambient temperature before $t = 0$, that is, before we enter the warm room. T_{a_2} is the ambient temperature in the warm room. Since we assume that the temperature sensor always sees the temperature at time $t - t_{lag}$ instead of the actual temperature at time t, Equation 3.5 describes the ambient temperature as seen by the temperature sensor. In Equation 3.1, $T_a(t)$ corresponds to the body temperature, T_b. This means that for $t < t_{lag}$ we have

$$T_1(t) = T_{a_1} - (T_{a_1} - T_0) \cdot e^{-r \cdot t} \tag{3.6}$$

and, for $t \geq t_{lag}$

$$T_2(t) = T_{a_2} - (T_{a_2} - T_1(t_{lag})) \cdot e^{-r \cdot (t - t_{lag})} \tag{3.7}$$

The parameters in the last two equations have the same interpretations as above in Equation 3.1. Note that $T_1(t_{lag})$ has been used as the initial temperature in Equation 3.7 since we are shifting from T_1 to T_2 at time t_{lag}, and $T_1(t_{lag})$ is the actual temperature at that time. Note also that $t - t_{lag}$ appears in the exponent of

Equation 3.7 to make sure that we have $T_2(t_{\text{lag}}) = T_1(t_{\text{lag}})$. The overall model can now be written as

$$T(t) = \begin{cases} T_1(t) & t < t_{\text{lag}} \\ T_2(t) & t \geq t_{\text{lag}} \end{cases} \tag{3.8}$$

3.2.3.4 Comparing Model and Data

Some of the parameters of the last equations can be estimated based on Figure 3.2b and the corresponding data in Room.csv:

$$T_0 \approx 18.5 \tag{3.9}$$

$$t_{\text{lag}} \approx 2.5 \tag{3.10}$$

$$T_{a_2} \approx 21 \tag{3.11}$$

In principle, the remaining parameters (T_{a_1} and r) can now either be determined by heuristic arguments as was done above in the context of Equation 3.1, or by nonlinear regression methods as described in Section 2.4. However, before this is done, it is usually efficient to see if reasonable results can be achieved using hand-tuned parameters. In this case, a hand tuning of the remaining parameters T_{a_1} and r shows that no satisfactory matching between Equation 3.8 and the data can be achieved, and thus any further effort (e.g. nonlinear regression) would be wasted. Figure 3.4 shows a comparison of Equation 3.8 with the data of Figure 3.2b based on the hand-fitted values $T_{a_1} = 16.7$ and $r = 0.09$. The figure has been produced using the *Maxima* code RoomExp.mac and the data room.csv in the book software. Looking at RoomExp.mac, you will note that the if...then command is used to implement Equation 3.8 (see *Maxima*'s help pages for more information on this and other "conditional execution" commands).

RoomExp.mac computes a coefficient of determination $R^2 = 92.7\%$, reflecting the fact that data and model are relatively close together. Nevertheless, the result

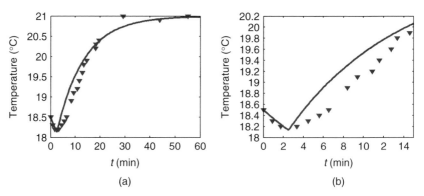

(a)   (b)

Fig. 3.4 (a) Comparison of Equation 3.8 (line) with the data of Figure 3.2b (triangles) using $T_{a_1} = 16.7$ and $r = 0.09$. (b) Same picture on a different scale, showing a dissimilarity between model and data.

is unsatisfactory since the qualitative pattern of the model curve derived from Equation 3.8 differs from the data. This is best seen if model and data are plotted for $t < 14$ as shown in Figure 3.4b. As can be seen, there is a sharp corner in the model curve, which is not present in the data. Also, the model curve is bent upward for $t > 3$ while the data points are bent downward there. Although the coefficient of determination is relatively high and although we might hence be able to compute reasonable temperature predictions based on this model, the qualitative dissimilarity of the model and the data indicates that the modeling approach based on Equation 3.8 is wrong. As it was mentioned in Section 1.2.2, the qualitative coincidence of a model with its data is an important criterion in the validation of models.

3.2.3.5 Validation Fails – What Now?

We have to reject our first idea of how temperature memory could be included into the model. Admittedly, it was a very simple idea to assume that the sensor sees "old" temperatures $T_a(t - t_{lag})$ shifted by a constant time t_{lag}. One of the reasons why this idea was worked out here is the simple fact that it led us to a nice example of a model rejected due to its qualitative dissimilarity with the data. Equation 3.8 also is a nice example showing that one cannot always distinguish in a strict sense between phenomenological and mechanistic models. On the one hand, it is based on the phenomenological model of temperature adaption, Equation 3.1. On the other hand, Equation 3.1 has been used here in a modified form based on our mechanistic considerations regarding the temperature memory of the system. As was already mentioned in Section 1.5, models of this kind lying somewhere between phenomenological and mechanistic models are also called *semiempirical* or *gray-box* models.

Before going on, let us spend a few thoughts on what we did so far in terms of the modeling and simulation scheme (Note 1.2.3). Basically, our systems analysis above led us to the conclusion that our model needs some kind of temperature memory. Equation 3.8 corresponds to the modeling step of Note 1.2.3, the simulation and validation steps correspond to Figure 3.4. After the validation of the model failed, we are now *back in the systems analysis step*. Principally, we could go now into a more detailed study of the internal mechanics of the temperature sensor. We could, for example, read technical descriptions of the sensor, hoping that this might lead us on the right path. But this would probably require a considerable effort and might result into unnecessarily sophisticated models. Before going into a more detailed modeling of the sensor, it is better to ask if there are other, simple hypotheses that could be used to explain the temperature memory of the system.

Remember that Note 1.2.2 says that the simplest model explaining the data is the best model. If we find such a simple hypothesis explaining the data, then it is the best model of our data. This holds true even if we do not know that this hypothesis is wrong, and even if the data could be correctly explained *only* based on the temperature sensor's internal mechanics. If both the models – the wrong model based on the simple hypothesis and a more complex model based on the temperature sensor's internal mechanics – explain the data equally and indistinguishably well, then we should, for all practical purposes, choose the simple model – at least based on the data, in the absence of any other indications that it is a wrong model.

3.2.3.6 A Different Way to Explain the Temperature Memory

Fortunately, it is easy to find another, simple hypothesis explaining the temperature memory of the system. In contrast to our first hypothesis above ("temperature sensor sees old temperatures"), let us now assume that the qualitative difference between the data in the body temperature and alarm clock examples (Figures 3.1a and 3.2b) is *not* a consequence of differences in the internal mechanics of the temperature sensors, but let us assume that the temperature sensors used in both the examples work largely the same way. If this is true, then the difference between the data in the body temperature and alarm clock examples must be related to differences in the construction of the clinical thermometer and the alarm clock as a whole, namely to differences in their construction related to temperature measurements. There is indeed an obvious difference of this kind: when the clinical thermometer is used, the temperature sensor is in direct contact with the body temperature that is to be measured. In the alarm clock, on the other hand, the temperature sensor sits somewhere inside, not in direct contact with the ambient temperature that is to be measured. This leads to the following.

> **Hypothesis:**
> The temperature memory of the alarm clock is physically realized in terms of the temperature of its immediate surroundings inside the alarm clock, for example, as the air temperature inside the alarm clock or as the temperature of internal parts of the alarm clock immediately adjacent to the temperature sensor.

To formulate this idea in mathematical terms, we need one or more state variable(s) expressing internal air temperature or the temperatures of internal parts immediately adjacent to the temperature sensor. Now it was emphasized several times that we should start with the simplest approaches. The simplest thing that one can do here is to use an *effective internal temperature* T_i, which can be thought of as some combination of internal air temperature and the temperature of relevant internal parts of the alarm clock. A more detailed specification of T_i would require a detailed investigation of the alarm clock's internal construction, and of the way in which internal air and internal parts' temperatures affect the temperature sensor. This investigation would be expensive in terms of time and resources, and it would hardly improve the results, which is achieved below based on T_i as a largely unspecified "black box quantity".

> **Note 3.2.4 (Effective quantities)** Effective quantities expressing the cumulative effects of several processes (such as T_i) are often used to achieve simple model formulations.

Introducing T_i as a new state variable in Model A, we obtain *Model B* as an improved conceptual model of the alarm clock (Figure 3.3). As a next step, we now need mathematical statements (the M of the mathematical model (S, Q, M) to be developed) that can be used to compute the state variables. Since ODEs are required here, we will go on with this example at the appropriate place below (Section 3.4.2). It will turn out that Model B explains the data in Figure 3.2b very well.

3.2.3.7 Limitations of the Model

Remember that as a mechanistic modeler you are a "systems archaeologist", uncovering the internal system mechanics from data similar to an archaeologist who derives subsurface structures from ground-penetrating radar data. Precisely in this "archaeological" way, Figure 3.3 was derived from the data in Figure 3.2b by our considerations above. Our starting point was given by the data in Figure 3.2b with the puzzling initial decrease in the temperatures, an effect that obviously "wants to tell us something" about internal system mechanics. Model B now represents a hypothesis of what the data in Figure 3.2b might tell us about the internal mechanics of the alarm clock during temperature measurement.

Note that Model B is a really brutal simplification of the alarm clock as a real system (Figure 3.2a), similar to our brutal simplification of the system "car" in Section 1.1. In terms of Model B, nothing remains of the initial complexity of the alarm clock except for the three temperatures T_s, T_i, and T_a. It must be emphasized that Model B represents an *hypothesis* about what is going on inside the alarm clock during temperature measurement. It may be a right or wrong hypothesis. The only thing we can say with certainty is that Model B probably represents the simplest *hypothesis* explaining the data in Figure 3.2b. Model B may fail to explain more sophisticated temperature data produced with the alarm clock, and such more sophisticated data might force us to go into a more detailed consideration of the alarm clock's internals; for example, into a detailed investigation of the temperature sensor, or into a more detailed modeling of the alarm clock's internal temperatures, which Model B summarizes into one single quantity T_i.

As long as we are concerned with the data only in Figure 3.2b, we can be content with the rather rough and unsharp picture of the alarm clock's internal mechanics provided by Model B. The uncertainty remaining when mechanistic models such as Model B are developed from data is a principal problem that cannot be avoided. A mechanistic model represents a hypothesis about the internal mechanics of a system, and it is well known that it is, as a matter of principle, impossible to prove a scientific hypothesis based on data [95]. Data can be used to show that a model is wrong, but they can never be used to prove its validity. From a practical point of view, this is not a problem since we can be content with a model as long as it explains the available data and can be used to solve our problems.

3.3
General Idea of ODE's

3.3.1
Intrinsic Meaning of π

Simplicity is a characteristic of good mathematical models (Note 1.2.2), but also a characteristic of good science in general (a fact that does not really surprise us since good science must of course be based on good mathematical models ...). A scientific result is likely to be well understood if it can be expressed in simple terms. For example, if there is someone who wants to know about the meaning of π, you could begin with a recitation like this

$$\pi = 3.14159265358979323846264338327950288419716939937510\ldots \quad (3.12)$$

If that someone is a clever guy, he might be able to do a similar recitation after some time. But, of course, he would not understand the meaning of π, even if he could afford the time to recite every digit. Your next idea may be to use formulas such as [96]

$$\pi = 4 \cdot \sum_{n=0}^{\infty} \frac{(-1)^n}{2n+1} \quad (3.13)$$

Probably, your listener would be happy that it is not so much effort to memorize π this way. But, of course, he still would not understand. He would not understand until you say a simple thing: "π is the surface area of a circle with radius 1." He would understand it this way because π is treated in the right setting here, namely in terms of geometry. π can be treated in terms of algebra as above, but its intrinsic meaning belongs to the realms of geometry, not algebra. This is reflected by the fact that you have to tell comparatively long and complicated stories about π in terms of algebra (Equations 3.12 and 3.13 above), while much more can be said by a very simple statement in terms of geometry. The example may seem trivial, but its practical importance can hardly be underestimated. To understand things, we should always try to find a natural setting for our objects of interest.

3.3.2
e^x Solves an ODE

With this idea in mind, let us reconsider the body temperature example from the last section. Remember that the data in Figure 3.1a were described using the function

$$T(t) = T_b - (T_b - T_0) \cdot e^{-r \cdot t} \quad (3.14)$$

The main ingredient of this function is the exponential function. Now, similar to above, let us ask a naive question: What is the exponential function? If you never have thought about this, you may start with similar answers as above. For example, the exponential function is the power function e^t having the Euler number

$$e = 2.71828182845904523536028747135266249775724709369959\ldots \quad (3.15)$$

as its base. Or, you may use a formula similar to Equation 3.13 [17]:

$$e^x = \sum_{n=0}^{\infty} \frac{x^n}{n!} \qquad (3.16)$$

Undoubtedly, answers of this kind would prove that you are knowing a lot more than those students who believe that "the exponential function is the e^x key on my pocket calculator". But you never would be able to understand the intrinsic meaning of the exponential function this way. As above, algebra just is not the right setting if you want to understand e^t.

The right setting for an understanding of the exponential function are ODEs, just as geometry is the right setting to understand π. This is related with a fact that most readers may remember from calculus: The derivative of e^t is e^t, that is, e^t and its derivative coincide. In terms of ODEs, this is expressed as follows: e^t solves the ODE

$$y' = y \qquad (3.17)$$

with initial condition

$$y(0) = 1 \qquad (3.18)$$

3.3.3
Infinitely Many Degrees of Freedom

Let us say a few words on the meaning of the last two equations (precise definitions are given in Section 3.5). At a first glance, Equation 3.17 looks similar to many other algebraic equations that the reader will already have encountered during his mathematical education. There is an unknown quantity y in this equation, and we have to find some particular value for y such that Equation 3.17 is satisfied. There is, however, an important difference between Equation 3.17 and standard algebraic equations: the unknown y in Equation 3.17 is a *function*. To solve Equation 3.17, y must be replaced by a function $y(t)$ which satisfies Equations 3.18 and 3.17 for *every* $t \in \mathbb{R}$. Note that y' is the derivative of y with respect to the independent variable, t in this case. Any of the common denotations of derivatives can be used when writing down differential equations. In particular, $\dot{y}$ is frequently used to denote a time derivative.

The fact that *functions serve as unknowns* of differential equations makes their solution a really challenging task. A simple analogy may explain this. Consider, for example, a simple linear equation such as

$$3x - 4 = 0 \tag{3.19}$$

which everyone of us can easily solve for x. Now you know that the number of unknowns and equations can be increased arbitrarily. For example,

$$2x - 8y = 4$$
$$7x - 5y = 2 \tag{3.20}$$

is a system of two linear equations in the two unknowns x and y, and this generalizes to a system of n linear equations in n unknowns as follows:

$$a_{11}x_1 + a_{12}x_2 + \cdots + a_{1n}x_n = b_1$$
$$a_{21}x_1 + a_{22}x_2 + \cdots + a_{2n}x_n = b_2$$
$$\vdots \tag{3.21}$$
$$a_{n1}x_1 + a_{n2}x_2 + \cdots + a_{nn}x_n = b_n$$

To solve the last system of equations, the n unknowns $x_1, x_2, \ldots, x_n$ must be determined. Although there are efficient methods to solve equations of this kind (e.g. *Maxima's* `solve` command could be used, see also [17]), it is obvious that the computational effort increases beyond all bounds as $n \to \infty$. Now remember that the unknown of a differential equation is a function. At a first glance, it may seem that we, thus, have "less unknowns" in an equation such as (3.17) compared to Equation 3.21: y as the only unknown of Equation 3.17, and $x_1, x_2, \ldots, x_n$ as the n unknowns of Equation 3.21. But now ask yourself how many numbers $x_1, x_2, \ldots, x_n$ you would need to describe a function $y = f(x)$ on some interval $[a,b]$. In fact, you would need *infinitely many* numbers, as you would have to recite all the values of that function for those infinitely many values of the independent variable x in the interval $[a,b]$. This is related to the fact that a real function $y = f(x)$ on some interval $[a,b]$ is said to have *infinitely many degrees of freedom* on its domain of definition. If we solve a differential equation, we thus effectively solve an equation having infinitely many unknowns corresponding to the solution functions degrees of freedom. The solution of differential equations is indeed one of the most challenging tasks in mathematics, and a substantial amount of the scientific work in mathematics has been devoted to differential equations since many years.

3.3.4
Intrinsic Meaning of the Exponential Function

It was said that to solve Equation 3.17, we need to replace the unknown y by a function $f(t)$ such that a valid equation results. Choosing $f(t) = e^t$ and using $f'(t) = e^t$, Equation 3.17 turns into

$$e^t = e^t \tag{3.22}$$

which means that e^t indeed solves Equation 3.17 in the sense described above. Note that $f(t) = c \cdot e^t$ would also solve Equation 3.17, where $c \in \mathbb{R}$ is some constant. This kind of nonuniqueness is a general characteristic of differential equations, usually overcome by imposing appropriate additional *initial conditions* or *boundary conditions*. Equation 3.18 provides an example of an initial condition (boundary conditions are discussed further below). It is not really hard to understand why this condition is needed here to make the solution of Equation 3.17 unique. Remember how we motivated the use of differential equations above (Note 3.1.1): they provide us with a means to formulate equations in terms of the rates of change of the quantities we are interested in. From this point of view, we can say that Equation 3.17 fixes the rate of change for the quantity y in a way that it always equals the current value of y. But it is of course obvious that we cannot derive the absolute value of a quantity from its rate of change. Although your bank account may have the same interest rates as the bank account of Bill Gates, there remains a difference that can be made precise in terms of the concrete values of those bank accounts at some particular time, which is exactly what is meant by an initial condition.

Equation 3.18 is an initial condition for the ODE (Equation 3.17) since it prescribes the initial value of y at time $t = 0$. $f(t) = e^t$, our solution of Equation 3.17, obviously also satisfies Equation 3.18, and it thus solves the so-called initial value problem comprising Equations 3.17 and 3.18. Thus, we can now answer the above question "What is the exponential function?" as follows:

> The exponential function is the real function $f(t)$ which coincides everywhere with its derivative and satisfies $f(0) = 1$.

This statement characterizes the exponential function in a simple and natural way, similarly simple and natural as the characterization of π in Section 3.3.1. Note that unique solvability of Equations 3.17 and 3.18 is assumed here – this is further discussed in Section 3.5. There is indeed no simpler way to explain the exponential function, and the complexity of algebraic statements such as Equations 3.15 and 3.16 emanates from that simple assertion above just as the algebraic complexity of π emanates from the assertion that "π is the surface area of a circle having radius 1". As an alternative, less mathematical formulation of the above "explanation of the exponential function", we could also say (neglecting the initial condition)

> The exponential function describes a process where the rate of change of a quantity of interest always equals the actual value of that quantity.

This second formulation makes it understandable why the exponential function appears so often in the description of technical processes. Technical processes involve rates of changes of quantities of interest, and the exponential function describes the simplest hypothesis that one can make regarding these rates of changes. This is the intrinsical meaning of the exponential function, and this is why it is so popular. Note that the exponential function can also be used in situations where the rate of change of a quantity of interest is a linear function of the actual value of that quantity (Sections 3.4.1 and 3.7.2.1).

3.3.5
ODEs as a Function Generator

Ordinary differential equations also provide the right setting for an understanding of many other transcendental functions. For example, the cosine function can be characterized as the solution of

$$y'' = -y \tag{3.23}$$

with the initial conditions

$$\begin{aligned} y(0) &= 1 \\ y'(0) &= 0 \end{aligned} \tag{3.24}$$

$\cos(t)$ satisfies Equation 3.23 in the sense described above since $\cos''(t) = -\cos(t)$ for $t \in \mathbb{R}$. Equation 3.24 is also obviously valid since $\cos(0) = 1$ and $\cos'(0) = -\sin(0) = 0$. Note that two initial conditions are required here, which is related to the fact that Equation 3.23 is a so-called second-order equation since it involves a second-order derivative. The *order of an ODE* is always the order of its highest derivative (precise definitions are given in Section 3.5).

> **Note 3.3.1 (Understanding transcendental functions)** Ordinary differential equations provide a natural framework for an understanding of the intrinsic meaning of functions that are frequently used to describe technical processes such as the exponential function and the trigonometric functions.

Beyond this, we will see that ODEs virtually serve as what might be called a *function generator* in the sense that they can be used to produce functions that do not belong to the classical zoo of functions such as the exponential and trigonometric functions. An example is the function describing the time dependence of cell biomass during wine fermentation (see Figure 3.20 in Section 3.10.2). This function cannot by described by any of the classical functions, but it can be described using a system of ODEs as it is explained below.

> **Note 3.3.2 (ODEs as function generator)** ODEs can be used to generate functions that do not belong to the classical zoo of functions such as the exponential and trigonometric functions.

3.4
Setting Up ODE Models

Based on the last section, the reader knows some basic facts about ODEs. As this section will show, this is already sufficient to set up and solve some simple ODE models. This is not to say, however, that you should not read the next sections. As Section 3.8 will show, things can go wrong when you do not know what you are doing, particularly when you solve ODEs using numerical methods.

3.4.1
Body Temperature Example

The models discussed in Section 3.2 are reformulated here in terms of ODE's. Let us begin with a reinvestigation of the body temperature data (Figure 3.1a). Above, we used the following phenomenological model of these data (compare Equation 3.1):

$$T(t) = T_b - (T_b - T_0) \cdot e^{-r \cdot t} \tag{3.25}$$

3.4.1.1   Formulation of an ODE Model

This formulation involves the exponential function, and in the last section we have learned that the intrinsic meaning of the exponential function becomes apparent only when it is expressed in terms of an ODE. We may, thus, hope to learn more about the background of this equation by reformulating it as an ODE. To achieve this, we need to express $T(t)$ somehow in terms of its derivative. So let us write down the derivative of $T(t)$:

$$T'(t) = r \cdot (T_b - T_0) \cdot e^{-r \cdot t} \tag{3.26}$$

This equation can already be viewed as an ODE, since it is an equation in the unknown $T(t)$ which involves a derivative of the unknown function. An appropriate initial condition would be

$$T(0) = T_0 \tag{3.27}$$

since T_0 is the initial temperature at time 0 (see the discussion of Equation 3.1 above). Mathematically, this ODE is easily solved by an integration of the right-hand side of Equation 3.26, using Equation 3.27 to determine the integration constant. This leads us back to Equation 3.25, without learning anything about the background of that equation. Does this mean it makes no sense to reformulate Equation 3.25 as an ODE?

Not at all. Remember the structure of the ODE characterizing the exponential function, Equation 3.17, where the unknown function appears in the right-hand side of the equation. The right-hand side of Equation 3.26 can also be expressed in terms of the unknown $T(t)$ as follows:

$$T'(t) = r \cdot (T_b - T(t)) \tag{3.28}$$

Together with the initial condition (3.27) we thus arrive at the following *initial value problem*:

$$T'(t) = r \cdot (T_b - T(t)) \tag{3.29}$$

$$T(0) = T_0 \tag{3.30}$$

3.4.1.2 ODE Reveals the Mechanism

Equations 3.29 and 3.30 now are an equivalent formulation of Equation 3.25 in terms of an ODE as desired. And, as we will see now, these equations indeed help us to understand the mechanism that generated the data in Figure 3.1a. Similar to our discussion of the exponential function (Section 3.3), the ODE formulation reveals the intrinsic meaning of Equation 3.25.

What Equation 3.29 obviously tells us is this: $T'(t)$, the temperature's rate of change, is proportional to the difference between the body temperature, T_b, and the actual temperature, $T(t)$. Since an almost perfect coincidence between Equation 3.25 and the data in Figure 3.1a was achieved above (Figure 3.1b), we may safely assume that it is this mechanism that generated the data in Figure 3.1a. Qualitatively, the proportionality expressed in Equation 3.29 makes good sense: Assume we are in a situation where $T(t) < T_b$, that is, where the temperature displayed by the clinical thermometer used in our example is smaller than the body temperature. Then we would expect an increase in $T(t)$, and indeed $T'(t)$, the rate of change of the temperature, will be positive since we have $T_b - T(t) > 0$ in this case. On the other hand, $T(t) > T_b$ implies a decrease in $T(t)$ by the same argument. The proportionality expressed by Equation 3.29 is also in good coincidence with our everyday experience: the higher the difference between $T(t)$ and T_b is, the higher is the temperature adjustment rate, $T'(t)$.

The *parameters of the model* can also be better understood based on the ODE formulation of the model. For example, the meaning of T_0 and T_b is much more evident from Equations 3.29 and 3.30 compared to Equation 3.25. Equation 3.30 immediately tells us that T_0 is the initial temperature, while we have to insert $t = 0$ into Equation 3.25 to see the same thing in the context of that equation. Equations 3.29 and 3.30 obviously just provide a much more natural formulation of the model. Regarding r, one can of course see from Equation 3.25 that this parameter controls the speed of temperature adjustment. Based on Equation 3.29, however, we can easily give a precise interpretation of this parameter as follows: Equation 3.29 implies

$$r = \frac{T'(t)}{T_b - T(t)} \tag{3.31}$$

Now since

$$T'(t) \approx \frac{T(t + \Delta t) - T(t)}{\Delta t} \tag{3.32}$$

for small Δt, we have

$$r \approx \frac{(T(t + \Delta t) - T(t))/(T_b - T(t))}{\Delta t} \tag{3.33}$$

This means that r expresses the percent rate of temperature adjustment per unit of time, expressed relative to the actual difference between sensor temperature and body temperature, $T_b - T(t)$. The unit of r is s^{-1}. For example, $r = 0.1$ would mean that we can expect a 10% reduction in the actual difference between sensor temperature and body temperature per unit of time (which will slightly overestimate the actual reduction in that difference in a given unit of time since this difference is continuously reduced). Interpretations of this kind are important because we should of course always know what we are doing, and which quantities we are dealing with. They are also practically useful since they, for example, help us to find reasonable a priori values of parameters. In this case, since r is a percent value, we know a priori that its value will probably be somewhere between 0 and 1.

> **Note 3.4.1 (Natural interpretations of parameters)**   While the parameters of phenomenological models are hardly interpretable tuning parameters in many cases, the parameters of mechanistic models do often have natural interpretations in terms of the system.

3.4.1.3   ODE's Connect Data and Theory

Equation 3.29 is also known as *Newton's law of cooling* [97]. Newton's law of cooling can be derived under special assumptions from the *heat equation*, which is the general law describing the variation of temperature in some given region in time and space (see [97] and Section 4.2 for more details). Precisely, Newton's law of cooling can be derived from the heat equation in cases where the surface conductance of the body under consideration is "much smaller" than the interior thermal conductivity of that body. In such cases, the temperature gradients within the body will be relatively small since the relatively high internal conductivity of the body tends to flatten out any higher temperature gradients. Now if there is little temperature variation inside the body, so-called lumped models of the body that neglect the variation of temperature in space, such as Newton's law of cooling, become applicable (compare Section 1.6.3). The *Biot number* expresses the ratio between surface conductance and interior thermal conductivity of a body [97]. If its value is less than 0.1, Newton's law of cooling is usually considered applicable.

In the light of this theory, our mechanistic model, thus, tells us even more about the "internal mechanics" of the clinical thermometer that generated the data in

Figure 3.1a. We have seen in Figure 3.1b that there is a good coincidence between these data and Newton's law of cooling (Equation 3.29). This means that we have a good reason to assume the validity of the assumptions on which this law is based, that is, we can assume a Biot number less than 0.1 for the clinical thermometer. In other words, the interior conductivity of the thermometers metal head can be expected to be much higher than its surface conductivity. Generally, what we see here is this:

> **Note 3.4.2 (Mechanistic models generate theoretical insights)** Mechanistic models make it easier to see connections with existing theories, and to make use of the results and insights of these theories.

3.4.1.4 Three Ways to Set up ODEs

Let us make a last point about the body temperature model. When we derived the ODE model above (Equations 3.29 and 3.30), we took the phenomenological Equation 3.25 as a starting point. Although the phenomenological Equation 3.25 fitted the data very well (Figure 3.1b), and although one might, thus, be tempted to say that there is no need to do anything beyond Equation 3.25, we have seen that it made good sense to derive a mechanistic ODE model from Equation 3.25. In doing this, we achieved a better understanding of Equation 3.25, and of the way in which the clinical thermometer generated the data in Figure 3.1a. The *derivation of an ODE model based on a phenomenological model*, however, is only one among several possibilities to set up an ODE model. Starting with the data in Figure 3.1a, we could also have used a *theoretical approach*, trying to understand the mechanisms generating these data based on a study of the relevant literature. This would have led us to a study of the heat equation and the Biot number as discussed above, and making the natural assumption that the interior thermal conductivity within the metal head of the clinical thermometer exceeds the surface conductivity at the surface where it touches the human body, we might have come up with Newton's law of cooling independently from Equation 3.25.

Another possibility to set up the ODE model would have been the *rate of change approach*. In the rate of change approach (compare Note 3.1.1), we would start with the data in Figure 3.1a. As a first step, we would have to identify and define the state variables needed to describe these data in terms of a model. As above, we would find there is one state variable (temperature), and as above we would designate it, for example, as $T(t)$. Now in the rate of change approach, the central question is

What are the rates of change of the state variables?

This question then would lead us to considerations similar to those made above in our discussion of Equation 3.29. Similar to above, we would ascertain that $T'(t)$, the temperature's rate of change, depends both on the sign and on the absolute value of $T_b - T(t)$ as described above. The simplest assumption that one can make in this case is a linear dependence of $T'(t)$ on $T_b - T(t)$ as expressed by Equation 3.29, and this would again lead us to the same ODE model as above.

> **Note 3.4.3 (Phenomenological, theoretical, and rate of change approach)** First-order ODEs can be set up using phenomenological equations, theoretical considerations, or by expressing the state variable's rate of change.

3.4.2
Alarm Clock Example

As a second example of setting up an ODE model, let us reconsider the alarm clock data (Figure 3.2b). In Section 3.2, a conceptual *Model B* was developed for these data (Figure 3.3). This model involves three state variables:
- $T_s(°C)$: the sensor temperature
- $T_i(°C)$: the effective internal temperature
- $T_a(°C)$: the ambient temperature.

3.4.2.1   A System of Two ODEs
In Section 3.4.1.4, three ways to set up ODEs have been introduced. We cannot (easily) use the *phenomenological approach* here since, in contrast to the body temperature model discussed above, no phenomenological model of the data in Figure 3.2b has been developed in Section 3.2. The *theoretical approach*, however, can be used similar to the discussion in Section 3.4.1, since the relationship between T_s and T_i as well as the relationship between T_i and T_a follows a similar logic as the relationship between T and T_b discussed in Section 3.4.1. Looking at T_s and T_i, we can say that, in the sense of Newton's law of cooling, T_i is the temperature of the "cooling environment" that surrounds the sensor described by T_s. From a theoretical point of view, Newton's law of cooling is the simplest way to describe this situation, which leads to

$$T_s'(t) = r_{si} \cdot (T_i(t) - T_s(t)) \tag{3.34}$$

This equation is exactly analogous to Equation 3.29 (an interpretation of r_{si} is given in the next section). As we have just said, this equation is the simplest way to describe the relationship between T_s and T_i, and it is, thus, appropriate here since we should always begin with simple approaches (Note 1.2.4). More sophisticated approaches can be used if we should observe substantial deviations between this model and the data. Now with the same reasoning as above, we can write down a corresponding equation relating T_i and T_a as follows:

$$T_i'(t) = r_{ia} \cdot (T_a - T_i(t)) \tag{3.35}$$

The last two equations could also have been established using the rate of change approach and a similar argumentation as it was made above for Equation 3.29.

Regarding Equation 3.34, for example, the rate of change approach would be based on the observations

- $T_s'(t)$ increases with $|T_i(t) - T_s(t)|$
- $T_i(t) - T_s(t)$ determines the sign of $T_s'(t)$.

Then, as above, the proportionality expressed by Equation 3.34 is the simplest way to mathematically express these observations. Denoting the temperatures of T_i and T_s at time $t = 0$ with T_{i0} and T_{s0}, respectively, the overall problem can now be written as an *initial value problem for a system of two ODEs* as follows:

$$T_s'(t) = r_{si} \cdot (T_i(t) - T_s(t)) \tag{3.36}$$

$$T_i'(t) = r_{ia} \cdot (T_a - T_i(t)) \tag{3.37}$$

$$T_s(0) = T_{s0} \tag{3.38}$$

$$T_i(0) = T_{i0} \tag{3.39}$$

Note that this problem involves not only two ODEs, Equations 3.36 and 3.37, but also *two corresponding initial conditions*, Equations 3.38 and 3.39. We may, thus, note here that the number of initial (or boundary) conditions required to solve ODEs increases not only with the order of the ODE (see the discussion in Section 3.3.5), but – naturally – also with the number of ODEs under consideration. Every first-order ODE needs its own initial or boundary condition – the "Gates" argument used in Section 3.3.4 applies to every single ODE.

3.4.2.2 Parameter Values Based on A priori Information

To solve Equation 3.36–3.39, we need to specify the following parameters: r_{si}, r_{ia}, T_a, T_{s0}, and T_{i0}. Based on the same argumentation used above for the parameter r in Section 3.4.1.2, the coefficients of proportionality in Equations 3.36 and 3.37 can be interpreted as follows:

- r_{si} (min^{-1}) is the percent rate of temperature adjustment between $T_s(t)$ and $T_i(t)$, expressed relative to the actual difference $T_i(t) - T_s(t)$
- r_{ia} (min^{-1}) is the percent rate of temperature adjustment between $T_i(t)$ and T_a, expressed relative to the actual difference $T_a - T_i(t)$

We, thus, know a priori that the values of these parameters are positive, probably somewhere between 0 and 1. Regarding T_a, the data in Figure 3.2b tell us that it is reasonable to set $T_a \approx 21\,°C$. To see this, you may also look into the raw data of this figure in the file room.csv, which you find in the book software. Regarding T_{s0}, Figure 3.2b or room.csv tells us that we should set $T_{s0} \approx 18.5$. Regarding T_{i0}, finally, it seems that we have a little problem here since we do not have any measurement values of this quantity. But remember that initial decrease in the sensor temperature in Figure 3.2b, and remember why we introduced T_i into Model B (Figure 3.3). T_i serves as the temperature memory of the alarm clock model: it memorizes the fact that the alarm clock was in a cold environment before $t = 0$.

Physically, this temperature memory is realized in terms of the relatively cold temperatures of the internal air inside the alarm clock or of certain parts inside the alarm clock that are immediately adjacent to the alarm clock. We had introduced T_i above as an "effective temperature" representing internal air temperature and internal parts' temperatures. As explained in Section 3.2.3, the temperature sensor "sees" only T_i instead of T_a. Now at time $t = 0$, when we enter the warm room, T_i is still colder than the actual sensor temperature, T_s, and this is why that initial decrease in the sensor temperature in Figure 3.2b is observed. In terms of this model, the initial decrease in the sensor temperature can, thus, only be explained if $T_{i0} < T_{s0}$, even $T_{i0} < \min_{t \geq 0} T_s(t)$. Looking into `room.csv`, it can be seen that this means $T_i < 18.2$. But we know even more. From Figure 3.2b or `room.csv` you can see that around $t \approx 2.5$ min we have $T'_s = 0$. In terms of Equation 3.36, this means $T_i(2.5) \approx T_s(2.5)$ or $T_i(2.5) \approx 18.2$ (using `Room.csv` again).

Our a priori knowledge of the parameters of Equations 3.36–3.39 can be summarized as follows:

- r_{si}, r_{ia}: percent values, probably between 0 and 1
- $T_a \approx 21\,°C$
- $T_{s0} \approx 18.5$
- T_{i0}: to be determined such that $T_i(2.5) \approx 18.2$.

3.4.2.3 Result of a Hand-fit

The criterion to be applied for the determination of more exact values of the parameters is a good coincidence between the model and the data. According to the nonlinear regression idea explained in Section 2.4, the parameters have to be determined in a way such that T_s as computed from Equations 3.36–3.39 matches the data in Figure 3.2b as good as possible. Note, however, that if we do not use the closed form solution of Equations 3.36–3.39 that is discussed in Section 3.7.3, the methods in Section 2.4 need to be applied in a slightly modified way here since T_s, which serves as the nonlinear regression function, is given only implicitly as the solution of Equations 3.36–3.39. This problem is addressed in Section 3.9. Here, we confine ourselves to a simple hand tuning of the parameters. This is what is usually done first after a new model has been created, in order to see if a good fit with the data can be obtained in principle. Figure 3.5 shows the result of Equations 3.36–3.39 obtained for the following hand-fitted parameter values: $r_{si} = 0.18, r_{ia} = 0.15, T_a = 21, T_{s0} = 18.5$, and $T_{i0} = 17$.

The figure was produced using the *Maxima* code `RoomODED.mac` and the data `room.csv` in the book software. This code is similar to `FeverExp.mac` discussed in Section 3.2.2.1, and it is based on Equations 3.199 and 3.200, a "closed form solution" of Equations 3.36–3.39, which is discussed in Section 3.7.3. Obviously, the figure shows a very good fit between T_s and the data. Note that the T_i curve in Figure 3.5 intersects the T_s curve exactly at its minimum, as it was required above. Note also that it is by no means "proved" by the figure that we have found the "right" values of the parameters. There may be ambiguities, that is, it might be true that a similarly perfect fit could be produced using an entirely different set of parameters. Also, you should note that even if there are no ambiguities of this kind, the parameter values were obtained from data, and thus are estimates in a statistical sense as it was discussed in Section 2.2.4.

3.4.2.4 A Look into the Black Box

Our result is a nice example of how mechanistic models allow us a look into the hidden internal mechanics of a system, that is, into the system viewed as a black box. Before we applied the model, we had no information about the dynamics of the temperatures inside the alarm clock. Based on Figure 3.5, we may now conjecture a pattern of the internal temperatures similar to the T_i curve in the figure, that is, starting with an initial temperature substantially below the starting temperature of the sensor (17.0 °C as compared to 18.5 °C), and reaching room temperature not before about 20 min. Below, we will see a number of similar applications where models allow us to "see" things that would be invisibly hidden inside the system black box without the application of mathematical models.

However, *careful interpretations* of what we are "seeing" in terms of models are necessary. We had deliberately chosen to say that the model allows us to *conjecture* a pattern of the internal temperatures. We have no proof that this pattern really describes the pattern of the alarm clock's internal temperature unless we really measure that temperature. In the absence of such data, the T_i pattern in Figure 3.5 is no more than a hypothesis about the alarm clock's internal temperature. Its validity is somewhere located between a mere speculation and a truth proven in terms of data. Certainly, we may consider it a relatively well-founded hypothesis as the T_s data are so nicely explained using our model. But we should know about its limitation.

Basically, mathematical models are a means to generate clever hypotheses about what is going on inside the system black box. You as the user of mathematical models should know that this is different from actually seeing the true "internal mechanics" of the black box. This is the reason why *experimental data will never become superfluous* as a consequence of the application of mathematical models or computer simulations. The author of this book frequently met experimentalists who seemed to believe that models and simulations would be a threat to their jobs. What everyone should learn is that modelers and experimentalists depend on each other, and that more can be achieved if they cooperate. No experimentalist

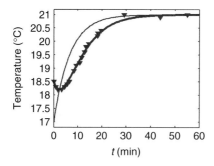

Fig. 3.5 T_s (thick line), T_i (thin line), and sensor temperature data.

needs to tremble with fear when the modelers come along with their laptops and mathematical equations. Every modeler, on the other hand, should tremble with fear if his model is insufficiently validated with data, and he should read the "Dont's of mathematical modeling" in Section 1.8 (particularly that "Do not fall in love with your model"...).

> **Note 3.4.4 (Importance of experimental data)** Mechanistic mathematical models can be viewed as a clever way to generate hypotheses about the inside of system black boxes, but this is different from actually seeing what is going on there. Experimental data provide the only way to validate hypotheses regarding black box systems.

3.5
Some Theory You Should Know

Based on the previous sections, the reader should now have a general idea of ODEs and how they can be used to formulate mathematical models. This section and the following sections (Sections 3.6–3.9) are intended to give you a more precise idea about the mathematical aspects of ODE problems and their various subtypes, and about what you are doing when you solve ODEs using computer programs, which is the typical case. You may skip this and the following sections at a first reading if you just want to get an idea of how ODEs can be used in various applications. In that case, you may choose to go on with the examples in Section 3.10. But as always in life, it is a good idea to have an idea of what one is doing, so do not forget to read this and the following sections.

3.5.1
Basic Concepts

Let us summarize what we already know about ODEs. The first ODE considered in Section 3.3.2 was

$$y' = y \tag{3.40}$$
$$y(0) = 1 \tag{3.41}$$

Note that this is an *abbreviated notation*. If not stated otherwise, $y' = y$ means $y'(t) = y(t)$ for $t \in \mathbb{R}$. The *order of an ODE* is the order of the highest derivative of the unknown function, so you see that Equation 3.40 is a first-order ODE. Equation 3.41 is the so-called initial condition, which is required to make (3.40) uniquely solvable. Both equations together constitute what is called an *initial value problem*. As stated above, the exponential function is the *solution* of (3.40) and (3.41) since both equations are satisfied for $t \in \mathbb{R}$ if we replace $y(t)$ and $y'(t)$ with e^t (remember that e^t and its derivative coincide).

Using a reformulation of the body temperature model, an ODE with a slightly more complex right-hand side was obtained in Section 3.4.1:

$$T'(t) = r \cdot (T_b - T(t)) \tag{3.42}$$

$$T(0) = T_0 \tag{3.43}$$

Comparing this with Equations 3.40 and 3.41, you see that we can of course use a *range of different notations*, and this is also what we find in the literature. You may write y, $y(t)$, T, $T(t)$ or anything else to designate the unknown function. A simple y such as in Equations 3.40 and 3.41 is often preferred in the theoretical literature on ODEs, while the applied literature usually prefers a notation which tells us a little more about the meaning of the unknown, such as the T in Equations 3.42 and 3.43, which designates temperature.

We also considered a *second-order ODE* in Section 3.3:

$$y'' = -y \tag{3.44}$$

$$y(0) = 1 \tag{3.45}$$

$$y'(0) = 0 \tag{3.46}$$

Here, Equation 3.44 is the ODE, while Equations 3.45 and 3.46 are the initial conditions, which are again required to assure unique solvability as before. Two initial conditions are required here since Equation 3.44 is a second-order ODE. Again, the system consisting of Equations 3.44–3.46 is called an *initial value problem,* and we saw in Section 3.3 that it is solved by the cosine function.

In Section 3.4.1, we considered the following *system of first-order ODEs*:

$$T'_s(t) = r_{si} \cdot (T_i(t) - T_s(t)) \tag{3.47}$$

$$T'_i(t) = r_{ia} \cdot (T_a - T_i(t)) \tag{3.48}$$

$$T_s(0) = T_{s0} \tag{3.49}$$

$$T_i(0) = T_{i0} \tag{3.50}$$

Here, Equations 3.47 and 3.48 are the ODEs while Equations 3.49 and 3.50 are the initial conditions. As was mentioned above in connection with these equations, every first-order ODE needs a corresponding initial condition (if no boundary conditions are imposed, see below).

First-order ODEs are in an *exceptional position* in that the majority of ODE models used in practice is based on first-order equations (which is also reflected by the examples treated in this book). This is not really surprising since, as discussed above, many ODE models just express the rates of change of its state variables, which can be naturally done in terms of a first-order ODE system. The alarm clock model discussed in Section 3.4.2 is a nice example of how you arrive at a first-order system based on rate of change considerations. The exceptional position of first-order ODEs is underlined by the fact that higher-order ODEs can be reformulated as first-order ODEs. For example, setting $z = y'$ in Equations 3.44–3.46, we get the following equivalent first-order formulation:

$$y' = z \tag{3.51}$$

$$z' = -y \tag{3.52}$$

$$y(0) = 1 \tag{3.53}$$

$$z(0) = 0 \tag{3.54}$$

3.5.2
First-order ODEs

To keep things simple, we will thus confine ourselves here to a formal definition of first-order ODEs. This can be done as follows [98]:

Definition 3.5.1 (First-order ODE) Let $\Omega \subset \mathbb{R}^2$, $F : \Omega \to \mathbb{R}$ a continuous function. Then,

$$y'(t) = F(t, y(t)) \tag{3.55}$$

is a *first-order ODE* in the unknown function $y(t)$. A function $y : [a, b] \to \mathbb{R}$ is called a *solution of the ODE* (3.55) if this equation is satisfied for every $t \in [a, b] \subset \mathbb{R}$.

Note that the definition implicitly assumes that $(t, y(t)) \in \Omega$ holds for all $t \in [a, b]$. In many practical cases, we will have $\Omega = \mathbb{R}^2$, that is, the ODE will be defined for any $t \in \mathbb{R}$ and for any $y \in \mathbb{R}$. An exception is, for example, the equation

$$y' = \frac{1}{y} \tag{3.56}$$

where the right-hand side is undefined for $y = 0$, which means we have to set $\Omega = \mathbb{R}^2 \setminus \mathbb{R} \times \{0\}$. Applying Definition 3.5.1 to the last equation, we have

$$F(t, y) = \frac{1}{y} \tag{3.57}$$

which is continuous on $\Omega = \mathbb{R}^2 \setminus \mathbb{R} \times \{0\}$ as required by the definition. Applied to (3.40), on the other hand, we have

$$F(t, y) = y \tag{3.58}$$

which is again a continuous function on $\Omega = \mathbb{R}^2$ as required. For Equation 3.42, we have instead

$$F(t, T) = r \cdot (T_b - T) \tag{3.59}$$

which is also obviously continuous on $\Omega = \mathbb{R}^2$ (remember that we can use any variable as the unknown of the ODE, and hence as the second argument of $\mp$!). We may, thus, conclude that Equations 3.40, 3.42 and 3.57 are indeed first-order ODEs in the sense of Definition 3.5.1. The last equations should have made it clear what is meant by $F(t, y(t))$ on the right-hand side of Equation 3.55. This term just serves as a placeholder for arbitrary expressions involving t and $y(t)$, which may of course be much more complex than those in the last examples.

Note that the solution concept used in the definition complies with our discussion of ODE solutions in the previous sections. For example, as discussed in Section 3.3, $y(t) = e^t$, viewed as a function $y : \mathbb{R} \to \mathbb{R}$, solves Equation 3.40 since this equation is satisfied for all $t \in \mathbb{R}$. In terms of Definition 3.5.1, this means we have used $[a, b] = [-\infty, \infty]$ here. Similarly, as discussed in Section 3.4.1, $T(t) = T_b - (T_b - T_0) \cdot e^{-r \cdot t}$, again viewed as a function $T : \mathbb{R} \to \mathbb{R}$, solves Equation 3.42 since this equation is satisfied for all $t \in \mathbb{R}$.

3.5.3
Autonomous, Implicit, and Explicit ODEs

Equations 3.40 and 3.42 are examples of so-called *autonomous ODEs* since their right-hand side does not depend on t explicitly. Note that there is, of course, an implicit t dependence in these equations via $y(t)$ and $T(t)$, respectively. In terms of Definition 3.5.1, we can say that the right-hand side of autonomous ODEs can be written as $F(y)$. Equation 3.42 would turn into a nonautonomous ODE, for example, if T_b, the room temperature, would be time dependent. Note that in this case $T(t) = T_b - (T_b - T_0) \cdot e^{-r \cdot t}$ would no longer solve Equation 3.42. If T_b is time dependent, you can verify that the derivative of this function would be

$$T'(t) = T'_b(t)(1 - e^{-rt}) + r(T_b(t) - T_0)e^{-rt} \tag{3.60}$$

which obviously means that Equation 3.42 is not satisfied.

3.5.4
The Initial Value Problem

As it was mentioned above, ODEs without extra conditions are not uniquely solvable. For example, Equation 3.40 is solved by $f(t) = ce^t$ for arbitrary $c \in \mathbb{R}$. We have seen that this can be overcome using the initial conditions for the unknown function, which leads to *initial value problems*. A formal definition can be given as follows:

Definition 3.5.2 (Initial value problem) Let $\Omega \subset \mathbb{R}^2$, $F : \Omega \to \mathbb{R}$ a continuous function and $y_0 \in \mathbb{R}$. Then,

$$y'(t) = F(t, y(t)) \tag{3.61}$$

$$y(a) = y_0 \tag{3.62}$$

is an *initial value problem* for the first-order ODE, Equation 3.61. A function $y : [a, b] \to \mathbb{R}$ solves this problem if it satisfies Equation 3.62 as well as 3.61 for every $t \in [a, b] \subset \mathbb{R}$.

Theoretically, it can be shown that this initial value problem is *uniquely solvable* if F is *Lipschitz continuous* [98–100]. Basically, this continuity concept limits F in how fast it can change. We will not go into a discussion of this concept here since unique solvability is no problem in the majority of applications, including the examples in this book. Equations 3.40–3.43 are obvious examples of initial value problems in the sense of Definition 3.5.2.

3.5.5
Boundary Value Problems

Characteristic of initial value problems is the fact that conditions on the unknown function are imposed for only one value of the independent variable. Problems where conditions are imposed at more than one point are called *boundary value problems*. Problems of this kind provide another possible way to select one particular solution among the many solutions of an ODE. As we will see in Chapter 4, boundary value problems are particularly important in partial differential equation models. This is related to the fact that many boundary value problems are formulated in terms of spatial coordinates, while, on the other hand, most differential equation models involving spatial coordinates are formulated using partial differential equations. If models involving spatial coordinate are written in terms of an ODE, they can frequently be generalized in terms of a PDE. For example, the ODE

$$T''(x) = 0 \tag{3.63}$$

describes the temperature distribution in a one-dimensional, homogeneous body that is obtained after a very long (in fact, infinite) time when you do not change the environment of that body. Since this temperature distribution is time independent, it is usually called the *stationary temperature distribution* of that body (compare Definition 1.6.2). You may imagine, for example, a *metal rod*, which is insulated against heat flow except for its ends where you keep it at constant temperature for a long time. Equation 3.63 is a special case of the *heat equation*, which provides a general description of the variation of temperature in time and space [97, 101]. The general form of this equation is

$$\frac{\partial T}{\partial t} = k \left(\frac{\partial^2 T}{\partial x^2} + \frac{\partial^2 T}{\partial y^2} + \frac{\partial^2 T}{\partial z^2} \right) \tag{3.64}$$

This equation is a PDE since it involves derivatives with respect to several variables, and it is discussed in more detail in Section 4.2. Assuming a one-dimensional body means that temperature variations in the y and z directions can be neglected (see the discussion of spatial dimensionality in Section 4.3.3), which means we have

$$\frac{\partial^2 T}{\partial y^2} = \frac{\partial^2 T}{\partial z^2} = 0 \tag{3.65}$$

in Equation 3.64. On the other hand, the stationarity of T leads to

$$\frac{\partial T}{\partial t} = 0 \tag{3.66}$$

Inserting Equations 3.65 and 3.66 into Equation 3.64, Equation 3.63 is obtained as desired.

Now let us be more specific about that metal rod. Consider a metal rod of length 1 m with constant temperatures $20°$ and $10\,°C$ at its ends. Mathematically, this means that Equation 3.63 turns into the problem

$$T''(x) = 0, \quad x \in [0, 1] \tag{3.67}$$
$$T(0) = 20 \tag{3.68}$$
$$T(1) = 10 \tag{3.69}$$

This is a boundary value problem in the sense explained above since the conditions accompanying the ODE are imposed at two different points. The mathematical problem (Equations 3.67–3.69) is easily solved based on the observation that Equation 3.67 is satisfied exactly by the straight lines. The boundary conditions (3.68) and (3.69) fix the endpoints of this straight line such that the solution becomes

$$T(x) = 20 - 10x \tag{3.70}$$

This equation describes a linear transition between the endpoint temperatures, and this is of course what one would naturally expect here since we have assumed a homogeneous metal rod. Looking at our derivation of Equation 3.70, it is obvious that this is the unique solution of the boundary value problem (Equations 3.67–3.69). Note that Equation 3.67 can be considered as the "second simplest" ODE. The only simpler ODE is this:

$$T'(x) = 0 \tag{3.71}$$

Equation 3.71 describes a straight line with slope zero, parallel to the x axis. It is again a nice demonstration of the fact that initial or boundary conditions are needed to obtain unique solution of ODEs. Without such a condition, Equation 3.71 is solved by the entire family of straight lines parallel to the x axis, that is, by all functions of the form

$$T(x) = c \tag{3.72}$$

for arbitrary $c \in \mathbb{R}$. Only after imposing an initial condition such as $T(0) = 4$, the c in Equation 3.72 is fixed and a unique solution of Equation 3.71 is obtained ($T(x) = 4$ in this case).

3.5.6
Example of Nonuniqueness

Let us go back to our consideration of boundary value problems. Unfortunately, in contrast to initial value problems, it cannot be proved in general that boundary value problems would lead us to unique solutions of an ODE. For example, consider

$$y''(x) = -y(x) \tag{3.73}$$

In Equations 3.44–3.46, we have looked at this equation as an initial value problem. Now without any extra conditions, the general solution of Equation 3.73 can be written as

$$y(x) = a \cdot \sin(x) + b \cdot \cos(x) \tag{3.74}$$

This means: For any values of $a, b \in \mathbb{R}$, Equation 3.74 is a solution of Equation 3.73 (see also the remarks on the solution of Equation 3.73 in Section 3.6). Now it depends on the particular boundary conditions that we impose whether we get a uniquely solvable boundary value problem or not. Let us first consider the problem

$$y''(x) = -y(x) \tag{3.75}$$
$$y(0) = 1 \tag{3.76}$$
$$y\left(\frac{\pi}{2}\right) = 0 \tag{3.77}$$

Inserting the general solution, Equation 3.74, into Equations 3.76 and 3.77, you can easily verify that $a = 0$ and $b = 1$ in Equation 3.74, which means that we get $y(x) = \cos(x)$ as the unique solution of this boundary value problem. On the other hand, you will find that the boundary value problem

$$y''(x) = -y(x) \tag{3.78}$$
$$y(0) = 0 \tag{3.79}$$
$$y(\pi) = 0 \tag{3.80}$$

is solved by any function of the form $y(x) = a \cdot \sin(x)$ ($a \in \mathbb{R}$), which means there is no unique solution of this problem. Finally, inserting the general solution, Equation 3.74, into conditions (3.82) and (3.83) of the boundary value problem

$$y''(x) = -y(x) \tag{3.81}$$

$$y(0) = 0 \tag{3.82}$$

$$y(2\pi) = 1 \tag{3.83}$$

you are led to the contradictory requirements $b = 0$ and $b = 1$, which means that there is no solution of this boundary value problem. This means that we cannot expect any general statement assuring unique solvability of boundary value problems, as it was possible in the case of initial value problems.

In the problems considered below as well as in our discussion of solution methods, we will confine ourselves to initial value problems. While it was indicated in the above discussion of the metal rod problem that boundary value problems can frequently be seen as special cases of problems that are naturally formulated in a PDE context, initial value problems, on the other hand, may be considered as the "natural" setting for ODEs, since time serves as the independent variable in the majority of applications (note that "initial value" immediately appeals to time as the independent variable). The fact that unique solvability can be proved only for initial value problems may be considered as an additional aspect of the naturalness of these problems. This is not to say, however, that boundary value problems over ODEs are unimportant. From a numerical point of view, boundary value problems over ODEs are particularly interesting: they involve numerical methods such as the shooting method and the multiple shooting method for the solution of ODEs, as well as global methods like finite differences or collocation methods [102]. All these, however, are beyond the scope of a first introduction into mathematical modeling.

3.5.7
ODE Systems

Note that not all initial value problems discussed above are covered by Definitions 3.5.1 and 3.5.2. Equations 3.47–3.50 or 3.51–3.54 involve systems of two ODEs and two state variables, while the above definitions refer only to one ODE and one state variable. This is solved in the usual way based on a vectorial notation. Using boldface symbols for all vectorial quantities, the vector versions of Definitions 3.5.1 and 3.5.2 read as follows:

Definition 3.5.3 (System of first-order ODEs) Let $\Omega \subset \mathbb{R}^{n+1}$, $\mathbf{F} : \Omega \to \mathbb{R}^n$ a continuous function. Then,

$$\mathbf{y}'(t) = \mathbf{F}(t, \mathbf{y}(t)) \tag{3.84}$$

is an *ODE* of first order in the unknown function $\mathbf{y}(t)$. A function $\mathbf{y} : [a, b] \subset \mathbb{R} \to \mathbb{R}^n$ is called a *solution of the ODE* (3.84) if this equation is satisfied for every $t \in [a, b]$.

Definition 3.5.4 (Initial value problem for a system of first-order ODEs) Let $\Omega \subset \mathbb{R}^{n+1}$, $\mathbf{F} : \Omega \to \mathbb{R}^n$ a continuous function and $\mathbf{y_0} \in \mathbb{R}^n$. Then,

$$\mathbf{y}'(t) = \mathbf{F}(t, \mathbf{y}(t)) \tag{3.85}$$

$$\mathbf{y}(a) = \mathbf{y_0} \tag{3.86}$$

is an *initial value problem* for the first-order ODE, Equation 3.85. A function $\mathbf{y} : [a, b] \to \mathbb{R}^n$ solves this problem if it satisfies Equation 3.86 as well as Equation 3.85 for every $t \in [a, b] \subset \mathbb{R}$.

As before, unique solvability of the initial value problem can be proved if $\mathbf{F}$ is Lipschitz continuous [99]. The interpretation of Equations 3.47–3.50 or 3.51–3.54 in terms of these definitions goes along the usual lines. For example, using the identifications

$$
\begin{aligned}
y_1 &= T_s \\
y_2 &= T_i \\
y_{01} &= T_{s0} \\
y_{02} &= T_{i0} \\
a &= 0 \\
F_1(t, \mathbf{y}(t)) &= r_{si} \cdot (y_2(t) - y_1(t)) \\
F_2(t, \mathbf{y}(t)) &= r_{ia} \cdot (T_a - y_2(t))
\end{aligned}
\tag{3.87}
$$

Equations 3.47–3.50 turn into

$$y_1'(t) = F_1(t, \mathbf{y}(t)) \tag{3.88}$$

$$y_2'(t) = F_2(t, \mathbf{y}(t)) \tag{3.89}$$

$$y_1(a) = y_{01} \tag{3.90}$$

$$y_2(a) = y_{02} \tag{3.91}$$

Using the vectors

$$\mathbf{y} = \begin{pmatrix} y_1 \\ y_2 \end{pmatrix}, \quad \mathbf{y_0} = \begin{pmatrix} y_{01} \\ y_{02} \end{pmatrix}, \quad \mathbf{y}' = \begin{pmatrix} y_1' \\ y_2' \end{pmatrix}, \quad \mathbf{F}(t, \mathbf{y}(t)) = \begin{pmatrix} F_1(t, \mathbf{y}(t)) \\ F_2(t, \mathbf{y}(t)) \end{pmatrix} \tag{3.92}$$

you finally arrive at the form of the initial value problem as in Definition 3.5.4 above. So, as usual with vectorial laws, after writing a lot of stuff you see that (almost) everything remains the same in higher dimensions.

3.5.8
Linear versus Nonlinear

Particularly when solving ODEs (Section 3.6), it is important to know whether you are concerned with a linear or nonlinear ODE. As usual, linear problems can be treated much easier, and nonlinearity is one of the main causes of trouble that you may have. You remember that linearity can be phrased as "unknowns may be added, subtracted, and multiplied with known quantities". Applying this to ODEs, it is obvious that a *linear ODE* is this:

$$y'(x) = a(x) \cdot y(x) + b(x) \tag{3.93}$$

This can be generalized to a *system of linear ODEs* as follows:

$$y'_i(x) = \sum_{j=1}^{n} a_{ij}(x) \cdot y_j(x) + b_i(x), \quad i = 1, \ldots, n \tag{3.94}$$

Using $\mathbf{y}(x) = (y_1(x), y_2(x), \ldots, y_n(x))$, $\mathbf{b}(x) = (b_1(x), b_2(x), \ldots, b_n(x))$ and

$$\mathbf{A}(x) = \begin{pmatrix} a_{11}(x) & a_{12}(x) & \cdots & a_{1n}(x) \\ a_{21}(x) & a_{22}(x) & \cdots & a_{2n}(x) \\ \vdots & \vdots & \cdots & \vdots \\ a_{n1}(x) & a_{n2}(x) & \cdots & a_{nn}(x) \end{pmatrix} \tag{3.95}$$

Equation 3.94 can be written in matrix notation as follows:

$$\mathbf{y}'(x) = \mathbf{A}(x) \cdot \mathbf{y}(x) + \mathbf{b}(x) \tag{3.96}$$

Most ODEs considered so far have been linear. For example, the ODE of the body temperature model (Equation 3.29 in Section 3.4.1.1),

$$T'(t) = r \cdot (T_b - T(t)) \tag{3.97}$$

can be reformulated as

$$T'(t) = -rT(t) + r \cdot T_b \tag{3.98}$$

which attains the form of Equation 3.93 if we identify $y = T$, $x = t$, $a(x) = -r$, and $b(x) = r \cdot T_b$. As another example, consider the ODE system of the alarm clock model (Equations 3.36 and 3.37 in Section 3.4.2.1)

$$T'_s(t) = r_{si} \cdot (T_i(t) - T_s(t)) \tag{3.99}$$
$$T'_i(t) = r_{ia} \cdot (T_a - T_i(t)) \tag{3.100}$$

Let us write this as

$$T_s'(t) = -r_{si} \cdot T_s(t) + r_{si} \cdot T_i(t) \tag{3.101}$$

$$T_i'(t) = -r_{ia} \cdot T_i(t) + r_{ia} \cdot T_a \tag{3.102}$$

Then, using $x = t$, $\mathbf{y} = (y_1, y_2) = (T_s, T_i)$, $\mathbf{b} = (b_1, b_2) = (0, r_{ia} \cdot T_a)$ and

$$\mathbf{A}(x) = \begin{pmatrix} -r_{si} & r_{si} \\ 0 & -r_{ia} \end{pmatrix} \tag{3.103}$$

Equation 3.96 is obtained from Equations 3.99 and 3.100, which means that Equations 3.99 and 3.100 are a linear system of ODEs.

A *nonlinear ODE* considered above is (compare Equation 3.57 in Section 3.5.2)

$$y' = \frac{1}{y} \tag{3.104}$$

This cannot be brought in the form of Equation 3.93 since it involves a division operation. As discussed above, we have of course to be careful regarding the domain of definition of this equations right-hand side, in contrast to the unrestricted domains of definition of the linear ODEs considered above, which illustrates our above statement that "nonlinearity is one of the main causes of trouble that you may have with ODEs". A number of nonlinear ODE models are discussed below, such as the predator–prey model (Section 3.10.1) and the wine fermentation model (Section 3.10.2).

3.6
Solution of ODE's: Overview

3.6.1
Toward the Limits of Your Patience

As it was already mentioned above, solving differential equations is a big challenge and a big topic in mathematical research since a long time, and it seems unlikely that mathematicians working in this field will loose their jobs in the near future. It is the significance of differential equations in the applications that drives this research effort. While the example applications considered so far referred to simple technical systems (see the body temperature, alarm clock, and steel rod examples), the simplicity of these examples is, on the other hand, a nice demonstration of the broad range of differential equation applications. They are needed not only for an investigation of sophisticated and complex systems; rather, as it was demonstrated in Section 3.3, differential equations are frequently a part of the true mathematical structure of a model as soon as exponential or other transcendental functions are used.

In Section 3.3, a comparison of the solution of linear systems of equations versus the solution of differential equations was used to give you an idea of what it is that makes differential equations hard to solve. As it was discussed there, it is the infinite dimensionality of these equations, that is, the fact that these equations ask for an *infinite-dimensional unknown*: a function. A great part of what is done by mathematicians working on the numerical solution of differential equations today can be described as an effort to deal with the infinite dimensionality of differential equations in an effective way. This is even more important when we are dealing with PDEs, since, as explained above, these equations ask for functions depending on several variables, that is, which are infinite dimensional in the sense explained above not only with respect to one independent variable (time in many cases), but also with respect to one or more other variables (such as spatial coordinates).

Particularly when solving PDEs, you will easily be able to explore the limits of your computer, even if you consider models of systems that do not seem to be too complex at a first glance, and even if you read this book years after this text is written, that is, at a time when computers will be much faster than today (since more complex problems are solved on faster computers). PDE models involving a complex coupling of fluid flow with several other phenomena, such as models of casting and solidification or climate phenomena, may require *several hours of computation time* or more, even on today's fastest computers (Section 4.6.8 and [103, 104]). PDE's may thus not only help us to explore the limits of our computers, but also of our patience . . .

3.6.2
Closed Form versus Numerical Solutions

There are two basic ways how differential equations can be solved, which correspond to Section 3.7 on closed form solutions and Section 3.8 on numerical solutions. A *closed form solution* – which is sometimes also called an *analytical solution* – is a solution of an equation that can be expressed as a formula in terms of "well-known" functions such as e^x and $\sin(x)$. All solutions of ODEs considered so far are closed form solutions in this sense. In the body temperature example above (Section 3.4.1), Equation 3.25 is a closed form solution of the initial value problem (Equations 3.29 and 3.30). In practice, however, most ODEs cannot be solved in terms of closed form solutions, frequently due to nonlinear right-hand sides of the equations. Approximate *numerical solutions* of such ODEs are obtained using appropriate numerical algorithms on the computer.

The borderline between closed form and numerical solutions is not really sharp in the sense that *well-known functions* such as e^x and $\sin(x)$ are of course also computed approximately using numerical algorithms, and also in the sense that it is a matter of definition what we call "well-known functions". For example, people working in probability and statistics frequently need the expression

$$\frac{2}{\sqrt{\pi}} \int_0^x e^{-t^2}\, dt \tag{3.105}$$

The integral in this formula cannot be obtained in closed form. But since this expression is needed so often in probability and statistics, it received its own name and is referred to as the *error function* erf (x) [105]. Using this function as a part of the "well-known functions", many formulas can be written in closed form in probability and statistics, which would not have been possible based on the usual set of "well-known functions".

From a modeling point of view, it is desirable to have closed form solutions since they tell us more about the system compared to numerical solutions. To see this, consider Equation 3.25 again, the closed form solution of the body temperature model, Equations 3.29 and 3.30:

$$T(t) = T_b - (T_b - T_0) \cdot e^{-r \cdot t} \tag{3.106}$$

In this expression, the effects of the various model parameters on the resulting temperature curve can be seen directly. In particular, it can be seen that the ambient temperature, T_b, as well as the initial temperature, T_0, affect the temperature essentially linear, while the rate parameter r has a strong nonlinear (exponential) effect on the temperature pattern. To make this precise, Equation 3.106 could also be used for a so-called *sensitivity analysis*, computing the derivatives of the solution with respect to its parameters. The expression of $T(t)$ in Equation 3.106 allows T to be viewed as a multidimensional function

$$T(T_0, T_b, r, t) = T_b - (T_b - T_0) \cdot e^{-r \cdot t} \tag{3.107}$$

where we can take, for example, the derivative with respect to r as follows:

$$\frac{\partial T(T_0, T_b, r, t)}{\partial r} = r \cdot (T_b - T_0) \cdot e^{-r \cdot t} \tag{3.108}$$

This is called the *sensitivity* of T with respect to r. On the basis of a Taylor expansion, it can be used to estimate the effect of a change from r to $r + \Delta r$ (Δr being small) as

$$T(T_0, T_b, r + \Delta r, t) \approx T(T_0, T_b, r, t) + \frac{\partial T(T_0, T_b, r, t)}{\partial r} \cdot \Delta r \tag{3.109}$$

A numerical solution of Equations 3.29 and 3.30, on the other hand, would give us the temperature curve for any given set of parameters T_0, T_b, r similar to an experimental data set, that is, it would provide us with a list of values (t_1, T_1), $(t_2, T_2), \ldots, (t_n, T_n)$ which we could then visualize, for example, using the plotting capabilities of *Maxima* (Section 3.8.1). Obviously, we would not be able to see any parameter effects based on such a list of data, or to compute sensitivities analytically as above. The only way to analyze parameter effects using numerical solutions is to compute this list of values (t_1, T_1), $(t_2, T_2), \ldots, (t_n, T_n)$ for several different values of a parameter, and then to see how it changes. Alternatively, parameter sensitivities could also be computed using appropriate numerical procedures. However, you would have these numerically computed sensitivities only at some particular points, and it is of course better and gives you more information on the system if you have a general formula like Equation 3.108. In situations where you cannot get a closed form solution for a mathematical model, it may, thus, be worthwhile to consider a simplified version of the model that can be solved in terms of a closed form solution. You will then have to trade off the advantages of the closed form solution against the disadvantages of considering simplified versions of your model only.

3.7
Closed Form Solutions

Although mathematical models expressed as closed form solutions are very useful as discussed above, an exhaustive survey of the methods that can be used to obtain closed form solutions of ODEs would be beyond the scope of a first introduction to mathematical modeling techniques. This section is intended to give the reader a first idea of the topic based on a discussion of a few elementary methods that can be applied to first-order ODEs. For anything beyond this, the reader is referred to the literature such as [100].

3.7.1
Right-hand Side Independent of the Independent Variable

Let us start our consideration of closed form solutions with the simplest ODE discussed above (Equation 3.71 in Section 3.5.5):

$$T' = 0 \tag{3.110}$$

3.7.1.1 General and Particular Solutions
Basically, we solved this equation "by observation", namely by the observation that straight lines parallel to the x-axis have the property that the slope vanishes everywhere as required by Equation 3.110. Here, "everywhere" of course refers to the fact that we are talking about solutions of Equation 3.110 over the entire set of real numbers. A more precise formulation of the problem imposed by Equation 3.110 would be

(P1): Find a function $T : \mathbb{R} \to \mathbb{R}$ such that $T'(x) = 0$ holds for all $x \in \mathbb{R}$.

In this formulation, it is implicitly assumed that the function T is differentiable such that $T'(t)$ can be computed everywhere in $\mathbb{R}$. We could be more explicit (and more precise) in this point using a specification of the space of functions in which we are looking for the solution:

(P2) Find a function $T \in C^1(\mathbb{R})$ such that $T'(x) = 0$ holds for all $x \in \mathbb{R}$.

Here, $C^1(\mathbb{R})$ is the set of all functions having a continuous first derivative on $\mathbb{R}$. In the theory of differential equations, particularly in the theory of PDEs, a precise consideration of the function spaces where the solutions of differential equations are sought for is of great importance. These functions spaces are used as a subtle measure of the differentiability or smoothness of functions. For example, the theory of PDEs leads to the so-called Sobolev spaces that involve functions that are not differentiable in the classical sense, but are, nevertheless, used to solve differential equations involving derivatives (see Section 4.7.1 for more details). All this, however, is beyond the scope of this first introduction into mathematical modeling techniques.

Note 3.7.1 (Notation convention) If we write down an equation such as Equation 3.110 with no further comments and restrictions, it will be understood that we are looking for a sufficiently differentiable function on all of $\mathbb{R}$. If there can be no misunderstanding, a formulation like Equation 3.110 will always be preferred to alternative formulations such as (P1) or (P2) above.

Now let us go back to the problem of solving Equation 3.110. As discussed in Section 3.5.5, the observation "straight lines parallel to the x axis solve Equation 3.110" leads to the following expression for the solution of Equation 3.110:

$$T(x) = c, \quad c \in \mathbb{R} \tag{3.111}$$

This is called the *general solution* of Equation 3.110 since it was derived from that equation without any extra (initial or boundary) conditions. Note that Equation 3.111 describes an entire *family of solutions* parameterized by c, that is, for every $c \in \mathbb{R}$ we get a *particular solution* of Equation 3.110 from the general solution (Equation 3.111). General solutions of differential equations are typically described by such families of solutions that are parameterized by certain parameters. We will see several more examples of general solutions below. As explained before, initial or boundary conditions are the appropriate means to pick out one particular solution from the general solution, that is, to fix one particular value of c in the case of Equation 3.111.

3.7.1.2 Solution by Integration

To say that we solved Equation 3.110 "by observation" is of course a very unsatisfactory statement, which cannot be generalized to more complex equations We obviously need computational procedures to obtain the solution. Regarding Equation 3.110, we can simply integrate the equation (corresponding to using the fundamental theorem calculus) which then leads directly to the general solution (Equation 3.111). The same technique can be applied to all ODEs having the simple form

$$T'(x) = f(x) \tag{3.112}$$

Assuming that $f(x)$ is defined and continuous on some interval $[a, b]$, the fundamental theorem calculus states [17] that

$$T(x) = \int_a^x f(s)\, ds + c, \quad c \in \mathbb{R}, \quad x \in [a, b] \tag{3.113}$$

solves Equation 3.112. Since f is assumed to be continuous (Definition 3.5.1), we know that the integral on the right-hand side of Equation 3.113 can be evaluated. Of course, continuity of the right-hand side of an ODE was required in Definition 3.5.1 exactly for this reason, that is, to make sure that solution formulas such as Equation 3.113 make good sense. In the sense explained above, Equation 3.113 is the general solution of Equation 3.112, and it is again a family of solutions parameterized by $c \in \mathbb{R}$. As before, particular solutions of Equation 3.112 are obtained by imposing an initial condition. From Equation 3.113, you see that $T(0) = c$, which shows that c is the initial value of T at $x = 0$, so you see that the value of c (and hence a particular case of Equation 3.113) is indeed fixed if you impose an initial condition.

3.7.1.3 Using Computer Algebra Software

You will agree that what we did so far did not really involve a great deal of deep mathematics. Anyone with that basic mathematical education assumed in this book will do all this easily by hand. Nevertheless, we have a good starting point here to see how a computer algebra software such as *Maxima* can be used to solve ODEs (Appendix C). As explained before, the *Maxima* procedures used in this book translate easily into very similar corresponding procedures in other computer algebra software that you might prefer. Let us begin with Equation 3.110. In terms of *Maxima*, this equation is written as

$$\text{'diff(T,x)=0;} \tag{3.114}$$

Comparing this with Equation 3.110, you see that the operator 'diff(.,x) designates differentiation with respect to x. To work with this equation in *Maxima*, it should be stored in a variable like this:

$$\text{eq: 'diff(T,x)=0;} \tag{3.115}$$

Now you can solve the ODE using the ode2 command as follows:

$$\text{ode2(eq,T,x);} \tag{3.116}$$

This command instructs *Maxima* to solve the equation eq for the function T which depends on the independent variable x. Figure 3.6a shows how this procedure is realized in a *wxMaxima* session. As can be seen, *Maxima* writes the result in the form

$$\text{T=\%c;} \tag{3.117}$$

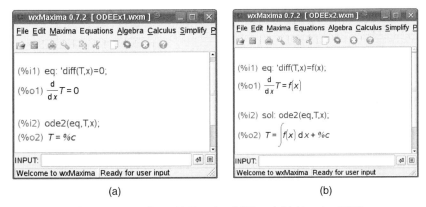

Fig. 3.6 *wxMaxima* sessions solving (a) Equation 3.110 and (b) Equation 3.112.

which is *Maxima*'s way to express our general solution in Equation 3.111. As you see, an expression such as %c in a *Maxima* output denotes a constant that may take on any value in $\mathbb{R}$. This example is a part of the book software (see the file ODEEx1.mac).

Figure 3.6b shows an analogous *wxMaxima* session for the solution of Equation 3.112. As the figure shows, Maxima writes the solution as

$$T = \int f(x)\, dx + \%c \tag{3.118}$$

This corresponds to the general solution derived above in Equation 3.113. Again, %c is an arbitrary real constant. *Maxima* expresses the integral in its indefinite form, which means that appropriate integration limits should be inserted by the "user" of this solution formula similar to those used in Equation 3.113. Note that the x in Equation 3.118 is just an integration variable, but it is *not* the independent variable on which the solution function T depends. Equation 3.113 clarifies this point. *Maxima*'s notation may be considered a little bit awkward here, but this is the usual notation for indefinite integrals, and it is not really a problem when you read it carefully.

Of course, *Maxima* can easily evaluate the integral in Equation 3.118 when you provide a concrete (integrable) expression for $f(x)$. For example, solving

$$T'(x) = x^2 \tag{3.119}$$

instead of Equation 3.112 and using the same procedure as above, *Maxima* produces the correct general solution

$$T = \frac{x^3}{3} + \%c \tag{3.120}$$

which corresponds to

$$T(x) = \frac{x^3}{3} + c, \quad c \in \mathbb{R} \tag{3.121}$$

in the usual mathematical notation (see ODEEx3.mac in the book software).

> **Note 3.7.2 (Computer algebra software solves ODEs)** Computer algebra software such as *Maxima* can be used to solve ODEs either in closed form or numerically. Closed form solutions are obtained in *Maxima* using the ode2 command (and/or the ic1, ic2, and desolve commands, see below) and numerical solutions using the rk command (see below).

3.7.1.4 Imposing Initial Conditions

So far we have seen how general solutions of ODEs can be computed using *Maxima*. The next step is to pick out particular solutions using initial or boundary conditions. *Maxima* provides the command ic1 to impose initial conditions in first-order equations and a corresponding command ic2 for second-order equations. Let us look at the following initial value problem for Equation 3.119:

$$T'(x) = x^2 \tag{3.122}$$

$$T(0) = 1 \tag{3.123}$$

Using the general solution, Equation 3.113, it is easily seen that the initial condition, Equation 3.123, leads us to $c = 1$ such that the solution of Equations 3.122 and 3.123 is

$$T(x) = \frac{x^3}{3} + 1 \tag{3.124}$$

In *Maxima*, this is obtained using the following lines of code (see ODEEx4.mac in the book software):

```
1:  eq:  ´diff(T,x)=x^2;
2:  sol: ode2(eq,T,x);
3:  ic1(sol,x=0,T=1);
```
(3.125)

As before, the "1:", "2:", "3:" at the beginning of each line are not a part of the code, but just line numbers which we will use for reference. The first two lines of this code produce the general solution (3.120) following the same procedure as above, the only difference being that the general solution is stored in the variable sol in line 2 for further usage. Line 3 instructs *Maxima* to take the general solution stored in sol, and then to set $T = 1$ at $x = 0$, that is, to impose the initial condition (3.123). *Maxima* writes the results of these lines as

$$T = \frac{x^3 + 3}{3} \tag{3.126}$$

which is the same as Equation 3.124 as required. If you want *Maxima* to produce this result exactly in the better readable form of Equation 3.124, you can use the expand command as follows:

```
1:  eq:  'diff(T,x)=x^2;
2:  sol: ode2(eq,T,x);
3:  ic1(sol,x=0,T=1);
4:  expand(%);
```
(3.127)

The expand command in line 4 is applied to "%". Here, "%" refers to the last output produced by *Maxima*, which is the result of the ic1 command in line 3. This means that the expand command in line 4 is applied to the solution of the initial value problem (Equation 3.126). Another way to achieve the same result would have been to store the result of line 3 in a variable, and then to apply the expand command to that variable, as it was done in the previous lines of the code. As you can read in the *Maxima* manual, the expand command splits numerators of rational expressions that are sums into their respective terms. In this case, this leads us exactly from Equation 3.126 to 3.124 as required.

As it was discussed above, the general form of Equation 3.122 is this:

$$T'(x) = f(x) \tag{3.128}$$

We have seen that Equation 3.118 is the general solution of this equation. This means that solving this kind of ODEs amounts to the problem of integrating $f(x)$. This can be done using the procedure described above, that is, by treating Equation 3.128 as before. On the other hand, we could also have used a direct integration of $f(x)$ based on *Maxima's* command for the integration of functions: integrate. For example, to solve Equation 3.119, we could have used the simple command

```
integrate(x^2,x);
```
(3.129)

See ODEEx5.mac in the book software. *Maxima* writes the result in the form

$$\frac{x^3}{3} \tag{3.130}$$

Except for the fact that *Maxima* skips the integration constant here, we are back at the general solution (Equation 3.113) of Equation 3.119. In cases where the integral of the right-hand side of Equation 3.128 cannot be obtained in closed form, the numerical integration procedures of *Maxima* can be used (see e.g. *Maxima's* QUADPACK package).

3.7.2
Separation of Variables

The simple differential equations considered so far provided us with a nice playground for testing *Maxima* procedures, but differential equations of this kind would of course hardly justify the use of computer algebra software. So let us go on now toward more sophisticated ODEs. Solving ODEs becomes interesting when the right-hand side depends on y. The simplest ODE of this kind is this:

$$y' = y \tag{3.131}$$

You remember that this was the first ODE considered above in Section 3.3. Again, we solved it "by observation", observing that Equation 3.131 requires the unknown function to coincide with its derivative, which is satisfied by the exponential function:

$$y(x) = c \cdot e^x, \quad c \in \mathbb{R} \tag{3.132}$$

Equation 3.132 is the general solution of Equation 3.131. Of course, it is again quite unsatisfactory that we got this only "by observation". We saw above that the fundamental theorem of calculus can be used to solve ODEs having the form Equation 3.112. Obviously, this cannot be applied to Equation 3.131. Are there any computational procedures that can be applied to this equation? Let us hear what *Maxima* says to this question. First of all, let us note that *Maxima* solves Equation 3.131 using the above procedure (see ODEEx7.mac in the book software):

```
1:  eq:  'diff(y,x)=y;
2:  ode2(eq,y,x);
```
$$\tag{3.133}$$

If you execute this code, *Maxima* expresses the solution as

$$y = \%c\%e^x \tag{3.134}$$

As before, %c designates an arbitrary constant. You may be puzzled by the second percent sign, but this is just a part of the Euler number which *Maxima* writes as %e. So we see that *Maxima* correctly reproduces Equation 3.132, the general solution of Equation 3.131. But remember that we were looking for a computational procedure to solve Equation 3.131. So far we have just seen that *Maxima* can do it *somehow*. To understand the computational rule used by *Maxima*, consider the following generalization of Equation 3.131:

$$y' = f(x) \cdot g(y) \tag{3.135}$$

Obviously, Equation 3.131 is obtained from this by setting $g(y) = y$ and $f(x) = 1$. Equation 3.128 is also obtained as a special case from Equation 3.135 if we set $g(y) = 1$. Now confronting *Maxima* with Equation 3.135, it must tell us the rule which it uses since it will of course be unable to compute a detailed solution like Equation 3.134 without a specification of $f(x)$ and $g(y)$. To solve Equation 3.135, the following code can be used (see ODEEx8.mac in the book software):

```
1:  eq:  'diff(y,x)=f(x)*g(y);
2:  ode2(eq,y,x);
```
$$\tag{3.136}$$

Maxima writes the result as follows:

$$\int \frac{1}{g(y)} \, dy = \int f(x) \, dx + \%c \tag{3.137}$$

This leads to the following note:

Note 3.7.3 (Separation of variables method) The solution $y(x)$ of Equation 3.135 satisfies

$$\int \frac{1}{g(y)} \, dy = \int f(x) \, dx + c, \quad c \in \mathbb{R} \tag{3.138}$$

In the literature, this method is often formally justified by writing Equation 3.135 as

$$\frac{dy}{dx} = f(x) \cdot g(y) \tag{3.139}$$

that is, by using Leibniz's notation of the derivative, and then by "separating the variables" on different sides of the equation as

$$\frac{dy}{g(y)} = f(x) \, dx \tag{3.140}$$

assuming $g(y) \neq 0$, of course. Then, a formal integration of the last equation yields the separation of variables method (Equation 3.138).

Note that if $g(y) = 1$, that is, in the case of Equation 3.128, we get

$$y = \int 1 \, dy = \int f(x) \, dx + c, \quad c \in \mathbb{R} \tag{3.141}$$

which means that we correctly get the general solution (3.113), which turns out to be a special case of Equation 3.138. Now let us look at Equation 3.131, which is obtained by setting $g(y) = y$ and $f(x) = 1$ in Equation 3.135 as observed above. Using this in Equation 3.138 and assuming $y \neq 0$, we get

$$\int \frac{1}{y} \, dy = \int 1 \, dx + c, \quad c \in \mathbb{R} \tag{3.142}$$

or

$$\ln|y| = x + c, \quad c \in \mathbb{R} \tag{3.143}$$

which leads by exponentiation to

$$|y| = e^{x+c}, \quad c \in \mathbb{R} \tag{3.144}$$

or, resolving the absolute value and denoting $\text{sgn}(y) \cdot e^c$ as a new constant c,

$$y = c \cdot e^x, \quad c \in \mathbb{R} \backslash \{0\} \tag{3.145}$$

Since $y = 0$ is another obvious solution of Equation 3.131, the general solution can be written as

$$y = c \cdot e^x, \quad c \in \mathbb{R} \tag{3.146}$$

3.7.2.1 Application to the Body Temperature Model

As another example application of Equation 3.138, let us derive the solution of the body temperature model (Equation 3.29 in Section 3.4.1):

$$T'(t) = r \cdot (T_b - T(t)) \tag{3.147}$$

To apply Equation 3.138, we first need to write the right-hand side of the last equation in the form of Equation 3.135, that is,

$$T' = f(t) \cdot g(T) \tag{3.148}$$

Since Equation 3.147 is an autonomous equation, we can set

$$f(t) = 1 \tag{3.149}$$
$$g(T) = r \cdot (T_b - T) \tag{3.150}$$

Equation 3.138 now tells us that the solution is obtained from

$$\int \frac{1}{r \cdot (T_b - T)} dT = \int 1 \, dx + c, \quad c \in \mathbb{R} \tag{3.151}$$

Using the substitution $z = T_b - T(t)$ in this integral and an analogous argumentation as above in Equations 3.142–3.146 the general solution is obtained as

$$T(t) = T_b + c \cdot e^{-rt}, \quad c \in \mathbb{R} \tag{3.152}$$

Applying the initial condition

$$T(0) = T_0 \tag{3.153}$$

one obtains

$$c = T_0 - T_b \tag{3.154}$$

and hence

$$T(t) = T_b + (T_0 - T_b) \cdot e^{-r \cdot t} \tag{3.155}$$

which is the solution obtained "by observation" in Section 3.4.1 (Equation 3.25).

3.7.2.2 Solution Using *Maxima* and *Mathematica*

Let us now see how Equation 3.147 is solved using *Maxima*, and let us first try to use the same code as it was used, for example, in Equation 3.136 (see ODEEx9.mac in the book software):

```
1:   eq:  'diff(T,t)=r*(T[b]-T);
2:   ode2(eq,T,t);
```
(3.156)

Maxima's result is as follows:

$$-\frac{\log(T - T_b)}{r} = t + \%c$$
(3.157)

This result is unsatisfactory for two reasons. First, it is an *implicit solution* of Equation 3.147 since it expresses the solution in an equation not solved for the unknown T. More seriously, note that the expression $\log(T - T_b)$ in Equation 3.157 is defined only for $T > T_b$, which means that this solution is valid only in the case $T > T_b$. When we introduced the body temperature model in Section 3.2.2, we were interested exactly in the reverse case, $T < T_b$. In the case of $T > T_b$, it is easy to derive Equation 3.152 from Equation 3.157 by exponentiation as before. Nevertheless, we are definitely at a point here where we encounter *Maxima's* limits for the first time. Using the commercial *Mathematica*, Equation 3.147 is solved without problems using a code very similar to program 3.156:

```
1:   eq=T'[t]==r*(Tb-T[t]);
2:   DSolve[eq,T[t],t]
```
(3.158)

Mathematica's result is

$$\{\{T[t] \rightarrow Tb + e^{-rt}C[1]\}\}$$
(3.159)

which is exactly the general solution (Equation 3.152). Note that without going into details of *Mathematica* notation, the "→" in Equation 3.159 can be understood as an equality sign "=" here, and note also that *Mathematica* writes the arbitrary real constant as C[1] (instead of %c in *Maxima*).

Does this mean we should recommend using the commercial *Mathematica* instead of *Maxima*? The answer is a definite "no and yes". You will see in a minute that although *Maxima's* ode2 command fails here, there is a way to solve Equation 3.147 in *Maxima* using a different command. This is the "no". On the other hand, undoubtedly, *Mathematica* is more comfortable here since it solves all ODEs considered so far with one single command (DSolve). Also, comparing *Mathematica's* and *Maxima's* capabilities to solve ODE, you will see that *Mathematica* solves a lot more ODEs. This is the "yes". The true answer lies in between. It depends on your needs. If you constantly need to solve computer algebra problems beyond *Maxima's* scope in your professional work, then it can be economically advantageous to pay for *Mathematica*, at least until your problems will be covered by improved versions of *Maxima* or other open-source computer algebra software.

At least within the scope of this book where we want to learn about the principles and ideas of computer algebra software, *Maxima's* capabilities are absolutely satisfactory. This is nicely illustrated by a comparison of the *Maxima* code 3.156 with the *Mathematica* code 3.158. Except for different names of the ODE solving commands and different ways to write derivatives and equations, the structure of these two codes is very similar, and we can safely say that after you have learned about *Maxima* in this book, you will not have any big problems in using *Mathematica* or other computer algebra software.

Now it is time to say how to solve Equation 3.147 in *Maxima*. *Maxima* provides a second command solving ODEs called `desolve`. This command is restricted to linear ODEs (including systems of linear ODEs), and it can be applied here since Equation 3.147 is indeed linear (see the discussion of linearity in Section 3.5.8 above). The following code can be used (see `ODEEx10.mac` in the book software):

$$
\begin{array}{ll}
\texttt{1: eq: 'diff(T(x),x)=r*(T[b]-T(x));} \\
\texttt{2: desolve(eq,T(x));}
\end{array}
\qquad (3.160)
$$

This is again very similar to the two codes above (programs 3.156 and 3.158). Note that in contrast to the `ode2` command used in Equation 3.156, `desolve` wants you to write down the dependence of the unknown on the independent variable in the form $T(x)$. Equation 3.160 produces the following result:

$$
T(x) = (T(0) - T_b)\, \%e^{-rx} + T_b \qquad (3.161)
$$

which is equivalent with Equation 3.155.

3.7.3
Variation of Constants

As was discussed above in Section 3.5, Equation 3.147 is no longer solved by Equation 3.152 if we assume that the room temperature T_b is time dependent. In that case, we can write Equation 3.147 as

$$
T'(t) = r \cdot (T_b(t) - T(t)) \qquad (3.162)
$$

The separation of variables method (Note 3.7.3) is not applicable to this equation since we simply cannot separate the variables here. This method would require the right-hand side of Equation 3.162 to be written in the form

$$
T'(t) = f(t) \cdot g(T) \qquad (3.163)
$$

but there is no way how this could be done since the right-hand side of Equation 3.162 expresses a sum, not a product. The general form of Equation 3.162 is that of a linear ODE in the sense of Equation 3.93 (see the discussion of linear ODEs in Section 3.5.8):

$$y'(x) = a(x) \cdot y(x) + b(x) \tag{3.164}$$

As before, we ask *Maxima* to tell us how this kind of equation is solved. Analogous to Equation 3.156, the following *Maxima* code can be used (see `ODEEx11.mac` in the book software):

```
1:  eq: 'diff(y,x)=a(x)*y+b(x);
2:  ode2(eq,y,x);
```
(3.165)

In *Maxima*, this yields

$$y = \%e^{\int a(x)\,dx}\left(\int b(x) \cdot \%e^{-\int a(x)\,dx}\,dx + \%c\right) \tag{3.166}$$

which is known as the

Note 3.7.4 (Variation of constants method) The solution $y(x)$ of Equation 3.164 satisfies

$$y = e^{\int a(x)\,dx}\left(\int b(x) \cdot e^{-\int a(x)\,dx}\,dx + c\right), \quad c \in \mathbb{R} \tag{3.167}$$

This method – which is also known as the *variation of parameters method* – applies to an inhomogeneous ODE such as Equation 3.164, and it can be generalized to inhomogeneous systems of ODEs such as Equation 3.96. Remember from your linear algebra courses that the term "inhomogeneous" refers to a situation where we have $b(x) \neq 0$ in Equation 3.164. The case $b(x) = 0$ can be treated by the separation of variables method as described above. As in the case of the separation of variables method (Note 3.7.3), the variation of constants method does not necessarily lead us to a closed form solution of the ODE. Closed form solutions are obtained only when the integrals in Equation 3.167 (or Equation 3.138 in the case of the separation of variables method) can be solved in closed form, that is, if the result can be expressed in terms of well-known functions. In our above applications of the separation of variables method, we got everything in closed form (e.g. the solution Equation 3.155 of Equation 3.147).

3.7.3.1 Application to the Body Temperature Model

Now let us see whether we get a closed form solution if we apply the variation of constants method to Equation 3.162. Comparing Equations 3.162 and 3.164, you see that we have

$$\begin{aligned} a(t) &= -r \\ b(t) &= r \cdot T_b(t) \end{aligned} \tag{3.168}$$

and hence

$$\int a(t)\, dt = \int -r\, dt = -rt \qquad (3.169)$$

Note that we do not need to consider an integration constant here since the $\int a(t)\, dt$ in the variation of constant method refers to one particular integral of $a(t)$ that we are free to choose. Using the last three equations, we get the following general solution of Equation 3.162:

$$T(t) = e^{-rt}\left(\int r \cdot T_b(t) \cdot e^{rt}\, dt + c\right), \quad c \in \mathbb{R} \qquad (3.170)$$

If this formula is correct, it should give us the general solution that we derived above in the case $T_b = \text{const}$, Equation 3.152. In this case, the last equation turns into

$$T(t) = e^{-rt}\left(T_b \cdot \int r \cdot e^{rt}\, dt + c\right), \quad c \in \mathbb{R} \qquad (3.171)$$

or

$$T(t) = e^{-rt}\left(T_b \cdot e^{rt} + c\right), \quad c \in \mathbb{R} \qquad (3.172)$$

which leads us back to Equation 3.152 as required. In the case where T_b depends on time, we get a closed form solution from Equation 3.170 only when the integral in that formula can be expressed in closed form. For example, let us assume a linearly varying room temperature:

$$T_b(t) = \alpha \cdot t + \beta \qquad (3.173)$$

such that Equation 3.170 becomes

$$T(t) = e^{-rt}\left(\int r \cdot (\alpha \cdot t + \beta) \cdot e^{rt}\, dt + c\right), \quad c \in \mathbb{R} \qquad (3.174)$$

The integral in this formula can be expressed in closed form:

$$\int r \cdot (\alpha \cdot t + \beta) \cdot e^{rt}\, dt = e^{rt}\left(\alpha t + \beta - \frac{a}{r}\right) \qquad (3.175)$$

This can be obtained by hand calculation, for example, using an integration of parts, or by using *Maxima's* `integrate` command (Section 3.7.1.4). Using Equations 3.174 and 3.175, we get

$$T(t) = \alpha t + \beta - \frac{\alpha}{r} + c \cdot e^{-rt}, \quad c \in \mathbb{R} \qquad (3.176)$$

as the general solution of Equation 3.162. If you like, insert $T(t)$ from Equation 3.176 and its derivative $T'(t)$ into Equation 3.162 to verify that this equation is really solved by Equation 3.176. Note that the case $T_b = \text{const}$ corresponds to $\alpha = 0$ and $T_b = \beta$ in Equation 3.173. If this is used in Equation 3.176, we get

$$T(t) = T_b + c \cdot e^{-rt}, \quad c \in \mathbb{R} \tag{3.177}$$

that is, we get Equation 3.152 as a special case of Equation 3.176 again. Using Equation 3.173, Equation 3.176 can be brought in a form very similar to Equation 3.177:

$$T(t) = T_b\left(t - \frac{1}{r}\right) + c \cdot e^{-rt}, \quad c \in \mathbb{R} \tag{3.178}$$

This means that we almost get the same solution as in the case $T_b = \text{const}$, except for the fact that the solution $T(t)$ "uses" an "old" body temperature $T_b(t - 1/r)$ instead of the actual body temperature $T_b(t)$. This is related to the fact that the system needs some time until it adapts itself to a changing temperature. The time needed for this adaption is controlled by the parameter r. In Section 3.4.1.2, we saw that r is the percent decrease of the temperature difference $T_b - T(t)$ per unit of time. As a consequence, large (small) values of r refer to a system that adapts quickly (slowly) to temperature changes. This corresponds with the fact that the solution formula (3.178) uses almost the actual temperature for a quickly adapting system, that is, when r is large, since we have $T_b(t - 1/r) \approx T_b(t)$ in that case.

> **Note 3.7.5 (Advantage of closed form solutions)** A discussion of this kind is what makes closed form solutions attractive, since it leads us to a deeper understanding of the effects that the model parameters (r in this case) have on the solution. No similar discussion would have been possible based on a numerical solution of Equation 3.162.

Note that there is still a constant c in all solutions of Equation 3.162 derived above. As before, this constant is determined from an initial condition (Section 3.7.1.4).

3.7.3.2 Using Computer Algebra Software

Equation 3.170, the general solution of Equation 3.162, could also have been obtained using *Maxima*'s ode2 command similar to above (see ODEEx12.mac in the book software):

```
1:  eq: 'diff(T,t)=r*(Tb(t)-T);
2:  ode2(eq,T,t);
```
(3.179)

The result is as follows:

$$T = \%e^{-rt}\left(r \int \%e^{rt} T_b(t)\, dt + \%c\right) \tag{3.180}$$

which is – in *Maxima* notation – the same as Equation 3.170. Likewise, Equation 3.176 can be obtained directly as the solution of Equation 3.162 in the case of $T_b(t) = \alpha t + \beta$ using the following code (see ODEEx13.mac in the book software):

```
1:  eq: 'diff(T,t)=r*(a*t+b-T);
2:  ode2(eq,T,t);
```

(3.181)

which leads to the following result in *Maxima*:

$$
T = \%e^{-rt}\left(\frac{a(rt - 1)\%e^{rt}}{r} + b\%e^{rt} + \%c \right)
$$

(3.182)

which coincides with Equation 3.176 if we identify $a = \alpha$ and $b = \beta$.

3.7.3.3 Application to the Alarm Clock Model

Now let us try to get a closed form solution for the alarm clock model (Section 3.4.2):

$$
T_s'(t) = r_{si} \cdot (T_i(t) - T_s(t))
$$

(3.183)

$$
T_i'(t) = r_{ia} \cdot (T_a - T_i(t))
$$

(3.184)

$$
T_s(0) = T_{s0}
$$

(3.185)

$$
T_i(0) = T_{i0}
$$

(3.186)

In general, it is, of course, much harder to derive analytical solutions for systems of ODEs. With the last examples in mind, however, it is easy to see that something can be done here. First of all, note that Equations 3.184 and 3.186 can be solved independently from the other two equations since they do not involve T_s. Equations 3.183 and 3.184 are partially decoupled in the sense that the solution T_i of Equation 3.184 affects Equation 3.183, but not vice versa. We will come back to this point in our next example below (Section 3.7.4). To solve Equations 3.184 and 3.186, it suffices to say that we have already done this before. Except for different notation, Equation 3.184 is equivalent with Equation 3.147. Equation 3.155 was derived as the solution of Equation 3.147, which is

$$
T_i(t) = T_a + (T_{i0} - T_a) \cdot e^{-r_{ia} \cdot t}
$$

(3.187)

in this case if Equation 3.186 is used. Inserting this into Equations 3.183–3.186, we come up with a single ODE as follows:

$$
T_s'(t) = r_{si} \cdot (T_a + (T_{i0} - T_a) \cdot e^{-r_{ia} \cdot t} - T_s(t))
$$

(3.188)

$$
T_s(0) = T_{s0}
$$

(3.189)

Except for different notation, again, Equation 3.188 is of the form expressed by Equation 3.162, which has been generally solved using the variation of constants method above, as expressed by Equation 3.170. Using this solution, we find that the general solution of Equation 3.188 is

$$T_s(t) = e^{-r_{si}t}\left(\int r_{si} \cdot (T_a + (T_{i0} - T_a) \cdot e^{-r_{ia} \cdot t}) \cdot e^{r_{si}t}\,dt + c\right), \quad c \in \mathbb{R} \quad (3.190)$$

Everything that remains to be done is the evaluation of the integral in Equation 3.190, and the determination of the constant in that equation using the initial condition, Equation 3.189. Similar to above, you can do this by hand calculation, by using *Maxima*'s integrate command, or you could also go back to Equations 3.188 and 3.189 and solve these using *Maxima*'s ode2 or desolve commands as shown in our previous examples. Instead of proceeding with Equation 3.190, let us go back to the original system (Equations 3.183–3.186), which can be solved directly using *Maxima*'s desolve command as follows (see ODEEx14.mac in the book software):

```
1:  eq1:  'diff(Ts(t),t)=rsi*(Ti(t)-Ts(t));
2:  eq2:  'diff(Ti(t),t)=ria*(Ta-Ti(t));
3:  desolve([eq1,eq2],[Ts(t),Ti(t)]);
```
(3.191)

Note that we cannot use *Maxima*'s ode2 command here since it is restricted to single ODEs, that is, it cannot be applied to ODE systems. desolve is applicable here since Equations 3.183 and 3.184 are linear ODEs in the sense explained in Section 3.5. The code (3.191) produces a result which (after a little simplification by hand) reads as follows:

$$T_s(t) = T_a + (T_{s0} - T_a)e^{-r_{si}t} + (T_{i0} - T_a) \cdot r_{si}\frac{e^{-r_{si}t} - e^{-r_{ia}t}}{r_{ia} - r_{si}} \quad (3.192)$$

$$T_i(t) = T_a + (T_{i0} - T_a)e^{-r_{ia}t} \quad (3.193)$$

3.7.3.4 Interpretation of the Result

A number of things can be seen in these formulas. First of all, note that the initial conditions, Equations 3.185 and 3.186, are satisfied since all the exponential terms in Equations 3.192 and 3.193 equal 1 for $t = 0$. Moreover, the formulas show that we have $T_s = T_i = T_a$ in the limit $t \to \infty$, since all exponential terms vanish as t reaches infinity. This reflects the fact that both the sensor and the effective internal temperature will of course equal the ambient temperature after a long time. You can also see that the solution behaves as expected in the case $T_{i0} = T_a$, that is, in the case when we start with no difference between internal and ambient temperatures. In this case, we would expect a constant $T_i(t) = T_a$, which is obviously true due to $T_{i0} - T_a = 0$ in Equation 3.193. In Equation 3.192, $T_{i0} = T_a$ implies

$$T_s(t) = T_a + (T_{s0} - T_a)e^{-r_{si}t} \quad (3.194)$$

which is (except for different notation) equivalent with Equation 3.155, the solution of the body temperature model with constant body temperature, Equation 3.147.

This is correct since Equation 3.147 expresses the form of Equation 3.183 in the case of $T_{i0} = T_a$, since T_i is time independent in that case. So you see again that a lot can be seen from a closed form solution such as Equations 3.192 and 3.193. Without graphical plots of T_s and T_i, Equations 3.192 and 3.193 already tell us about the behavior of the solution in special cases such as those discussed above, and we can see from this that the behavior of T_s and T_i in these special cases is reasonable. No a priori discussion of this kind would have been possible based on a numerical solution of Equations 3.183 and 3.184.

Note that Equation 3.192 also shows that we have a problem in the case of $r_{si} = r_{ia}$ when the denominator in the fraction becomes zero. To understand this point, we can ask *Maxima* for the result in the case $r = r_{si} = r_{ia}$. After some hand simplification, again, we get (compare ODEEx15.mac in the book software)

$$T_s(t) = T_a + (T_{s0} - T_a)e^{-rt} + (T_{i0} - T_a)rte^{-rt} \tag{3.195}$$

$$T_i(t) = T_a + (T_{i0} - T_a)e^{-rt} \tag{3.196}$$

Now it is just a matter of simple calculus to see that these two solutions are compatible. Denoting $r = r_{si}$, compatibility of Equation 3.192 with 3.195 requires

$$\lim_{r_{ia} \to r} \frac{e^{-rt} - e^{-r_{ia}t}}{r_{ia} - r} = te^{-rt} \tag{3.197}$$

which is indeed true. Based on the usual definition of the derivative, Equation 3.197 just expresses the fact that te^{-rt} is the derivative of e^{-rt} with respect to r. If you like, you can show this by hand calculation based on any of the methods used to compute limits in calculus [17]. Alternatively, you can use *Maxima's* limit function to obtain Equation 3.197 as follows [106]:

$$\texttt{limit((\%e\^{}(-r*t)-\%e\^{}(-(r+h)*t))/h,h,0);} \tag{3.198}$$

Note that the argument of limit in program 3.198 corresponds to the left-hand side of Equation 3.197 if we identify $h = r_{ia} - r$. The closed form solution of Equations 3.183–3.186 can now be summarized as follows:

$$T_s(t) = \begin{cases} T_a + (T_{s0} - T_a)e^{-r_{si}t} + (T_{i0} - T_a) \cdot r_{si} \frac{e^{-r_{si}t} - e^{-r_{ia}t}}{r_{ia} - r_{si}}, & r_{si} \neq r_{ia} \\ T_a + (T_{s0} - T_a)e^{-rt} + (T_{i0} - T_a)rte^{-rt}, & r = r_{si} = r_{ia} \end{cases} \tag{3.199}$$

$$T_i(t) = \begin{cases} T_a + (T_{i0} - T_a)e^{-r_{ia}t}, & r_{si} \neq r_{ia} \\ T_a + (T_{i0} - T_a)e^{-rt}, & r = r_{si} = r_{ia} \end{cases} \tag{3.200}$$

An example application of this closed form solution has been discussed in Section 3.4.2.3 (Figure 3.5).

3.7.4
Dust Particles in the ODE Universe

So far we have introduced the *separation of variables method* and the *variation of constants method* as methods to obtain closed form solutions of ODEs by hand calculation. You should know that there is a great number of other methods that could be used to solve ODEs by hand, and that cannot be discussed in a book like this which focuses on mathematical modeling. There is a lot of good literature you can refer to if you like to know more [98–100]. Beyond hand calculation, we have seen that closed form solutions of ODEs can also be obtained efficiently using computer algebra software such as *Maxima*.

It was already mentioned above that ODEs having closed form solutions are rather the exception than the rule. More precisely, those ODEs are something like dust particles in the "ODE universe". Usually, small changes of an analytically solvable ODE suffice to make it analytically unsolvable (corresponding to the fact that you do not have to travel a long way if you want to leave a dust particle toward space). For example, one may find it impressive that Equations 3.199 and 3.200 can be derived as a closed form solution of Equations 3.183–3.186. But consider the following equations, which have a structure similar to Equations 3.183–3.186, and which do not really look more complicated compared to these equations at a first glance:

$$x' = x(a - by) \tag{3.201}$$

$$y' = -y(c - dx) \tag{3.202}$$

This is the so-called Lotka–Volterra model, which is discussed in Section 3.10.1. $x(t)$ and $y(t)$ are the unknowns in these equations, t is time, and a, b, c are real constants. Note that we have used a notation here that is frequently used in practice, with no explicit indication of the fact that x and y are unknown functions. This is implicitly expressed by the fact that x and y appear with their derivatives in the equations. Also, in the absence of any further comments, the usual interpretation of Equation 3.202 automatically implies that a, b, c, d are real constants.

In a first, naive attempt we could follow a similar procedure as above, trying to solve this ODE system using a code similar to Equation 3.191 (see ODEEx16.mac in the book software):

```
1:  eq1:  ´diff(x(t),t)=x(t)*(a-b*y(t));
2:  eq2:  ´diff(y(t),t)=-y(t)*(c-d*x(t));          (3.203)
3:  desolve([eq1,eq2],[x(t),y(t)]);
```

However, this code produces no result in *Maxima*. Trying to understand why this is so, you may remember that it was said above that desolve applies only to *linear* ODEs. As it was explained in Section 3.5, a linear ODE system has the form of Equation 3.94 or 3.96, that is, the unknowns may be added, subtracted, or multiplied by known quantities. In Equations 3.201 and 3.202, however, the unknowns $x(t)$ and $y(t)$ are multiplied, which means that these equations are nonlinear and hence *Maxima*'s `desolve` command is not applicable. We also discussed *Maxima*'s `ode2` command above, which is indeed applicable to nonlinear ODEs, but unfortunately not to ODE systems like Equations 3.201 and 3.202. Fortunately, on the other hand, this is no problem since `ode2` could not solve these equations even if it would be applicable, since the solution of Equations 3.201-and 3.202 cannot be expressed in closed form.

Comparing the structure of Equations 3.201 and 3.202 with the analytically solvable system (3.183)–(3.186), one can say that there are two main differences: Equations 3.201 and 3.202 are nonlinear, and they involve a stronger coupling of the ODEs. As it was already noted above, Equation 3.184 can be solved independently of Equation 3.183, while in the system of Equations 3.201 and 3.202 each of the two equations depends on the other equation.

> **Note 3.7.6 (Why most ODEs do not have closed form solutions)** Nonlinearity and a strong coupling and interdependence of ODEs are the main reasons why closed form solutions cannot be achieved in many cases.

3.8
Numerical Solutions

If there is no closed form solution of an ODE, you know from our discussion in Section 3.6 that the way out is to compute approximate *numerical solutions* of the ODE using appropriate numerical algorithms on the computer. Since our emphasis is on modeling aspects, we will not go into a detailed explanation of numerical algorithms here, just as we did not provide an exhaustive discussion of analytical methods in the previous section. This section is intended to make you familiar with some general ideas that are applied in the numerical solution of ODEs, and with appropriate software that can be used to solve ODEs.

Some people have a general preference for what might be called *closed form mathematics* in the sense of Section 3.7, that is, for a kind of mathematics where everything can be written down using "well-known expressions" as explained in Section 3.6.2. These people tend to believe that in what might be called the *house of mathematics*, numerical mathematics has its place at some lower level, several floors below closed form mathematics. When they write down quantities such as $\sqrt{2}$ or π in their formulas, they do not spend too much thoughts on it, and they believe that numerical mathematics is far away. *If* they would spend a few thoughts on their formulas, they would realize that numerical mathematics is everywhere, and that it gazes at us through most of the closed form formulas that we may write down. It is easy to write down transcendental numbers such as π or $\sqrt{2}$, but from a practical point of view, these symbols are just a wrapping around the numerical algorithms that must be used to obtain numerical approximations of these numbers that can really be used in your computations. These symbols denote something that is entirely unknown until numerical algorithms are applied that shed at least a little bit of light into the darkness (e.g. a few millions of π's digits). In the same sense, a general initial value problem (Definition 3.5.2) such as

$$y'(x) = F(x, y(x)) \tag{3.204}$$
$$y(0) = a \tag{3.205}$$

denotes an entirely unknown function $y(x)$, and ODE solving numerical algorithms help us to shed some light on this quantity.

3.8.1
Algorithms

Although a detailed account of numerical algorithms solving ODEs is beyond the scope of this book, let us at least sketch same basic ideas here that you should know when you are using these algorithms in practice. Referring to Equations 3.204–3.205, it is a priori clear that, as it is the case with π, an ODE solving numerical algorithm will never be able to describe the solution of these equations in its entirety. Any numerical algorithm can compute only a finite number of unknowns, and this means that a numerical approximation of the unknown $y(x)$ will necessarily involve the computation of $y(x)$ at some given points $x_0, x_1, \ldots, x_n$ ($n \in \mathbb{N}$), that is, the result will be of the form $y_0, y_1, \ldots, y_n$, where the y_i are numerical approximations of $y(x_i)$, that is, $y_i \approx y(x_i)$ ($i = 0, 1, \ldots, n$).

3.8.1.1 The Euler Method
The simplest and most natural thing one can do here is the so-called Euler method. This method starts with the observation that Equations 3.204 and 3.205 already tell us about two quantities of concern. Assuming $x_0 = 0$, we have

$$y_0 = y(x_0) = a \tag{3.206}$$
$$y'(x_0) = F(x_0, y(x_0)) = F(x_0, y_0) \tag{3.207}$$

This means we know the first point of the unknown function, (x_0, y_0), and we know the derivative of the function at this point. You know from calculus that this can be used for a linear approximation of $y(x)$ in the vicinity of x_0. Specifically, if we assume that $y(x)$ is continuously differentiable, a Taylor expansion of $y(x)$ around x_0 gives

$$y(x_0 + h) \approx y_0 + h \cdot y'(x_0) \tag{3.208}$$

or, using (3.207),

$$y(x_0 + h) \approx y_0 + h \cdot F(x_0, y_0) \tag{3.209}$$

These approximations are valid "for small h", that is, they become better as h is decreased toward zero. Now assuming constant and sufficiently small differences $h = x_i - x_{i-1}$ for $i = 1, 2, \ldots, n$, Equation 3.209 implies

$$y(x_1) \approx y_0 + h \cdot F(x_0, y_0) \tag{3.210}$$

The Euler method uses this expression as the approximation of $y(x)$ at x_1, that is,

$$y_1 = y_0 + h \cdot F(x_0, y_0) \tag{3.211}$$

Now (x_1, y_1) is an approximate point of $y(x)$ at x_1 and $F(x_1, y_1)$ is an approximate derivative of $y(x)$ at x_1, so we see that the same argument as before can be repeated and gives

$$y_2 = y_1 + h \cdot F(x_1, y_1) \tag{3.212}$$

as an approximation of $y(x)$ at x_2. This can be iterated until we arrive at y_n, and the general formula of the Euler method, thus, is

$$y_i = y_{i-1} + h \cdot F(x_{i-1}, y_{i-1}), i = 1, 2, \ldots, n \tag{3.213}$$

Note that the h in this equation is called the *stepsize* of the Euler method. Many of the more advanced numerical methods – including R's `lsoda` command that is discussed below – use nonconstant, *adaptive stepsizes*. Adaptive stepsizes improve the efficiency of the computations since they use a finer resolution (corresponding to small stepsizes) of the solution of the ODE in regions where it is highly variable, while larger stepsizes are used otherwise.

Numerical methods solving differential equations (ODEs as well as PDEs) such as the Euler method are also known as *discretization methods*, since they are based on a discrete reformulation of the originally continuous problem. Equations 3.204 and 3.205, for example, constitute a continuous problem since the unknown $y(x)$ depends continuously on the independent variable, x. On the other hand, the Euler method led us to a discrete approximation of $y(x)$ in terms of the numbers $y_0, \ldots, y_n$, that is, in terms of an approximation of $y(x)$ at a few, discrete times $x_0, \ldots, x_n$.

3.8.1.2 Example Application

Let us test the Euler method for the initial value problem

$$y'(x) = y(x) \tag{3.214}$$
$$y(0) = 1 \tag{3.215}$$

As we have seen in Section 3.3.2, the closed form solution of this problem is

$$y(x) = e^x \tag{3.216}$$

Thus, we are in a situation here where we can compare the approximate solution of Equations 3.214 and 3.215 derived using the Euler method with the exact solution (Equation 3.216), which is a standard procedure to test the performance and accuracy of numerical methods. A comparison of Equations 3.204 and 3.214 shows that we have $F(x, y) = y$ here, which means that Equation 3.213 becomes

$$y_i = y_{i-1} + h \cdot y_{i-1}, i = 1, 2, \ldots, n \qquad (3.217)$$

Using Equation 3.215, the entire iteration procedure can be written as follows:

$$y_0 = 1 \qquad (3.218)$$
$$y_i = (1 + h) \cdot y_{i-1}, i = 1, 2, \ldots, n \qquad (3.219)$$

To compute the approximate solution of Equations 3.214 and 3.215 for $x \in [0, 4]$, the following *Maxima* code can be used (compare `Euler.mac` in the book software):

```
1:   h:1;
2:   n:4/h;
3:   y[0]:1;
4:   for i:1 thru n step 1 do
5:       y[i]:(1+h)*y[i-1];
```
(3.220)

After the initializations in lines 1 and 2, the remaining lines are in direct correspondence with the iteration equations (3.218)–(3.219). Note that "`:`" serves as the *assignment operator* in *Maxima* [106], that is, a command such as line 1 in program 3.220 assigns a value to a variable (h receives the value 1 in this case). Many other programming languages including *R* (Appendix B) use the equality sign " $=$ " as assignment operator. In *Maxima*, the equality sign " $=$ " serves as the *equation operator*, that is, it defines unevaluated equations that can then, for example, be used as an argument of the `solve` command (line 4 of program 1.7 in Section 1.5.2). The `for...thru` command in line 4 iterates the command in line 5 for $i = 1, 2, \ldots, n$, that is, that command is repeatedly executed beginning with $i = 1$, then $i = 2$, and so on. See [106] for several other possibilities to formulate iterations using *Maxima*'s do command.

The code (3.220) produces the y_i values shown in Table 3.1. In the table, these values are compared with the corresponding values of the exact solution of Equations 3.214 and 3.215 (Equation 3.216). As can be seen, the y_i values deviate substantially from the exact values, e^{x_i}. This is due to the fact that a "small h" was assumed in the derivation of the Euler method above. Better approximations are achieved as h is decreased toward zero. This is illustrated in Figure 3.7a. The line in the figure is the exact solution, and it is compared with the result of program 3.220 obtained for $h = 1$ (points) and $h = 0.25$ (triangles). As can be seen, the approximation obtained for $h = 0.25$ is much closer to the exact solution. Still better approximations are obtained as h is further decreased toward zero – test this yourself using the code (3.220).

Table 3.1 Result of the *Maxima* code (3.220), y_i, compared with the values of the exact solution (Equation 3.216), e^{x_i}.

i	0	1	2	3	4
x_i	0	1	2	3	4
y_i	1	2	4	8	16
e^{x_i}	1	2.72	7.39	20.09	54.6

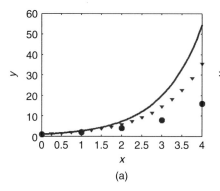
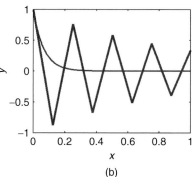

(a)   (b)

Fig. 3.7 (a) Solution of Equation 3.214 and 3.215: exact solution (Equation 3.216) (line), Euler method with $h = 1$ (points), and Euler method with $h = 0.25$ (triangles). Figure produced using Euler.mac. (b) Thin line: closed form solution (Equation 3.227). Thick line: numerical solution of Equations 3.225 and 3.226 obtained using the Euler method, Equations 3.228 and 3.229. Figure produced using Stiff.mac.

3.8.1.3 Order of Convergence

There is a great number of other numerical procedures solving ODEs which cannot be discussed in detail here. An important criterion to select an appropriate method among these procedures is the order of convergence, that is, how fast each procedure converges toward the exact solution if the stepsize h is decreased toward zero. The Euler method's order of convergence can be estimated fairly simple. Let us write down Equation 3.213 for $i = 1$:

$$y_1 = y_0 + h \cdot F(x_0, y_0) \tag{3.221}$$

Using the above notation and the exact solution $y(x)$ of Equation 3.204 and 3.205, this can be written as follows:

$$y_1 = y(0) + hy'(0) \tag{3.222}$$

Now y_1 is an approximation of $y(h)$, which can be expressed using a Taylor expansion as follows:

$$y(h) = y(0) + hy'(0) + \frac{1}{2}h^2y''(0) + O(h^3) \tag{3.223}$$

where $O(h^3)$ summarizes terms proportional to h^3. The error introduced in Equation 3.222 is the difference $y(h) - y_1$, which is obtained from the last two equations as

$$y(h) - y_1 = \frac{1}{2}h^2 y''(0) + O(h^3) \tag{3.224}$$

In this expression, $1/2h^2 y''(0)$ is the dominant term for sufficiently small h. Thus, we see that the error introduced by a single step of the Euler method (Equation 3.221) is proportional to h^2. Now suppose you want to obtain the solution of Equations 3.204 and 3.205 on an interval $[0, L]$. If you use the stepsize h, you will need $n = L/h$ Euler steps. Since the error in each Euler step is proportional to h^2 and the number of Euler steps needed is proportional to $1/h$, it is obvious that the total error will be proportional to h. Therefore, the Euler method is said to be a first-order method. Generally, an algorithm solving an ODE is said to be a *kth order method* if the total error is proportional to h^k. In practice, higher-order methods are usually preferred to the Euler method, such as the fourth order *Runge–Kutta method* which is used in *Maxima's* rk function. In this method, the total error is proportional to h^4, which means that it converges much faster to the exact solution compared with the Euler method. The use of *Maxima's* rk function is explained below. This function, however, has its limitations. For example, it does not provide an automatic control of the approximation error, uses a constant stepsize, and may fail to solve so-called stiff ODEs.

3.8.1.4 Stiffness

Stiff ODEs are particularly difficult to solve in the sense that numerical approximations may oscillate wildly around the exact solution and may require extremely small stepsizes until a satisfactory approximation of the exact solution is obtained. As an example, consider the initial value problem

$$y'(x) = -15 \cdot y(x) \tag{3.225}$$
$$y(0) = 1 \tag{3.226}$$

Using the methods in Section 3.7, the closed form solution is easily obtained as

$$y(x) = e^{-15x} \tag{3.227}$$

Let us apply the Euler method to get the numerical solution. Analogous to Equations 3.218 and 3.219, the iteration procedure is

$$y_0 = 1 \tag{3.228}$$
$$y_i = (1 - 15h) \cdot y_{i-1}, i = 1, 2, \ldots, n \tag{3.229}$$

The *Maxima* code Stiff.mac in the book software compares the numerical solution obtained by these formulas with the closed form solution, Equation 3.227. The result is shown in Figure 3.7b. You can see those wild oscillations of the numerical solution around the exact solution mentioned above. These oscillations become much worse as you increase the stepsize h (try this using $\mathtt{Stiff.mac}$), and they obviously reflect by no means the qualitative behavior of the solution. Stiffness is not such an unusual phenomenon and may occur in all kinds of applications. It is frequently caused by the presence of different time scales in ODE systems, for example, if one of your state variables changes its value slowly on a time scale of many years, while another state variable changes quickly on a time scale of a few seconds. There are methods specifically tailored to treat stiff ODEs such as the *backward differentiation formulas* (BDF) method which is also known as the *Gear method* [107]. This method is a part of a powerful and accurate software called $\mathtt{lsoda}$ which is a part of *R*'s $\mathtt{odesolve}$ package.

$\mathtt{lsoda}$ ("livermore solver for ordinary differential equations") has been developed by Petzold and Hindmarsh at California's Lawrence Livermore National Laboratory [108, 109]. It features an automatic switching between appropriate methods for stiff and nonstiff ODEs. In the nonstiff case, it uses the Adams method with a variable stepsize and a variable order of up to 12th order, while the BDF method with a variable stepsize and a variable order up to fifth order is used for stiff systems. Compared to *Maxima*'s $\mathtt{rk}$ function, $\mathtt{lsoda}$ is a much more professional software and it will be used below as the main tool for the numerical solution of ODEs. From now on, we will use *Maxima*'s $\mathtt{rk}$ and *R*'s $\mathtt{lsoda}$ functions without any more reference to the numerical algorithms that work inside. Readers who want to know more on numerical algorithms solving ODEs are referred to an abundant literature on this topic, for example, [107, 110].

3.8.2
Solving ODE's Using *Maxima*

Let us consider the body temperature model again, which was introduced in Section 3.2.2 as follows:

$$T(t) = T_b - (T_b - T_0) \cdot e^{-r \cdot t} \tag{3.230}$$

In Section 3.4.1, an alternative formulation of this model as an initial value problem was derived:

$$T'(t) = r \cdot (T_b - T(t)) \tag{3.231}$$

$$T(0) = T_0 \tag{3.232}$$

As it was seen in Section 3.4.1, Equation 3.230 solves Equations 3.231 and 3.232. This means that when we solve Equations 3.231 and 3.232 numerically using *Maxima*'s $\mathtt{rk}$ function, we will be able to assess the accuracy of the numerical solution by a comparison with Equation 3.230. Exactly this is done in the *Maxima* program $\mathtt{FeverODE.mac}$ in the book software. In this program, the solution of

Equations 3.231 and 3.232 is achieved in the following lines of code.

```
1:  eq: 'diff(T,t) = r*(T[b]-T);
2:  sol: rk(rhs(eq),T,T[0],[t,aa,bb,h]);
```
(3.233)

In line 1, the ODE (3.231) is defined as it was done above in Section 3.7. In that section, the ODE was then solved in closed form using *Maxima*'s ode2 and desolve commands. In line 2 of program 3.233, *Maxima*'s rk command is used to solve the ODE using the fourth order Runge–Kutta method as discussed above. As you see, rk needs the following arguments:

- the right-hand side of the ODE, rhs(eq) in this case
  (rhs(eq) gives the right-hand side of its argument);
- the variable the ODE is solved for, T in this case;
- the initial value of that variable, T[0] in this case;
- a list specifying the independent variable (t in this case), the limits of the solution interval (aa and bb), and the stepsize (h).

The result of the rk command in line 2 of program 3.233 is stored in a variable sol, which is then used to plot the solution in FeverODE.mac. Of course, all constants in program 3.233 need to be defined before line 2 of that code is performed: just see how program 3.233 is embedded into the *Maxima* program FeverODE.mac in the book software. Note that the right-hand side of the ODE can of course also be entered directly in the rk command, that is, it is not necessary to define eq and then use the rhs command as in Equation 3.233. The advantage of defining eq as in Equation 3.233 is that it is easier to apply different methods to the same equation (e.g. ode2 needs the equation as defined in Equation 3.233, see above). Using the same constants as in Figure 3.1b and a stepsize $h = 0.2$, FeverODE.mac produces the result shown in Figure 3.8. As can be seen, the numerical solution matches the exact solution given by Equation 3.230 perfectly well.

3.8.2.1 Heuristic Error Control

In the practical use of rk, one will of course typically have no closed form solution that could be used to assess the quality of the numerical solution. In cases where you have a closed form solution, there is simply no reason why we should use rk. But, how can the quality of a numerical solution be assessed in the absence of a closed form solution? The best way to do this is to use a code that provides you with parameters that you can use to control the error, such as *R*'s lsoda function, which is explained in Section 3.8.3. In *Maxima*'s rk command, the only parameter that can be used to control the error is the stepsize h. As it was discussed above, ODE solving numerical algorithms such as the Runge–Kutta method implemented in the rk function have the property that the numerical approximation of the solution

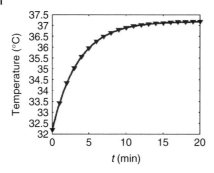

Fig. 3.8 Line: numerical solution of Equations 3.231 and 3.232 using *Maxima's* rk command and a stepsize $h = 0.2$. Triangles: exact solution (Equation 3.230). Figure produced using FeverODE.mac.

converges to the exact solution of the ODE as the stepsize h approaches zero. So you should use a small enough stepsize if you want to get good approximations of the exact solution. On the other hand, your computation time increases beyond all bounds if you are really going toward zero with your stepsize. A heuristic procedure that can be used to make sure that you are not using too big stepsizes is as follows:

Note 3.8.1 (Heuristic error control)
1. Compute the numerical solutions y_h and $y_{h/2}$ corresponding to the stepsizes h and $h/2$.
2. If the difference between y_h and $y_{h/2}$ is negligibly small, the stepsize h will usually be small enough.

But note that this is a heuristic only and that you better should use a code such as *R's* lsoda that provides a control of the local error (see below). Note also that it depends of course on your application what is meant by "negligibly small". Applied to Equations 3.231 and 3.232, you could begin to use FeverODE.mac with a big stepsize such as $h = 6$, which produces Figure 3.9a Then, halving this stepsize to $h = 3$ you would note that you get the substantially different (better) result in Figure 3.9b, and you would then go on to decrease the stepsize until your result would constantly look as that of Figure 3.8, regardless of any further reductions of h.

3.8.2.2 ODE Systems
Maxima's rk command can also be used to integrate systems of ODEs. To see this, let us reconsider the alarm clock model. In Section 3.4.2, the alarm clock model was formulated as an initial value problem as follows:

$$T_s'(t) = r_{si} \cdot (T_i(t) - T_s(t)) \tag{3.234}$$

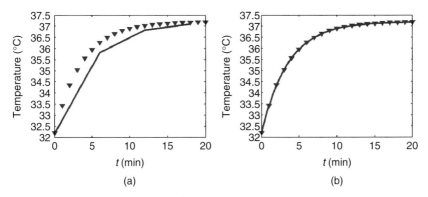

Fig. 3.9 (a) Line: numerical solution of Equations 3.231 and 3.232 using *Maxima*'s rk command and a stepsize $h = 6$. Triangles: exact solution (Equation 3.230). (b) Same plot for $h = 3$. Plots generated using FeverODE.mac.

$$T_i'(t) = r_{ia} \cdot (T_a - T_i(t)) \tag{3.235}$$

$$T_s(0) = T_{s0} \tag{3.236}$$

$$T_i(0) = T_{i0} \tag{3.237}$$

In Section 3.7.3.4, the following closed form solution of Equations 3.234–3.237 was derived:

$$T_s(t) = \begin{cases} T_a + (T_{s0} - T_a)e^{-r_{si}t} + (T_{i0} - T_a) \cdot r_{si}\dfrac{e^{-r_{si}t} - e^{-r_{ia}t}}{r_{ia} - r_{si}}, & r_{si} \neq r_{ia} \\ T_a + (T_{s0} - T_a)e^{-rt} + (T_{i0} - T_a)rte^{-rt}, & r = r_{si} = r_{ia} \end{cases} \tag{3.238}$$

$$T_i(t) = \begin{cases} T_a + (T_{i0} - T_a)e^{-r_{ia}t}, & r_{si} \neq r_{ia} \\ T_a + (T_{i0} - T_a)e^{-rt}, & r = r_{si} = r_{ia} \end{cases} \tag{3.239}$$

Again, the closed form solution Equations 3.238 and 3.239 can be used to assess the quality of the numerical solution of Equations 3.234–3.237 obtained using *Maxima*'s rk command. The numerical solution of Equations 3.234–3.237 is implemented in the *Maxima* program RoomODE.mac in the book software. Within this code, the following lines compute the numerical solution:

```
1:  eq1:´diff(T[s],t)=r[si]*(T[i]-T[s]);
2:  eq2:´diff(T[i],t)=r[ia]*(T[a]-T[i]);
3:  sol: rk([rhs(eq1),rhs(eq2)],
            [T[s],T[i]],[T[s0],T[i0]],[t,aa,bb,h]);
```
(3.240)

Note that everything is perfectly analogous to program 3.233, except for the fact that there are two equations eq1 and eq2 instead of a single equation eq, and that rk needs a list of two right-hand sides, state variables, and initial conditions in this case. On the basis of the parameter values that were used to produce Figure 3.5

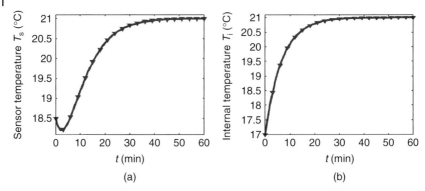

Fig. 3.10 (a) Line: numerical approximation of T_s based on Equations 3.234–3.237 using *Maxima*'s rk command and a stepsize $h = 0.5$. Triangles: exact solution (Equation 3.238). (b) Line: numerical approximation of T_i based on Equations 3.234–3.237 using *Maxima*'s rk command and a stepsize $h = 0.5$. Triangles: exact solution (Equation 3.239).

and a stepsize of $h = 0.5$, RoomODE.mac produces the result shown in Figure 3.10. Again, we get a perfect coincidence between the numerical and the closed form solution.

3.8.3
Solving ODEs Using *R*

Using the last two examples again, we will now explain how the same numerical solutions can be obtained using *R* and lsoda. As it was already emphasized above, lsoda is a much more professional software for the solution of ODEs (compared with *Maxima*'s rk command), and it will be the main instrument to solve ODEs in the following sections. Details about the numerical algorithms implemented in lsoda may be found in [108, 109]. Let us start with the body temperature example again, and let us again compare the closed form solution, Equation 3.230, with the numerical solution of Equations 3.231 and 3.232. To compute the numerical solution using *R*, we use the program ODEEx1.r in the book software.

3.8.3.1 Defining the ODE
Program 3.241 shows the part of ODEEx1.r that defines the differential equation 3.231. Comparing this with the *Maxima* code 3.233 where the ODE is defined in a single line (line 1), you see that the *Maxima* code is simpler and more intuitive (but, unfortunately, also limited in scope as explained above). To simplify the usage of the *R* programs discussed in this book as much as possible, the parts of the programs where user input is needed are highlighted as it can be seen in program 3.241.

```
 1:  dfn <-
 2:     function(t, y, p)
 3:     {
 4:  ### User: names of state variables
 5:  ###############################
 6:        T=y[1]
 7:  ### User: differential equation          (3.241)
 8:  ###############################
 9:        dTdt=r*(Tb-T)
10:  ### User: list of derivatives
11:  ###############################
12:        list(c(dTdt))
13:     }
```

As a whole, the code 3.241 defines a function dfn that gives the right-hand side of the ODE and is used later in ODEEx1.r to solve the ODE. In the general case, this function assumes a system of ODEs of the form (compare Definition 3.5.3)

$$\mathbf{y}'(t) = \mathbf{F}(t, \mathbf{y}(t), \mathbf{p}) \qquad (3.242)$$

where $\mathbf{y}(t) = (y_1(t), y_2(t), \ldots, y_n(t))^t$ expresses the n dependent variables of the ODE and $\mathbf{p} = (p_1, p_2, \ldots, p_s)^t$ is a vector of parameters on which the right-hand side of the ODE may depend ($n, s \in \mathbb{N}$). We do not use $\mathbf{p}$ here, but it is important in the estimation of parameters of the ODE from experimental data (Section 3.9). The arguments of dfn in program 3.241 are in obvious correspondence with the arguments of $\mathbf{F}(t, \mathbf{y}(t), \mathbf{p})$ in Equation 3.242. The components of y and p are accessed as y[1], y[2],... and p[1], p[2],... respectively, in R. Square brackets are used generally in R to access array components [45].

In lines 4–6 of program 3.241, you set names for the dependent variables that are then used in lines 7–9 to define the ODE. Lines 4–6 can be omitted if you formulate the ODE using y[1], y[2] etc. as the dependent variables. Note, however, that your code will be better readable if you are using the original names of your dependent variables instead of y[1], y[2] etc. Line 9 defines the ODE in obvious correspondence with Equation 3.231. Line 12 of program 3.241 is the last statement within the function body of dfn (all statements enclosed in brackets between lines 3 and 13), and it is, thus, the function value produced by dfn. dfn is expected to produce a list of all right-hand sides of the ODE system, which is a mere list(c(dTdt)) in this case and which would read, for example, list(c(dTdt,dSdt)) if our ODE would involve a second state variable S (see the examples below). Note that the concatenate operator c(...) is used in R to build vectors from its arguments [45].

3.8.3.2 Defining Model and Program Control Parameters

Now you know how to define an ODE in R – let us go on looking through ODEEx1.r. The next thing you can do is the definition of a *closed form solution*, if available, and if you would like to compare the numerical result with a closed form solution. In the body temperature example, we have Equation 3.230 as a closed form solution of Equations 3.231 and 3.232. Program 3.243 shows the part of ODEEx1.r that defines the closed form solution, Equation 3.230. As you see, the closed form solution function is denoted as AnSol in ODEEx1.r. AnSol corresponds to $T(t)$ in Equation 3.230, and in R it can be invoked in the form AnSol(t). The right-hand side of Equation 3.230 is in immediate correspondence with the expression you see in line 3 of 3.243.

```
1:  ### User: closed form solution (if available)
2:  #############################################          (3.243)
3:  AnSol=function(t) Tb-(Tb-T0)*exp(-r*t)
```

After this, there is a section **Parameters of the model** in ODEEx1.r where you do exactly what this section title suggests. In the body temperature model, there are three parameters to be defined (r, T_0, and T_b) and looking into ODEEx1.r you will see that exactly the same values as in the *Maxima* program FeverODE.mac are used here.

In the **Program control parameters** section of ODEEx1.r, you begin with a selection of the plots you want to see by setting the appropriate logical variables. After this, the start and end times aa and bb are defined, which limit the interval in which you want to see the numerical solution. If applicable, you can set the number of points of the analytical solution in the plot using nAnSol, and you can define a data file in csv-format in the variable DataFile (e.g. if you want to compare the numerical solution with data). Using nTime, you can prescribe the number of intermediate points between aa and bb where you want lsoda to compute numerical approximations (this corresponds to n in the Euler method, see Section 3.8.1.1). Basically, nTime controls the smoothness of the numerical solution curve that you see in the plots, since this curve is a polygon connecting the points of the solution that have been computed by the numerical algorithm. Note that nTime assumes an equidistant distribution of the intermediate points between aa and bb. You can easily change this if you like by looking at the way in which nTime is used later in ODEEx1.r. It is important that you do not confuse the distance between numerical approximation points controlled by nTime with the stepsize h discussed above. h is automatically controlled by lsoda so as to satisfy given local error bounds (see below), while nTime just controls the amount of output you need.

3.8.3.3 Local Error Control in lsoda

After this you set the absolute and relative local error tolerances, atolDef and rtolDef. As it was mentioned above, the fact that these error tolerances can be prescribed using lsoda is an important reason why you should better use lsoda instead of *Maxima's* rk function in your everyday work, although the use of *Maxima* is simpler as we have seen. Suppose we are in n dimensions, that is, we have $\mathbf{y}(t) = (y_1(t), y_2(t), \ldots, y_n(t))^T$ in Equation 3.242. Assume that the solver has just performed one step of its numerical algorithm leading to a numerical approximation $\tilde{\mathbf{y}}(t) = (\tilde{y}_1(t), \tilde{y}_2(t), \ldots, \tilde{y}_n(t))^T$ at some time t. You may imagine one step of the Euler method here that advances the numerical solution from a previous time $t - h$ to the current approximation at time t. Above it was demonstrated how the error in one step of the Euler method can be estimated, see Equation 3.224. This kind of per-step-error is called the *local error* of the numerical algorithm. It must be distinguished from the *global error* of the numerical method, which is the difference between the numerical approximation of the solution of the ODE and the exact solution of the ODE, both considered over the entire interval of interest (the difference being measured in some appropriate norm that we do not need to discuss here).

Using methods appropriate for the numerical algorithms used in lsoda, the *local* error in one step of this algorithm can be estimated similar to Equation 3.224. For each of the n components of $\tilde{\mathbf{y}}(t)$, lsoda computes such an estimated local error, which we denote e_i. Then, the local errors e_i are required to satisfy

$$e_i \leq \mathrm{rtol}_i \cdot |\tilde{y}_i(t)| + \mathrm{atol}_i, \quad i = 1 \ldots n \tag{3.244}$$

Here, rtol_i and atol_i are the relative and absolute local error tolerances that you specify in ODEEx1.r using atolDef and rtolDef. For example, assume $n = 2$, and assume you want to have $\mathrm{atol}_1 = 10^{-2}$ and $\mathrm{atol}_2 = 10^{-4}$. Then, you can do this by setting

$$\begin{array}{ll} 1\text{:} & \text{\# absolute tolerance} \\ 2\text{:} & \text{atolDef=c(1e-2,1e-4)} \end{array} \tag{3.245}$$

in the "program control parameters" section of ODEEx1.r. On the other hand,

$$\begin{array}{ll} 1\text{:} & \text{\# absolute tolerance} \\ 2\text{:} & \text{atolDef=1e-3} \end{array} \tag{3.246}$$

would give $\mathrm{atol}_1 = 10^{-3}$ and $\mathrm{atol}_2 = 10^{-3}$. rtol_i can be controlled in the same way. Basically, lsoda uses Equation 3.244 to optimize the stepsize, that is, the stepsize is chosen as large as possible such that Equation 3.244 is still satisfied.

3.8.3.4 Effect of the Local Error Tolerances

Various strategies of local error control can be realized by an appropriate choice of atolDef and rtolDef. Consider one particular solution component $i \in \{1 \ldots n\}$. Setting $\mathrm{rtol}_i = 0$ would transform Equation 3.244 into

$$e_i \leq \mathrm{atol}_i, \quad i = 1 \ldots n \tag{3.247}$$

Choosing atol$_i = 10^{-3}$ would then mean that a step of the numerical algorithm inside lsoda is accepted only if the absolute value of the estimated local error of solution component i is below 10^{-3}. Setting atol$_i = 0$, on the other hand, would transform Equation 3.244 into

$$e_i \leq \text{rtol}_i \cdot |\tilde{y}_i(t)|, \quad i = 1 \ldots n \tag{3.248}$$

In this case, choosing rtol$_i = 10^{-3}$ would mean that a step of the numerical algorithm inside lsoda is accepted only if the local error of solution component i is smaller than 10^{-3} when expressed *relative* to the size of the numerical approximation of solution component i, $|\tilde{y}_i|$. Equations 3.247 and 3.248 correspond to (purely) *absolute error control* and (purely) *relative error control*, respectively, while Equation 3.244 with rtol$_i \neq 0$ and atol$_i \neq 0$ expresses a mixed absolute and relative error control. Such a mixed error control basically corresponds to a relative error test when the solution component $|\tilde{y}_i(t)|$ is much larger than atol$_i$, while it approximates an absolute error test when $|\tilde{y}_i(t)|$ is smaller than atol$_i$.

3.8.3.5 A Rule of Thumb to Set the Tolerances

This is the theory – but how should atol$_i$ and rtol$_i$ be chosen in a practical situation? Naively, one might conjecture that it is a good idea to choose these tolerances as small as possible. But this is a bad idea since Equation 3.244 would then force the numerical algorithms inside lsoda to use extremely small stepsizes, which would increase the computational time beyond all bounds and, more seriously, which might increase the global error in an uncontrollable way since much more computational steps would be needed until the numerical approximation of the solution of the ODE would be obtained.

If you want to know more about the various kinds of errors in numerical algorithms, we recommend [111] for a nice discussion of "errors and uncertainties in computations" or the comments on "error, accuracy and stability" in [92]. Two important kinds of error discussed there are the **roundoff error** and the **truncation error** or **approximation error**. Basically, the truncation or approximation error is related to the fact that any numerical algorithm can only perform a finite number of steps. Remember the discussion of the Euler method in Section 3.8.1.1, which evaluates the right-hand side of the ODE only at a finite number of x values, and which is based on formula (3.208), a truncated version of the originally infinite Taylor series. The roundoff error, on the other hand, arises because computers store floating-point numbers using only a finite number of digits, which means, for example, that a number such as 1/3 is stored as 0.3333 if we assume a computer storing four decimal digits. If you then use 0.3333 instead of the exact 1/3 in your computations, it is obvious that this and any other arithmetical operation on such a computer may introduce a roundoff error in the order of 10^{-4}. These errors add up with the number of steps performed in your numerical algorithm, and this is the reason why atol$_i$ and rtol$_i$ should be chosen with care (not too small). As a guideline, we recommend to use the following "rule of thumb" from [112]:

Note 3.8.2 (Rule of thumb to set the tolerances)   If m is the number of significant digits required for solution component y_i, set $rtol_i = 10^{-m+1}$. Set $atol_i$ to the value at which $|y_i|$ is essentially insignificant.

Let us apply this to the **body temperature example**. In this example, there is only one dependent variable (T in Equation 3.231), and hence we have $n = 1$, which means we can omit the index at $atol_i$ and $rtol_i$ and simply write atol and rtol instead. The body temperature example refers to a clinical thermometer that works in a range roughly between 30 and 42 °C, and which displays one digit behind the decimal point. Therefore, it is clear that we need at least three significant digits here. In terms of the above "rule of thumb", this means $m = 3$ and $rtol = 10^{-2}$. Regarding atol, values below 10^{-1} can be considered as "essentially insignificant" since only one digit behind the decimal point is displayed by the clinical thermometer. In terms of the "rule of thumb" this means we should set $atol = 10^{-2}$. These settings of atol and rtol can be viewed as the largest values of these tolerances that may give us the desired result with sufficient accuracy.

However, you should remember that atol and rtol limit the *local error* of the algorithm. As explained above, what really counts is the *global error*, that is, the final difference between the exact solution of the ODE and the numerical approximation. In each step of the numerical algorithm, the local errors add up to the global error. If we are lucky, the signs of the local errors produced in each step of the numerical algorithms may be randomly distributed and thus cancel out, giving a global error in a similar order of magnitude as the local errors. Fortunately, this is frequently the case, that is, we may expect the global errors to be in a similar order of magnitude as the local errors in many cases. But it may also happen that the local errors add up to give a much larger global error. You get a fairly reliable idea about the size of the global error by using the heuristic procedure explained above in our discussion of *Maxima's* rk command (Note 3.8.1). There it was recommended to compute numerical solutions for two stepsizes h and $h/2$, and then to compare the results. The same can be done here by reducing the tolerances, for example, by an order of magnitude. If the differences between the numerical solutions obtained in this way are negligible, you have a good reason to assume a negligible global error, and hence that your choice of the tolerances was acceptable.

3.8.3.6   **The Call of lsoda**
Remember that we are still discussing the R code ODEEx1.r in the book software. The error tolerances rtolDef and atolDef are the last settings in the "Program control parameters" section of ODEEx1.r. If you go through the rest of the code, you will find several other places where the user needs to do some smaller adjustments, such as the specification of column names in the data file, setting initial values for lsoda using appropriate variable names, and adjusting the plots. Those places are indicated by commentaries beginning with "# User:...", such as lines 1–2 in program 3.243. As a last point in our discussion of ODEEx1.r, let us have a look

at the call of lsoda within that code, since this will be important, for example, in our discussion of parameter estimation in ODEs in Section 3.9:

```
1:  out = as.data.frame(
2:  lsoda(
3:  c(T=T0)
4:  ,seq(aa,bb,length=nTime)
5:  ,dfn
6:  ,c()
7:  , rtol= rtolDef
8:  , atol= atolDef
9:  ))
```

$$(3.249)$$

Note that we have left out the user comments here for brevity. In line 1 of 3.249, the output of lsoda is stored as a "data frame" since this is the form needed further below in ODEEx1.r to produce the plots (see [45] for more on the data frame concept). Line 3 corresponds to Equation 3.232 of the body temperature model, that is, here you supply the initial values of the state variables. Line 4 defines a vector of times where lsoda is required to generate numerical approximations of the state variable(s), that is, of the temperature $T(t)$ in the case of the body temperature model, Equations 3.231 and 3.232. The seq command in line 4 generates an array of nTime equally spaced times between aa and bb. You can replace this by any other vector of times that you might want to use. In Section 3.10.2, for example, a code Fermentation.r is discussed where a sophisticated, nonequally spaced time vector is used in the call of lsoda. Line 5 tells lsoda about the function that generates the right-hand sides of the ODE, which is dfn in our case (Section 3.8.3.1). In line 6 of 3.249 you can add parameters required by dfn, which will be used in the parameter estimation problems treated in Section 3.9. In that section, other optional parameters of lsoda such as hmax are also discussed, which can be set by adding a line such as

```
,hmax=0.001
```

e.g. between lines 8 and 9 of 3.249.

3.8.3.7 Example Applications

If you now set PlotStateAna=TRUE and PlotStateData=TRUE in the "Program control parameters" section of ODEEx1.r and then execute the code e.g. using the source command as described in Appendix B, you get the result shown in Figure 3.11. Figure 3.11a shows a comparison of the numerical solution of the **body temperature model** (Equations 3.231 and 3.232) with the appropriate closed form solution (Equation 3.230). The numerical and closed form solutions match perfectly well, as it was achieved previously using *Maxima*'s rk command (Figure 3.8). Looking at ODEEx1.r you will find that rtolDef=1e-4 and atolDef=1e-4 have been used for this result. Remember that the "rule of thumb" discussed above led us to rtolDef=1e-2 and atolDef=1e-2. The actual, smaller values of the

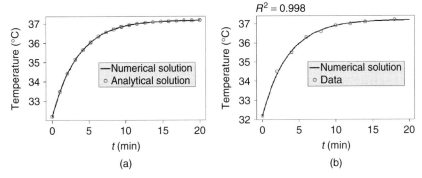

Fig. 3.11 (a) Line: numerical solution of Equations 3.231 and 3.232 computed with ODEEx1.r. Circles: exact solution (Equation 3.230). (b) Line: numerical solution of Equations 3.231 and 3.232 computed with ODEEx1.r. Circles: data Fever.csv.

tolerances have been chosen based on an application of the heuristic procedure discussed above, that is, by taking rtolDef=1e-2 and atolDef=1e-2 as a starting point, and then comparing the result with the result obtained using tolerance values one order of magnitude smaller, and so on, until a further reduction of the tolerances did not affect the result. Figure 3.11b shows a comparison of the numerical solution of Equations 3.231 and 3.232 with the data in Fever.csv similar to Figure 3.1b above which was produced using the *Maxima* program FeverExp.mac. Figure 3.11b reports an R^2 value of 0.998, which coincides with the value computed by FeverExp.mac.

As a second example of using R to solve ODEs, consider the *alarm clock model* (Equations 3.234–3.237). To solve these equations based on R, we use the program ODEEx2.r in the book software. The general structure of this program is identical with ODEEx1.r as discussed above, the only difference being the fact that we are concerned with a system of two ODEs here. In terms of the dfn function, this is expressed as follows (compare with 3.241):

```
 1:  dfn <-
 2:   -function(t, y, p)
 3:    {
 4:  ### User: names of state variables
 5:  ##############################
 6:      Ts=y[1]
 7:      Ti=y[2]
 8:  ### User: differential equation
 9:  ##############################
10:      dTsdt=rsi*(Ti-Ts)
11:      dTidt=ria*(Ta-Ti)
12:  ### User: list of derivatives
```

$$(3.250)$$

```
13:   ################################
14:       list(c(dTsdt,dTidt))
15:   }
```

As in 3.241, the code begins with a definition of the state variables in lines 6–7. In this case, there are two state variables, which are denoted as y[1] and y[2] (y[3], y[4], ... in cases with more than two variables). In lines 6–7, these state variables are named as Ts and Ti corresponding to T_s and T_i in Equations 3.234 and 3.235. Lines 10–11 define the ODEs (Equations 3.234 and 3.235). In line 14, finally, the list of derivatives computed in lines 10–11 is returned as the result of the dfn function. Note that in line 12 of 3.241 this was a list of length 1 since only one ODE was solved there. This list must always have as many elements as the number of ODEs in the ODE system to be solved. As before, the definition of dfn is followed by the definition of the closed form solution (Equations 3.238 and 3.239 in this case). Since the parameters used in ODEEx2.r satisfy $r_{si} \neq r_{ia}$, Equations 3.238 and 3.239 lead us to

```
1:  ### User: closed form solution (if available)
2:  ################################
3:      AnSolTs=function(t) Ta+(Ts0-Ta)*exp(-rsi*t)
        +(Ti0-Ta)*rsi*(exp(-rsi*t)
        -exp(-ria*t))/(ria-rsi)
4:      AnSolTi=function(t) Ta+(Ti0-Ta)*exp(-ria*t)
```
(3.251)

Compared with Equation 3.243 where there was only one closed form solution AnSol, you see that two closed form solutions AnSolTs and AnSolTi are defined here, corresponding to Equations 3.238 and 3.239. In the "Parameters of the model" section of ODEEx2.r, the parameters are set to the same values as in the *Maxima* code RoomODE.mac discussed above. In the "Solution of the ODE" section of ODEEx2.r, you have to set the initial values as follows:

```
1:  ### User: set initial values
2:  ################################
3:      c(Ts=Ts0,Ti=Ti0)
```
(3.252)

In ODEEx1.r, the corresponding code is

```
1:  ### User: set initial values
2:  ################################
3:      c(T=T0)
```
(3.253)

So you see that you have to provide a list of initial values here with as many elements as the number of ODEs in your ODE system. In the last part of ODEEx2.r where the plots are defined, we need two plots here comparing a numerical solution with a closed form solution, while only one such plot was needed in ODEEx1.r. Comparing ODEEx1.r and ODEEx2.r, you will see that this was achieved just by

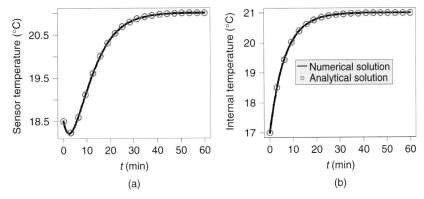

Fig. 3.12 (a) Line: numerical approximation of T_s based on Equations 3.234–3.237. Circles: exact solution (Equation 3.238). (b) Line: numerical approximation of T_i based on Equations 3.234–3.237. Circles: exact solution (Equation 3.239). Plots generated using ODEEx2.r.

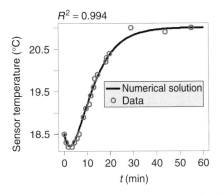

Fig. 3.13 Line: Numerical approximation of T_s based on Equations 3.234–3.237, obtained using ODEEx2.r. Circles: data from room.csv.

using two copies of the relevant part of ODEEx1.r, one of these copies plotting Ts compared with AnSolTs and the other one plotting Ti compared with AnSolTi. Figures 3.12 and 3.13 show the result produced by ODEEx2.r. Compare Figure 3.12 with 3.10, which shows the same result as it was obtained using *Maxima*'s rk function. As before, there is a perfect coincidence between the numerical solution and the closed form solution. Figure 3.13 shows a comparison of the numerical approximation with the data in room.csv similar to Figure 3.5 (see the discussion of these data in Section 3.2.3).

We have seen now how ODEs can be solved using *Maxima* rk or *R*'s lsoda commands. Although the procedure in *Maxima* is simpler, we have seen that it is better to use *R*'s lsoda in your everyday work since this is a much more professional software that provides, for example, a local error control and a number of other options which you do not have when using *Maxima*. If you want to apply R's lsoda to your own ODE systems, you can use the codes ODEEx1.r and ODEEx2.r discussed in this section as a starting point, editing these codes appropriately (for example, you will then have to insert your ODE into the dfn function as discussed above). All examples in the following sections are treated based on R's lsoda command. Some of these examples use extended version of ODEEx1.r and ODEEx2.r, using new options of R and lsoda. For example, in the discussion of the wine fermentation model (Section 3.10.2), a maximum stepsize hmax is prescribed in lsoda to make sure that external supplies of nitrogen are correctly used in the computations.

3.9
Fitting ODE's to Data

In Section 3.8, it was shown how ODEs can be solved using *Maxima* or R. One of the most important advantages of using R lies in the fact that within R the solution of ODEs can be coupled with R's unsurpassed statistical capabilities. This is of particular relevance when you are facing the important problem of fitting an ODE to experimental data. Consider, for example, Figure 3.13, which shows a comparison of the alarm clock model (Equations 3.234–3.237) with the data in room.csv. This figure was obtained by hand tuning of the parameters in Equations 3.234–3.237, that is, by hand tuning of r_{si}, r_{ia}, T_{i0}, and T_a (using the initial sensor temperature in room.csv for T_{s0}) until a good coincidence between the measurement data in room.csv and T_s as computed from Equations 3.234–3.237 was obtained. Besides the fact that such a hand tuning of the parameters of on ODE can be tedious and time consuming, it has the disadvantage that you do not get any information on the statistical quality of the parameters estimated in this way. For example, it would be nice to have confidence intervals around the parameter estimates that would tell us about the precision of these estimates, similar to the confidence intervals in our above discussion of nonlinear regression (Section 2.4).

3.9.1
Parameter Estimation in the Alarm Clock Model

Reconsidering the alarm clock model and data (Section 3.8.2.2), let us try to make it better now. First of all, note that we have already discussed the parameter estimation problem in Section 2.4. It was shown there how R's nls function can be used to statistically estimate the parameters $\mathbf{p} = (p_1, \dots, p_s)$ ($s \in \mathbb{N}$) of a nonlinear function $y = f(x, \mathbf{p})$ in a way that minimizes the distance between this function and a given data set $(x_1, y_1), \dots, (x_n, y_n)$. All we have to do here is to observe that the discussion in Section 2.4 applies to our situation. In the alarm clock model, $T_s(t)$ is the nonlinear function we want to fit to the data in room.csv using the parameter vector $\mathbf{p} = (r_{si}, r_{ia}, T_{i0}, T_a)$. Using the notation of Section 2.4, we can write $T_s(t)$ as $T_s(t, \mathbf{p})$. The only difference between our situation and the discussion in Section

2.4 lies in the fact that the functions considered there were all given as explicit formulas (i.e. in closed form), while $T_s(t, \mathbf{p})$ is given implicitly as the solution of an ODE system (Equations 3.234–3.237). But remembering our discussion in Section 3.3, you will note that this is a marginal difference – even functions such as the exponential functions are solutions of ODEs by definition, although it is common to use symbols to write them down "explicitly".

This means that all we have to do here is a *coupling of two methods* already discussed above: a coupling of lsoda as a means to solve ODEs (which will give us the function $T_s(t, \mathbf{p})$ in the alarm clock example) with nls as a means to estimate the parameters of the ODE from data. This is realized in the R program ODEFitEx1.r in the book software. The general structure of this code was derived from ODEEx2.r, and beyond this it uses R's nls function similar to NonRegEx1.r (Section 2.4.2). ODEFitEx1.r estimates the parameter r_{si} using the alarm clock ODE system (Equations 3.234–3.237) and the data in room.csv. We will just discuss those parts of ODEFitEx1.r which go beyond the codes NonRegEx1.r and ODEEx2.r that are discussed in detail in Sections 2.4.2 and 3.8.3.7.

3.9.1.1 Coupling lsoda with nls

Like ODEEx2.r, ODEFitEx1.r starts with a section where a function dfn is defined that describes the ODE system. Since ODEFitEx1.r refers to the alarm clock example, the definition of dfn is almost identical with program 3.250, the corresponding part of ODEEx2.r, except for the fact that r_{si}, the parameter to be estimated, must be supplied by the user as follows:

```
1:  ### User: estimated parameters
2:  ###############################              (3.254)
3:     rsi=p[1]
```

In contrast to ODEEx1.r and ODEEx2.r, a setting like $p = a$ defined in the subsequent *Parameters of the model section* of ODEFitEx1.r may have two meanings:
- If p is one of the parameters to be estimated, then a will be used as the initial value supplied to the algorithm doing the parameter estimation (nls).
- If p is not estimated, than a will be the fixed parameter value used by the ODE solving algorithm (lsoda).

In ODEFitEx1.r, this means that rsi=1 sets an initial value for r_{si} which is then estimated by nls, while the other parameter values defined in the "Parameters of the model" section are constants used in lsoda.

In the *Program control parameters section* of ODEFitEx1.r, there are a few new options:
- A logical variable NonlinRegress; if set to TRUE, nls will be invoked to estimate parameters, otherwise the ODE system will be solved treating all parameters in the "Program control parameters" section of ODEFitEx1.r as constants.

- A logical variable PrintConfidence; if set to TRUE, a
  confidence interval will be computed as described in Section
  2.4.2; note that if several parameters are estimated, this may
  take some time.
- A logical variable TraceProc; if set to TRUE, you will see the
  output produced by nls.
- A variable nlsTolDef which defines a numerical tolerance
  used by nls. Try to increase nlsTolDef from its default
  value 10^{-3} if nls does not converge.

The *Nonlinear regression section* of ODEFitEx1.r begins with a definition of the
nonlinear function dfn1 that is used by nls to estimate the parameters (again, we
leave out comments for brevity):

```
 1: dfn1 <- function(tt,rsi)
 2: {
 3: out <- lsoda(
 4: c(Ts=Ts0,Ti=Ti0)
 5: ,tData
 6: ,dfn
 7: ,c(rsi[1])
 8: ,rtol=rtolDef
 9: ,atol=atolDef
10: )
11: c(out[,"Ts"])
12: }
```
(3.255)

As discussed above, ODEFitEx1.r estimates the parameter r_{si} of the function
$T_s(t, r_{si})$, where T_s is determined from the alarm clock ODE system (Equations
3.234–3.237). It is exactly this function $T_s(t, r_{si})$ that is described by the function
dfn1 defined in program 3.255. Line 1 defines the dependence of dfn1 on time (tt)
and on the parameters to be estimated, which is rsi in our example. This list must
be edited by the user if you wish to estimate other parameters, or more parameters.
Looking at the definition of dfn1 in lines 3–11, you see that dfn1 mainly involves
a call of lsoda which solves Equations 3.234–3.237. The solution produced by
lsoda is stored in a variable out. In line 11, the part of out corresponding to the
numerical solution of T_s is returned as the result of the function.

The *call of lsoda* in lines 3–10 is similar to the call of lsoda in ODEEx1.r
and ODEEx2.r (Section 3.8.3.6). First, you have to provide the initial values in line
4. In line 5, you provide a vector with the times for which lsoda is required to
produce numerical approximations of the state variables. This is set to tData since
this variable is used in ODEFitEx1.r to denote the times corresponding to your
experimental data, and nls expects dfn1 to return the values of the state variables
exactly at these times. The rest of program 3.255 is self-explanatory.

After this, R's nonlinear regression procedure `nls` is invoked in the same way as it was discussed above in Section 2.4. Within `nls`, the nonlinear function is defined as `Ts~dfn1(t,rsi)`, which means that the function $T_s(t, r_{si})$ is indeed treated as if it was one of those explicitly given nonlinear function that were discussed in Section 2.4. The `nls` function hence does not "see" that `dfn1` involves a solution of an ODE system using `lsoda`. The rest of the code is similar to `NonRegEx1.r` (Section 2.4.2) and `ODEEx2.r` (Section 3.8.3.7). Using `ODEFitEx1.r` as a template to solve your own problems, just edit it as required at all the places highlighted by "`#{User:...}`".

3.9.1.2 Estimating One Parameter

In a typical usage of the parameter estimation procedure in `ODEFitEx1.r`, we first have to find an appropriate starting value for the parameter to be estimated (Note 2.4.1). We may know such a starting value a priori, for example, from the literature or based on theoretical considerations. `ODEFitEx1.r` estimates the r_{si} parameter of the alarm clock model, and we know from the discussion in Section 3.4.2 that this is a percent value and hence a number that can be expected to be in a range between 0 and 1. This means we could set, for example, $r_{si} = 0.5$ and hope that `ODEFitEx1.r` will be able to estimate r_{si} based on this starting value, which is indeed the case. If you do not get parameter estimates in this way, that is, if `nls` does not converge based on starting values obtained in this way, you can set `NonlinRegress=FALSE` and then try to optimize your starting values by hand. Setting `NonlinRegress=FALSE` in `ODEFitEx1.r` gives the result in Figure 3.14a, which shows the numerical solution obtained for $r_{si} = 0.5$. You see the deviation between the numerical solution and the data in the figure, and you can then gradually change r_{si} until you have reduced these deviations. After this, you can set

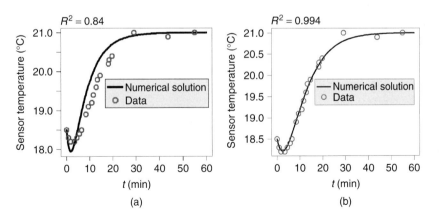

Fig. 3.14 (a) Line: numerical approximation of T_s based on Equations 3.234–3.237, obtained using `ODEFitEx1.r` with $r_{si} = 0.5$ and `NonlinRegress = FALSE`. Circles: data from `Room.csv`. (b) Same plot using `NonlinRegress = TRUE`.

NonlinRegress=TRUE again, hoping that nls will converge using the improved starting value of your parameter.

In this case, nls converges without problems using $r_{si} = 0.5$ as a starting value, which gives an almost perfect coincidence between model and data, see Figure 3.14b. In the R Console, the result is displayed as follows:

```
Estimated Coefficients:
       Estimate  Std. Error   t value      Pr(>|t|)
rsi 0.1800331 0.006866096  26.22059 2.206382e-16
...
Confidence Intervals:
    2.5%        97.5%
0.1666662 0.1947174
```

This means that nls estimates $r_{si} \approx 0.1800331$, and the 95% confidence interval is [0.1666662, 0.1947174]. As it was explained in Section 2.4, this means that this interval covers the unknown "true" value of r_{si} with a probability of 95%.

3.9.1.3 Estimating Two Parameters

ODEFitEx2.r estimates two parameters of the alarm clock model: r_{si} and r_{ia}. It was derived by an appropriate editing of ODEFitEx1.r, which is done in a minute – you just have to follow the "#User:..." instructions in ODEFitEx1.r. The necessary changes basically amount to an appropriate extension of parameter lists such as the replacement of 3.254 by

```
1:  ### User: estimated parameters
2:  ###############################
3:      rsi=p[1]
3:      ria=p[2]
```
(3.256)

Using $r_{si} = r_{ia} = 0.5$ as starting values (based on the above considerations), ODEFitEx2.r produces the following result in the R Console:

```
Estimated Coefficients:
       Estimate  Std. Error   t value      Pr(>|t|)
rsi 0.2151446 0.021490774  10.01102 8.780805e-09
ria 0.1358535 0.006333558  21.44980 2.868112e-14
...
Confidence Intervals:
           2.5%       97.5%
rsi 0.1786028 0.2608507
ria 0.1248202 0.1494279
...
Correlations:
           rsi           ria
```

```
rsi  1.0000000 -0.9255568
ria -0.9255568  1.0000000
```

The plot generated by ODEFitEx2.r is almost identical with Figure 3.14b, with a (very) slightly improved R^2 value of 0.995 (compared to 0.994 in the figure). Looking at the confidence intervals, you see that r_{ia} is estimated more precisely compared to r_{si}. r_{si}'s confidence interval is substantially larger compared to the one obtained above using ODEFitEx1.r. Remember from our discussion in Section 2.4 that the precision of the parameter estimates that can be achieved using nls depends on a number of factors. As it was said there, high correlations between your estimated parameters may indicate a bad experimental design, that is, your data may not provide the information that would be necessary to get sharp estimates of all parameters. In the above nls output, you see that there is indeed a high correlation between r_{si} and r_{ia}. It makes sense to assume a bad experimental design here since both parameters refer to the dynamics of T_i that has not been measured.

When you are solving your own parameter estimation problems over ODEs using the above procedure, it may happen that the *convergence regions* of the parameters are quite small, that is, it may happen that the numerical method converges only when you provide initial estimates of the parameters that are very close to the solution of the estimation problem. This is related to the fact that our above procedure is an *initial value approach* in the sense that all evaluations of the nonlinear function dfn1 by the numerical procedure in R's nls function require the solution of an initial value problem over the entire period under consideration. If you provide initial parameter estimates far away from the solution of the problem, this initial value problem may become unsolvable as discussed in [41]. From a practical point of view, we would of course prefer large convergence regions where the parameter estimates are obtained based on rough initial estimates of the parameters. So you should note that there are numerical methods which provide larger convergence regions, such as the *boundary value approach* developed by Bock [41, 113].

3.9.1.4 Estimating Initial Values

Our last example regarding the alarm clock model is ODEFitEx3.r, which estimates the following parameters of the alarm clock model: r_{ia}, T_a, and T_{i0}. The new aspect here is the fact that one of the parameters is an initial value of a state variable: T_{i0}. Initial values are treated a little bit different. They are not a part of the definition of dfn, which means they do not appear in the list of arguments of dfn. In particular, they do not need to be listed within dfn as in 3.256. It is obvious that it would not make sense to use initial values as arguments of dfn since dfn just gives the right-hand side of the ODE system, which does not depend on initial values of the state variables. However, estimated initial values must of course appear in the list of arguments of dfn1, that is, of the nonlinear function used by nls. In contrast to 3.255, the definition of dfn1 in this example, thus, looks as follows (again, all comments have been skipped):

```
1:  dfn1 <- function(tt,ria,Ta,Ti0)
```

```
 2:  {
 3:  out <- lsoda(
 4:  c(Ts=Ts0,Ti=Ti0[1])
 5:  ,tData
 6:  ,dfn
 7:  ,c(ria[1],Ta[1])
 8:  ,rtol=rtolDef
 9:  ,atol=atolDef
10:  )
11:  c(out[,"Ts"])
12:  }
```

(3.257)

ria and Ta are the parameters appearing as arguments of dfn, and they are supplied to dfn in line 7. Note that within dfn1, parameters must be written in the form ria[1], Tia[1], and Ti0[1] as before. Line 4 of program 3.257 defines Ti0[1] to be the initial value of Ti. Executing ODEFitEx3.r and using $r_{si} = 0.21$ (which approximates the estimate obtained in ODEFitEx2.r above), you get the following result in the R console:

```
Estimated Coefficients:
        Estimate  Std. Error    t value      Pr(>|t|)
ria   0.1376461 0.006992767   19.68406 3.881604e-13
Ta   21.0312734 0.044997657  467.38597 2.261465e-36
Ti0  16.9315415 0.113471697  149.21379 6.049605e-28

...
Confidence Intervals:
            2.5%       97.5%
ria   0.1240176   0.1525932
Ta   20.9394784  21.1264181
Ti0  16.6911157  17.1592343

...
Correlations:
          ria          Ta         Ti0
ria   1.0000000 -0.7211535 -0.8219467
Ta   -0.7211535  1.0000000  0.3582791
Ti0  -0.8219467  0.3582791  1.0000000
```

Except for relatively high correlations between some of the parameters (which can be interpreted similar as in Section 2.4.2 above), this is a good result in the sense that the confidence intervals of the parameters are relatively small.

3.9.1.5 Sensitivity of the Parameter Estimates

The above result must be interpreted with care, since it depends on our assumed value of $r_{si} = 0.21$. Of course, it would be natural to estimate r_{si}, too, but it turns out that nls does not converge in this case (edit ODEFitEx2.r appropriately and

Table 3.2 ODEFitEx3.r: Parameter estimates for r_{ia}, T_a, and T_{i0} and corresponding R^2 values (T_s compared with Room.csv) for different assumed values of r_{si}.

r_{si}	r_{ia}	T_a	T_{io}	R^2
0.14	0.2025652	21.0363632	16.2035026	0.996
0.21	0.1376461	21.0312734	16.9315415	0.996
0.3	0.1080993	21.0672026	17.2185061	0.994

try this yourself). As it was discussed above, we just do not have enough data to estimate so many parameters simultaneously.

This is also underlined by Table 3.2, which shows what happens with the parameter estimates if the assumed value of r_{si} is varied between 0.14 and 0.3. For all of these values, the resulting plots of T_s against the data Room.csv are almost indistinguishable, that is, each of these plots shows an almost perfect coincidence of the data with T_s, which is also reflected by the constantly high R^2 values in the table. Looking at the estimated parameters, one can say that only the estimate of T_a seems largely independent of the choice of r_{si}. The fact that T_a can be estimated so well is not surprising since in the alarm clock model, T_a is the temperature attained asymptotically by T_s for $t \to \infty$, and looking at the data, for example, in Figure 3.14 it is quite obvious that this asymptotic limit is very well characterized by the data. The other two parameters estimated by ODEFitEx3.r, r_{ia} and T_{i0}, on the other hand, depend on the dynamics of T_i that has not been measured, and so it is not surprising that these two parameters can be estimated with much less precision from the data. In Table 3.2, this is reflected by the fact that the estimates of r_{ia} and T_{i0} show a substantial dependence on the assumed value of r_{si}. If one particular value of r_{si} could be fixed on the basis of some a priori knowledge, for example, based on the literature data, this would be no problem. But, in the absence of such information, we must conclude that more experimental data, particularly on the dynamics of T_i, are required before we can hope to estimate all parameters of the alarm clock model with sufficient precision.

3.9.2
The General Parameter Estimation Problem

So far we have discussed the estimation of parameters in the alarm clock model, and we have seen how this can be done using software. The procedure described above applies to all situations where there are measurement data for one of the state variables of the ODE system only. In the general case, we may have simultaneous data for several state variables. Remember the above discussion of the alarm clock model where it was said that in order to estimate all parameters, we would need data for the dynamics of T_i in addition to the data of T_s in room.csv. To generalize the above procedure in this sense, it is useful to consider a general formulation of the parameter estimation problem over ODEs. Let us assume an initial value problem for an ODE system in the sense of Definition 3.5.4:

$$\mathbf{y}'(t) = \mathbf{F}(t, \mathbf{y}(t), \mathbf{p}) \tag{3.258}$$

$$\mathbf{y}(a) = \mathbf{y}_0 \tag{3.259}$$

where $\mathbf{y}(t) = (y_1(t), \ldots, y_n(t))^t$. Beyond Definition 3.5.4, it is assumed here that the right-hand side of the ODE depends on a vector of parameters $p = (p_1, \ldots, p_s)^t$ (same assumption as in Section 3.9.1). Now for $i = 1, \ldots, n$, let t_{ij} denote times where we have a measurement value $\tilde{y}_{ij}$ of state variable y_i ($j = 1, \ldots, m_i$). Similar to the discussion in Section 2.4, the parameters are now determined by minimizing an appropriate residual sum of squares. Denoting the solution of Equations 3.258 and 3.259 with $y_i(t, \mathbf{p})$ ($i = 1, \ldots, n$), $\mathbf{p}$ is determined by minimizing

$$G(\mathbf{p}) = \sum_{i=1}^{n} \sum_{j=1}^{m_i} w_{ij}(y_i(t_{ij}, \mathbf{p})) - \tilde{y}_{ij})^2 \tag{3.260}$$

where w_{ij} ($i = 1, \ldots, n, j = 1, \ldots, m_i$) are appropriate weighting factors [41]. Weighting factors different from 1 can be used, for example, in cases where your measurement data cover several orders of magnitude (see Section 3.10.2.9 or an example involving herbicide concentrations in soils discussed in [41]).

3.9.2.1 One State Variable Characterized by Data

It is exactly the problem of minimizing Equation 3.260 that has been solved by R's nls function in the alarm clock example. In that example, we have two state variables $y_1 = T_s$ and $y_2 = T_i$, which means we have $n = 2$. The fact that we have measurement data only for y_1 in this example means that the part of Equation 3.260 referring to $i = 2$ can be skipped, which leads to

$$G(\mathbf{p}) = \sum_{j=1}^{m_1}(y_1(t_{1j}, \mathbf{p}) - \tilde{y}_{1j})^2 \tag{3.261}$$

where we have set $w_{1j} = 1$ ($j = 1, \ldots, m_1$) since no weighting has been used in the alarm clock example. (Formally, the skipping of the $i = 2$ terms in Equation 3.260 could also have been achieved by setting $w_{2j} = 0$ for $j = 1, \ldots, m_2$.) Of course, the last equation is better written as

$$G(\mathbf{p}) = \sum_{j=1}^{m}(y_1(t_j, \mathbf{p}) - \tilde{y}_j)^2 \tag{3.262}$$

with obvious interpretations of m, t_j, and $\tilde{y}_j$. Now when you are using the software explained above, you should know where you find the ingredients of the last equation in the program. Referring to ODEFitEx1.r, the vectors $\mathbf{t} = (t_1, \ldots, t_m)^t$ and $\tilde{\mathbf{y}} = (\tilde{y}_1, \ldots, \tilde{y}_m)^t$ correspond to the vectors tData and TData in ODEFitEx1.r, which are read there directly from room.csv before they are used in the call of nls. The numerical approximations of y_1 at the times **t**, that is, the vector $(y_1(t_1, \mathbf{p}), \ldots, y_1(t_m, \mathbf{p}))^t$ is returned within ODEFitEx1.r as a result of the dfn1 function. See the code of this function in 3.255. Line 11 of that code corresponds to $(y_1(t_1, \mathbf{p}), \ldots, y_1(t_m, \mathbf{p}))^t$. Note that this holds true since in line 5 of 3.255, the data vector tData prescribes the times at which numerical approximations of the ODE system are computed, and this corresponds exactly to the times in $(y_1(t_1, \mathbf{p}), \ldots, y_1(t_m, \mathbf{p}))^t$.

3.9.2.2 Several State Variables Characterized by Data

Let us now generalize this to the case where we have data of several state variables y_i of the ODE system. In the wine fermentation model (Section 3.10.2), we have measurement data for three state variables S, E, and X. Using the above notation and proceeding analogously to the above discussion of ODEFitEx1.r, we will need the following quantities:

$$\mathbf{t} = (t_{11}, \ldots, t_{1m_1}, t_{21}, \ldots, t_{2m_2}, t_{31}, \ldots, t_{3m_3})^t \tag{3.263}$$

$$\tilde{\mathbf{y}} = (\tilde{y}_{11}, \ldots, \tilde{y}_{1m_1}, \tilde{y}_{21}, \ldots, \tilde{y}_{2m_2}, \tilde{y}_{31}, \ldots, \tilde{y}_{3m_3})^t \tag{3.264}$$

$$\mathbf{y}_p = (\mathbf{y}_{p1}, \mathbf{y}_{p2}, \mathbf{y}_{p3})^t \tag{3.265}$$

where for $i = 1, 2, 3$

$$\mathbf{y}_{pi} = (y_i(t_{i1}, \mathbf{p}), \ldots, y_i(t_{im_i}, \mathbf{p})), \tag{3.266}$$

The wine fermentation model is realized in the R program Fermentation.r, which has been derived from ODEFitEx1.r and is discussed in Section 3.10.2. Here we just take a short look at the parameter estimation part of that code, in order to see how the three quantities in Equations 3.263–3.265 are used in that code. Regarding **t**, it was said above in the discussion of ODEFitEx1.r that this quantity is used at two places: in the definition of dfn1 to prescribe the times at which dfn1 is required to produce numerical approximations of the state variables and in the call of R's nls function that is used to minimize the residual sum of squares (Equation 3.260). Looking at the definition of dfn1 in Fermentation.r, you will see that a time vector called tInt is used. If you check the definition of tInt a few lines upward in Fermentation.r, you will note that tInt involves more time points than those given by the vectors tS, tE, and tX, which correspond immediately to **t**. This is done for technical reasons related with the addition of nitrogen as discussed in Section 3.10.2. If you look at the last line in the function body of dfn1 in Fermentation.r

```
c(out[tSind "S"],out[tEind,"E"],out[tXind,"X"])          (3.267)
```

you see that the index vectors tSind, tEind, and tXind are used to make sure that dfn1 returns the approximations of the state variables corresponding to tS, tE, and tX (and hence corresponding to **t**) as required. Program 3.267 is what is returned by dfn1, and it corresponds directly to $\mathbf{y}_p$ in Equation 3.265 above. In the call of `nls` in `Fermentation.r`, the data vectors **t** and **ỹ** are supplied in the following line:

$$,\text{data=data.frame(y=c(S,E,X),x=c(tS,tE,tX))} \tag{3.268}$$

Here `c(tS,tE,tX)` corresponds directly to **t** in Equation 3.263 and `c(S,E,X)` corresponds to **ỹ** in Equation 3.264. This ends our discussion of how `nls` is applied in `Fermentation.r` to treat the case where several state variables are characterized by data. Starting, for example, with `ODEFitEx1.r` as a template, you may proceed along these lines to treat your own problems involving several state variables characterized by data. All other aspects of `Fermentation.r` and its results are discussed in Section 3.10.2.

3.9.3
Indirect Measurements Using Parameter Estimation

Note the elegance of the parameter estimation method discussed above. It can be viewed as an indirect measurement procedure that can be used in situations where the direct measurement of a quantity of interest is either impossible or too expensive in terms of time and money. In the case of the wine fermentation model, this applies to the parameters k (specific death constant of yeast) and N_0 (initial nitrogen concentration). As discussed in Section 3.10.2, N_0 is hard to measure because it refers to the yeast available nitrogen that cannot be easily characterized experimentally. Regarding k, the situation is different in the sense that it can be characterized experimentally but nobody did this so far, that is, there are no literature values that could be used. This means that if you want to use k in a model, you would normally have to perform an appropriate experiment that determines this parameter.

In such a situation where two parameters of a model you want to use are unavailable for different reasons, the parameter estimation provides an elegant way to determine these parameters from your available data. In the case of the wine fermentation model, this means that instead of performing specific experiments for the determination of the unknown parameters, you just use data of the overall fermentation process such as the sugar, ethanol and yeast biomass data discussed in Section 3.10.2, and then you use the mathematical procedure of parameter estimation to get approximate values of your unknown parameters from the data. Parameter estimation problems are *inverse problems* in the sense explained in Section 1.7.3. As it is explained there, the inverse character of parameter estimation lies in the fact that the process of parameter estimation does not aim at the determination of the output of your model (i.e. of its state variables) from the given input data (time interval and parameters); rather, you are going the "inverse" direction, asking for parameters of your model based on the given output data.

Note 3.9.1 (Indirect measurements) Suppose you want to determine a parameter that cannot be measured directly, and suppose that you have some data characterizing your system. Then, the parameter estimation procedures described in this section (and the underlying regression techniques described in Chapter 2) can often be used to estimate the parameter from the available data.

3.10
More Examples

ODEs can be applied in many, if not in most situations where your data describe the evolution of a quantity of interest over time. The following examples are intended to give you an idea of the wide applicability of this method. They will also be used to explain some useful concepts beyond the theory discussed above, such as the discussion of ideas of the theory of dynamical systems in Section 3.10.1 or of the concept of compartmental modeling in Section 3.10.3.

3.10.1
Predator–Prey Interaction

Population sizes of animals often have an oscillatory nature, that is, they increase and decrease periodically over time. In some cases, this can be explained by predator–prey interactions. Consider two animal species, a predator species and a prey species serving as the food of the predator. Let x and y denote their respective population sizes, measured, for example, as the number of species per square kilometer. If the *prey population* x is sufficiently large, the *predator population* y will increase since there is enough food available. The increase in the predator population y, however, will decrease the prey population x until the predator population finally runs short of food. Hence, the decrease in the prey population x will be followed by a decrease in the predator population y, until y is reduced so much that the prey population x increases again, which will be followed by a subsequent increase in y, and so on. If you sketch x and y in this way, you will find that one can expect periodical curves having the same period length, but a little bit shifted in time, that is, the maximum of x being followed by the maximum of y some time later.

3.10.1.1 Lotka–Volterra Model
In 1926, Volterra investigated fish population data exhibiting this kind of dynamics, and he used the following model to explain the data [114–116]:

$$x'(t) = (r - ay(t)) \cdot x(t) \tag{3.269}$$
$$y'(t) = (bx(t) - m) \cdot y(t) \tag{3.270}$$

In these equations, a, r, b and m are parameters (real constants) that are discussed below. Since the equations were independently found by Lotka and Volterra [115, 117], they are known as the *Lotka–Volterra model*.

To understand this model, let us consider some *special cases*. In the absence of the predator ($y = 0$), we have

$$x'(t) = r \cdot x(t) \tag{3.271}$$

Using the methods of Section 3.7, you can easily show that this is solved by

$$x(t) = x_0 \cdot e^{rt} \tag{3.272}$$

where $x_0 = x(0)$. Hence, the above model assumes an exponential increase in the prey population in the absence of the prey. Applying the same reasoning to the case $x = 0$, you find that the model assumes an exponential decrease in the predator population in the absence of preys:

$$y(t) = y_0 \cdot e^{-mt} \tag{3.273}$$

where $y_0 = y(0)$. Particularly, the exponential increase in the prey population is certainly a wrong assumption if the prey population becomes large enough. But this is irrelevant here since the model refers to the situation $y \neq 0$ where the prey population is limited by the presence of the predator. The growth rates r and m in Equations 3.269 and 3.270 can be interpreted similar to the interpretation of the parameter r of the body temperature model in Section 3.4.1.2:

- r, the growth rate of the prey, expresses the percent increase of x per unit of time, expressed relative to the actual size of x; a typical unit would be day^{-1}
- m, the death rate of the predator, expresses the percent decrease of y per unit of time, expressed relative to the actual size of y; again, a typical unit would be day^{-1}

Equations 3.269 and 3.270 show that the growth rate of the prey population is assumed to be reduced proportionally to the size of the predator population, while the death rate of the predator is reduced proportionally to the size of the prey population. This is governed by the parameters a and b with the following interpretations:

- a expresses the decrease in the growth rate of the prey per unit of y; a typical unit would be day^{-1}
- b expresses the increase in the growth rate of the predator per unit of x; a typical unit would be day^{-1}

A closed form solution of Equations 3.269 and 3.270 cannot be obtained (try it...), although one can at least derive an implicit equation characterizing the solution, see [114, 116]. Figure 3.15a shows an example where Equations 3.269 and 3.270 have been solved numerically using the R program Volterra.r in the book software. Volterra.r has been obtained by an appropriate editing of ODEEx2.r

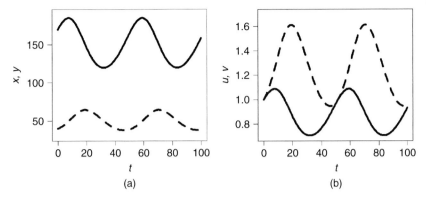

Fig. 3.15 (a) Prey (line) and predator (dashed line) population sizes obtained using `Volterra.r` and $r = 0.1$, $m = 0.15$, $a = 0.002$, $b = 0.001$, $x_0 = 170$, and $y_0 = 40$. (b) Same plot in nondimensional form obtained using `VolterraND.r` and $\tilde{r} = 0.1$, $\tilde{m} = 0.15$, $\tilde{a} = 0.08$, $\tilde{b} = 0.17$, $u_0 = 1$, and $v_0 = 1$.

(Section 3.8.3.7). Note that the curves in Figure 3.15a follow exactly that "shifted" periodical pattern that has been conjectured above.

3.10.1.2 General Dynamical Behavior

As it was said above, Volterra found the ODE system Equations 3.269 and 3.270 when he tried to understand certain oscillatory fish population data. He found a good coincidence between these ODEs and his data, and one could say that in this way these ODEs did what they where expected to do. But it is in fact a big advantage of mathematical modeling using ODEs that this is not necessarily the endpoint of the analysis. When a good coincidence between an ODE and data is found, we have a good reason to believe that these ODEs capture essential aspects governing the dynamics of the system. Then it makes sense to perform a theoretical investigation of the general dynamical behavior of the ODE system, that is, of the behavior of the system for all kinds of initial and parameter values, because in this way we may hope to learn about the general dynamical behavior of the real system that produced the data. For example, if we are investigating fish population data such as Volterra, we may be interested to learn about conditions that would increase the population size of a particular fish species beyond some acceptable level.

An analysis of the general dynamical behavior of an ODE system can be performed based on the *theory of dynamical systems*, which provides methods to understand and classify the patterns of dynamical behavior that solutions of ODE systems may have. An extensive discussion of these methods is beyond the scope of this book. You may find a detailed analysis of the dynamical behavior of the Lotka–Volterra model in [114, 116]. We will confine ourselves here to a discussion of two aspects of the analysis of the dynamical behavior of ODEs, which is something like a minimal knowledge you should have of this kind of analysis: the formulation of an ODE (or PDE) in dimensionless form and the phase plane plot.

3.10.1.3 Nondimensionalization

As to the first of these points, note that exactly the same picture as in Figure 3.15a would have been obtained using, for example, $x_0 = 1700$ and $y_0 = 400$ together with an appropriate scaling of the other parameters of the model. This means that if we want to classify the dynamical behavior of an ODE, we should first try to get rid of these scaling issues that just change the numbers at the axes of our plots, but that do not affect the qualitative dynamical behavior of the solution. This is done by bringing the *ODE in dimensionless form*. Referring to Equations 3.269 and 3.270, this can be done as follows. Let x_r, y_r, and t_r be reference values of x, y, and t (the appropriate choice of these values is discussed below). Now define u, v, and τ as follows:

$$u = \frac{x}{x_r} \tag{3.274}$$

$$v = \frac{y}{y_r} \tag{3.275}$$

$$\tau = \frac{t}{t_r} \tag{3.276}$$

All quantities defined in these equations are dimensionless since they are all expressed as fractions involving two quantities having the same dimensions. Since dimensions enter Equations 3.269 and 3.270 only through x, y, and t, we get rid of all dimensions in these equations if we substitute u, v, and τ into these equations using Equations 3.274–3.276. The result is

$$\frac{du}{d\tau} = (rt_r - at_r y_r v)u \tag{3.277}$$

$$\frac{dv}{d\tau} = (bt_r x_r u - mt_r)v \tag{3.278}$$

or

$$\frac{du}{d\tau} = (\tilde{r} - \tilde{a}v)u \tag{3.279}$$

$$\frac{dv}{d\tau} = (\tilde{b}u - \tilde{m})v \tag{3.280}$$

using the identifications $\tilde{r} = rt_r$, $\tilde{a} = at_r y_r$, $\tilde{b} = bt_r x_r$, and $\tilde{m} = mt_r$. Note that the structure of the last two equations is identical with the structure of Equations 3.269 and 3.270, and that all quantities appearing in Equations 3.279 and 3.280 are indeed dimensionless (verify this for $\tilde{r}, \tilde{a}, \tilde{b}, \tilde{m}$).

Now it is time to talk about the proper *choice of the reference values* x_r, y_r, and t_r in Equations 3.274–3.276. Principally, these quantities can be chosen arbitrarily. In some cases, your application may suggest "natural" reference values that can be used. t_r is usually set to the appropriate unit of time corresponding to the time scale on which the state variables of the ODE are observed, which means that you would set $t_r = 1$ day (or week, or year) in most models involving population dynamics. In the absence of other "natural" reference values, the reference values

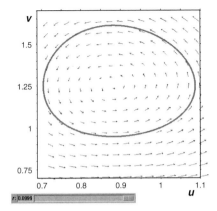

Fig. 3.16 Phase plot of the result in Figure 3.15b, generated using VolPhase.mac. The line is the curve $(u(t), v(t))$ and the arrows show the vector field $(du/d\tau, dv/d\tau)$. Note the slider below that can be used to change the value of r interactively.

for the state variables of the ODE are often set to the initial values of the ODE, which corresponds to $x_r = x_0$ and $y_r = y_0$ in our case.

Figure 3.15b shows the solution of Equations 3.279 and 3.280 using parameter values exactly corresponding to those used in Figure 3.15a. This figure was obtained using VolterraND.r in the book software. In contrast to Figure 3.15a, the axes in Figure 3.15b are in dimensionless units, which means that Figure 3.15b summarizes the dynamical behavior of the ODE for a great number of different situations. For example, depending on your setting of t_r, the values on the t axis may refer to seconds, days, years, and so on. In the same way, the values on the u, v axis may refer to the number of individuals, but they may also refer to any other units which you choose by setting x_r and y_r. Analyzing ODEs in this way it is much easier to get a picture of its overall dynamical behavior. See Murray [114] for a more detailed analysis of the dynamical behavior of the Lotka–Volterra equations. After introducing natural choices of the reference values x_r, y_r and t_r, Murray reduces Equations 3.279 and 3.280 to a form which involves only one parameter. An analysis of this system shows, for example, that the Lotka–Volterra equations are *structurally instable* in the sense that for certain initial conditions, small perturbations of these initial conditions or parameters can have large effects on the solution, which limits the practical usefulness of this model as discussed in [114].

3.10.1.4 Phase Plane Plots

For systems of two ODEs, the solution can also be plotted in a phase plane plot, which is particularly useful for an understanding of the overall dynamical behavior of the ODEs. Figure 3.16 shows a phase plane plot of the solution of Equations 3.279 and 3.280 using exactly the same parameters as in Figure 3.15b. In the *phase plane plot*, the coordinate axes correspond to the state variables u and v, and there is no time axis. This means that what you see in the phase plot is the curve $(u(t), v(t))$, drawn on some interval $[t_0, t_1]$. Comparing Figures 3.16 and 3.15b you will find that both figures indeed refer to the same solution of Equations 3.279 and 3.280. Figure 3.16 has been produced using the *Maxima* program VolPhase.mac. With some effort, R could also have been used to produce similar plots, but *Maxima* provides a really nice package called plotdf to produce phase plots that has been used in VolPhase.mac. The command producing the phase plot in this program is

```
 1:  plotdf(
 2:  [(r-a*v)*u,(b*u-m)*v]
 3:  ,[u,v]
 4:  ,[parameters, "r=0.1,m=0.15,a=0.08,b=0.17"]
 5:  ,[sliders,"r=0.07:0.1"]
 6:  ,[trajectory_at,1,1]
 7:  ,[tstep,0.1]
 8:  ,[nsteps,1000]
 9:  ,[u,0.69,1.1]
10:  ,[v,0.7,1.7]
11:  )$
```

$$(3.281)$$

In this code, line 2 defines the ODEs (3.279) and (3.280) via their right-hand sides. The rest of the code is self-explanatory. Note the definition of the slider element in line 5, which can be used to change the parameter value of r interactively (Figure 3.16). Several of such parameter sliders can be added to the plot in this way. Lines 7–8 define the interval $[0, T]$ for which the curve $(u(t), v(t))$ is plotted. We have $T = $ tstep $\cdot$ nsteps here which gives $T = 100$ based on the settings in 3.281, and this means that Figure 3.16 uses the same time interval as Figure 3.15b. tstep is the stepsize of the numerical algorithm. plotdf uses the Adams–Moulton method, but it can also be switched to a Runge–Kutta method (Section 3.8) with an adaptive stepsize [110]. If you use plotdf as above, you should choose tstep small enough, applying the heuristical procedures explained in Section 3.8 (Note 3.8.1).

The closed form of the trajectory in Figure 3.16 expresses the fact that the curve $(u(t), v(t))$ always goes along the same way, which means that this is indeed a *periodical solution*. It would not have been so easy to see this in the conventional plot, Figure 3.15b. The real benefit of the phase plot, however, lies in the fact that you can see the *effects of changes of the initial conditions* or of the parameters. If you move the parameter slider in Figure 3.16, you can see that the trajectory changes its size and position, and you can assess in this way the effects of parameter changes much better than it would have been possible using conventional plots such as Figure 3.15b. Try this yourself using VolPhase.mac. Effects of different initial condition can be studied particularly simple just by clicking into the plot. Every click produces a trajectory going through the initial condition at your mouse position. An example is shown in Figure 3.17a. This figure shows an interesting fact: you see that the amplitudes of the oscillations of the predator and prey populations go

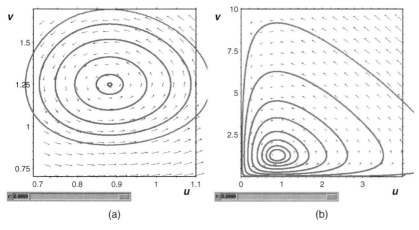

Fig. 3.17 (a) Phase plot as in Figure 3.16, including several other phase trajectories for different initial conditions. (b) Phase plot as in (a) but based on a different scaling, showing the instability of the Lotka–Volterra equations.

to zero as you approach a point in the center of the smallest, inner curve. This corresponds to a *singular point* of the ODEs, see the discussion in [114].

Figure 3.17b shows a similar plot using a different scaling of the axes, and a number of trajectories with initial conditions approaching the coordinate axes. This figure shows the *structural instability* of the Lotka–Volterra equations that was mentioned above (for a detailed analysis we refer to [114]): imagine you have initial conditions somewhere below the singular point infinitesimally close to the u axis of the plot, and imagine an infinitesimal perturbation of these initial conditions, which puts you on one of the neighboring trajectories. Then, comparing this neighboring trajectory with your original trajectory, you will find that you have large (i.e. noninfinitesimal) differences between these trajectories in regions of the phase space far away from the coordinate axes, as it can be seen in Figure 3.17b.

3.10.2
Wine Fermentation

Grape juice is transformed into wine in the fermentation process. This is a very complex process which is still not fully understood, and which involves the metabolization of sugar into ethanol by yeast cells [118, 119]. Loosely speaking, one can say the yeast in the fermenter "eats up" the sugar and excretes ethanol. If everything works well, the yeast cells will utilize most of the sugar in the fermenter during 7–10 days [120, 121]. It may happen, however, that the fermentation needs a much longer time (*sluggish fermentation*) or the fermentation may even stop before a sufficient amount of sugar is metabolized by the yeast cells (*stuck fermentation*). In an industrial setting, this kind of abnormalities can be very expensive, for example, in terms of extended processing times in the case of sluggish fermentations or in terms of microbial instabilities in the case of stuck fermentations that endanger the quality of the final product. This leads to the following problem:

Problem 1:
How can the fermentation process be controlled in a way that avoids sluggish or stuck fermentations?

3.10.2.1 Setting Up a Mathematical Model

To address this problem, the process engineer can try to tune a number of variables that have an impact on the way in which the fermentation proceeds. One of these control parameters is the *nitrogen concentration* in the fermenter which we denote $N(t)$ (gl^{-1}), where t is time (h^{-1}). Nitrogen is an important nutrient needed by the yeast cells (too low nitrogen levels can be a limiting factor of yeast cell growth). $N(t)$ refers to the total yeast available nitrogen concentration and includes various subtypes which we do not need to discuss here [121]. Another important control variable that is discussed here is the *temperature* $T(t)$ (K). Considering these two control variables, the process engineer has to answer the following question:

Q: How should $N(t)$ and $T(t)$ be adjusted during fermentation in order to avoid sluggish or stuck fermentations?

Referring to the definition of mathematical models as a triple (S, Q, M) consisting of a system S, a question Q and a set of mathematical statements M (Definition 1.4.1), this is the question we are going to ask here, and obviously S can be identified with the fermenter. What remains to be done is the formulation of the set of mathematical statements, M, which will be a system of ODEs. The above question Q tells us that $N(t)$ and $T(t)$ will be variables of our mathematical model, and since the question focuses on sluggish or stuck fermentations, it is obvious that we will also need to compute the *sugar concentration* $S(t)$ [gl^{-1}]. You can see here that the question Q we are asking is a really essential ingredient of a mathematical model that guides the formulation of the mathematical equations. Since we have said that the yeast cells "eat up" the sugar, it is obvious that we will not be able to compute the dynamics of $S(t)$ unless we have a variable expressing the (viable) *yeast cell concentration*. Let us denote this as $X(t)$ (gram biomass per liter). Now we have to formulate equations for these variables. In the modeling and simulation scheme (Note 1.2.3), this is at the heart of the systems analysis step, and this is the point where we have to refer to appropriate specialized literature. In this case, any book on fermentation technology can be used, which describes the processes that are involved in the metabolization of sugar and nitrogen into ethanol by the yeast cells [118, 119]. The result of such an analysis as well as a lot of background regarding the equations we are going to write now can be found in a model formulated by Blank [122]. Blank's model is an improved version of the fermentation model of Cramer *et al.* [121].

Table 3.3 State variables of the wine fermentation model.

State variable	Description	Unit
X	(Viable) yeast cell concentration	gram biomass per liter
N	(Yeast available) nitrogen concentration	gram nitrogen per liter
E	Ethanol concentration	gram ethanol per liter
S	Sugar concentration	gram sugar per liter

3.10.2.2 Yeast

Regarding $X(t)$, Cramer *et al.* use the following balance equation:

$$\frac{dX}{dt} = \mu \cdot X - k_d \cdot X \tag{3.282}$$

This means that the yeast cells grow proportionally to the actual yeast cell concentration X, the proportionality constant being the specific growth rate μ. At the same time, the yeast cells are inactivated or die proportionally to X, the proportionality constant being the death constant k_d. Note that you find all information regarding the state variables and parameters discussed here in Tables 3.3 and 3.4. If μ and k_d would be just constants, Equation 3.282 could be written as

$$\frac{dX}{dt} = \tilde{\mu} \cdot X \tag{3.283}$$

with $\tilde{\mu} = \mu - k_d$, which means that $X(t)$ would follow a simple exponential growth pattern. In reality, however, μ and k_d depend on a number of factors, and these dependencies must be described based on appropriate empirical data. In fact, the final predictive power of a fermentation model depends very much on the appropriate description of these empirical dependencies. Regarding μ, Cramer [121] uses the following expression:

$$\mu = \mu_{max} \cdot \frac{N}{K_N + N} \tag{3.284}$$

This means that the growth of the yeast cells is limited by nitrogen availability. The algebraic form of the right-hand side of this equation is frequently used to describe the rate of enzyme-mediated reactions, and it is known as the *Michaelis–Menten kinetics* term. Assuming a maximum rate V_{max} for some particular enzyme-mediated reaction and denoting the actual reaction rate and the substrate concentration with V and C, respectively, the Michaelis–Menten kinetics is usually written as

$$V = V_{max} \cdot \frac{C}{K_m + C} \tag{3.285}$$

Table 3.4 Parameters of the wine fermentation model.

Parameter	Description	Unit	Value	Source
μ	Specific growth rate	h^{-1}	(3.284)	[121]
μ_{max}	Maximum specific growth rate	h^{-1}	(3.286)	[122, 124, 125]
T	Temperature	K	Data	[122]
K_N	Monod constant for nitrogen	g nitrogen/l	0.01	[121]
k_d	Death constant	h^{-1}	(3.287)	[121]
k	Specific death constant	l/g ethanol//h	Unknown	
Y_{XN}	Stoichiometric yield coefficient of biomass on nitrogen	g biomass/g nitrogen	18	[127]
β	Specific ethanol production rate	g ethanol/g biomass/h	(3.289)	[121]
β_{max}	Maximum specific ethanol production rate	g ethanol/g biomass/h	(3.290)	[121,126]
$\beta_{max,24\,°C}$	Maximum specific ethanol production rate at 24 °C	g ethanol/g biomass/h	0.3	[121]
K_S	Michaelis–Menten-type constant for sugar	g sugar/l	10	[121]
Y_{ES}	Stoichiometric yield coefficient of ethanol on sugar	g ethanol/g sugar	0.47	[121]
X_0	Initial yeast concentration at $t = 0$	g biomass/l	0.2	fermentation.csv
E_0	Initial ethanol concentration at $t = 0$	g ethanol/l	0	fermentation.csv
S_0	Initial sugar concentration at $t = 0$	g sugar/l	205	fermentation.csv
N_0	Initial nitrogen concentration at $t = 0$	g nitrogen/l	Unknown	
N_{add}^i	ith nitrogen addition $(i = 1, \ldots, n)$	g nitrogen/l	0.03 $(i = 1,2)$	[122]
t_{add}^i	Time of ith nitrogen addition $(i = 1, \ldots, n)$	h	130 $(i = 1)$, 181 $(i = 2)$	[122]

where K_m is the *Michaelis–Menten constant* which corresponds to the substrate concentration that generates $V_{max}/2$ [123]. It can be easily derived from Equation 3.285 that $V \to V_{max}$ as $C \to \infty$.

Applying this to Equation 3.284, you see that μ_{max} is the maximum specific growth rate, and that μ approaches this maximum growth rate asymptotically as the available amount of nitrogen increases. The Michaelis–Menten constant K_N expresses the nitrogen concentration which corresponds to $\mu = \mu_{max}/2$. Note that it is not really surprising that such a fundamental relation from enzyme kinetics is used here since a number of enzyme-mediated reactions is involved in the metabolization of nitrogen by the yeast cells [124].

Equation 3.284 is appropriate under largely isothermal conditions. Since we want to consider the temperature as a (nonconstant) control variable here, the effects of the temperature on μ_{max} need to be taken into account, which can be done as follows [122, 124, 125]:

$$\mu_{max}(T) = 0.18 \cdot \exp\left(14\,200 \cdot \frac{T - 300}{300RT}\right) - 0.0054 \cdot \exp\left(121\,000 \cdot \frac{T - 300}{300RT}\right)$$
(3.286)

Here, $R = 8.314472\,\mathrm{J} \cdot \mathrm{K}^{-1} \cdot \mathrm{mol}^{-1}$ is the universal gas constant. The death constant k_d depends on the ethanol concentration $E(t)$. In [121], the following expression is used

$$k_d = k \cdot E$$
(3.287)

The last equations describe the dynamics of the viable yeast cell concentration, $X(t)$. In these equations, we needed the nitrogen concentration, $N(t)$, and the ethanol concentration, $E(t)$. This means we need equations characterizing $N(t)$ and $E(t)$. Let us begin with an ODE describing the dynamics of $E(t)$.

3.10.2.3 Ethanol and Sugar

Cramer describes the ethanol production rate proportional to the available amount of yeast cells, that is,

$$\frac{dE}{dt} = \beta X$$
(3.288)

where, similar to Equation 3.284, the specific ethanol production rate β depends on the available sugar concentration, $S(t)$:

$$\beta = \beta_{max} \cdot \frac{S}{K_S + S}$$
(3.289)

Again, the last two equations hold for essentially isothermal conditions, and a temperature dependence of β_{max} needs to be taken into account here. Following [122], we use an expression derived from data in [121, 126]:

$$\beta_{max}(T - 273.15) = \beta_{max,24\,°C} \cdot (0.00132 \cdot T^2 + 0.00987 \cdot T - 0.00781)$$
(3.290)

It remains to describe $N(t)$ and $S(t)$ that are used in the last equations. As discussed above, the dynamics of the sugar concentration $S(t)$ is intimately related with the ethanol production by the yeast cells. This is reflected by the fact that an ODE very similar to Equation 3.288 can be used to describe the sugar dynamics [121]:

$$\frac{dS}{dt} = -\frac{\beta}{Y_{ES}} X \tag{3.291}$$

In this equation, the coefficient Y_{ES} describes the stoichiometric bioconversion of sugar into ethanol.

3.10.2.4 Nitrogen

Remember that we wanted to use $N(t)$ as one of our control variables. In practice, any kind of fine tuning of $N(t)$ is hard to achieve. Even the measurement of nitrogen levels – which would be an essential prerequisite of controlling $N(t)$ levels – is a very difficult task [122, 124]. Nevertheless, nitrogen is an important variable in this process and people try to make sure that there is sufficient nitrogen in the fermenter. Since they usually do not know the actual nitrogen level $N(t)$, this is done largely in a "black box fashion", which means that nitrogen is added to the fermenter at times which have proven to be reasonable based on the experience made in previous fermentations. Let us denote with N_{add}^i the nitrogen concentrations added at times t_{add}^i ($i = 1, \ldots, n$). From the process engineers point of view, these are the nitrogen control variables, and they give rise to an essentially unknown dynamics of $N(t)$. One of the benefits of the model described here is that based on the nitrogen levels computed by the model, the process engineer can at least gain a qualitative idea of the effects that the nitrogen additions have on $N(t)$.

Note 3.10.1 (Qualitative optimization)   In a situation where an exact quantitative determination of a state variable cannot be achieved, mechanistic mathematical models often provide information about its qualitative behavior, which can be used for optimization. Such a "qualitative optimization" of a process is a definite step beyond pure "black box optimization" based on phenomenological models.

Since $N(t)$ is required in the above equations, it must be computed based on the nitrogen additions N_{add}^i at times t_{add}^i. To do this, we start with Cramer's nitrogen balance equation as follows:

$$\frac{dN}{dt} = -\frac{\mu}{Y_{XN}} \cdot X \tag{3.292}$$

This equation is based on the assumption that nitrogen is consumed proportional to the growth rate of yeast. The yield coefficient Y_{XN} is used to convert from nitrogen to yeast biomass. Now the nitrogen additions N_{add}^i at times t_{add}^i must be incorporated into the equations. Unfortunately, R's lsoda does not provide us with an option that would allow us to impose a discontinuous change of $N(t)$ by an amount of N_{add}^i at times t_{add}^i. Principally, this could be done by splitting the overall integration interval

$[0, T]$ into subintervals $[0, t_{add}^1]$, $[t_{add}^1, t_{add}^2]$, and so on, applying lsoda separately to each of the subintervals and using appropriate settings of the initial values that account for the nitrogen additions N_{add}^i. In our case, a more elegant and sufficiently accurate alternative is the assumption that the nitrogen additions are not given instantaneously, but at a constant rate over 1 h. This turns Equation 3.292 into

$$\frac{dN}{dt} = r(t) - \frac{\mu}{Y_{XN}} \cdot X \tag{3.293}$$

where

$$r(t) = \begin{cases} N_{add}^i & \text{if } t_{add}^i - \frac{1}{2} < t < t_{add}^i + \frac{1}{2}, \quad i = 1, \ldots, n \\ 0 & \text{else} \end{cases} \tag{3.294}$$

The overall model is a system of four ODEs that can be summarized as follows:

Wine fermentation model

$$\frac{dX}{dt} = \mu(T, N) \cdot X - k \cdot E \cdot X \tag{3.295}$$

$$\frac{dN}{dt} = r(t) - \frac{1}{Y_{XN}} \cdot \mu(T, N) \cdot X \tag{3.296}$$

$$\frac{dE}{dt} = \beta(T, S) \cdot X \tag{3.297}$$

$$\frac{dS}{dt} = -\frac{1}{Y_{ES}} \beta(T, S) \cdot X \tag{3.298}$$

We have used the notations $\mu(T, N)$ and $\beta(T, S)$ to emphasize the nonlinear dependence of these parameters on their arguments as discussed above.

3.10.2.5 Using a Hand-fit to Estimate N_0

Now let us see if the above model is capable to reproduce the concrete fermentation data in Figure 3.18. In this fermentation, the temperature has been varied between 12 and 18 °C by the process engineers (Figure 3.18), and nitrogen has been added at two times during the fermentation (see t_{add}^i and N_{add}^i in Table 3.4). As a result of these measures, the figure shows data of $X(t)$, $E(t)$ and $S(t)$. As it was mentioned above, the measurement of $N(t)$ is difficult and this is why no nitrogen data are available from this trial. Figure 3.18 was produced using the program *Fermentation.r* in the book software (set PlotData=TRUE in the "Program Control Parameters" section).

Fermentation.r will now also be used to solve the ODE system, Equations 3.295–3.298, and to estimate parameters of these equations from the data. This code has been derived by an appropriate editing of ODEFitEx1.r (Section 3.9.1). Some data handling aspects of Fermentation.r closely related to the parameter

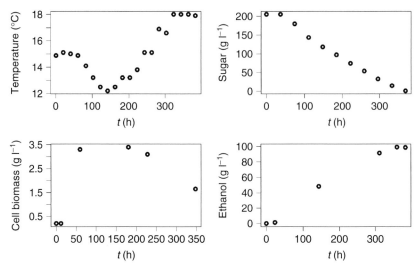

Fig. 3.18 Wine fermentation data from [122]. See `fermentation.csv` in the book software.

estimation procedure have already been discussed in Section 3.9.2.2. Other aspects of that code that are more closely related with the wine fermentation model are addressed below. See also the discussion of `ODEEx1.r` and `ODEEx2.r` in Section 3.8.3 for more information on the general way how you should work with this kind of codes. `Fermentation.r` is based on Equations 3.295–3.298 and uses the parameter values given in Table 3.4.

Looking at the table, you will note that there are *two unknown parameters*: k and N_0. N_0 is hard to measure because it refers to the yeast available nitrogen that cannot be easily characterized experimentally [122, 128]. Regarding k, [121] suggests a value of $k = 10^{-4}$ l/g ethanol/h for a temperature of 24 °C. This value could have been used here as a fixed value in the simulations, but [122] says it is better to estimate k from the data since this parameter is strongly temperature dependent, and the temperatures in the trial investigated here are substantially below 24 °C (Figure 3.18). Therefore, both N_0 and k are estimated from the data in `Fermentation.r` using the procedure explained in Section 3.9.

First of all, starting values of k and N_0 are needed as initial values for the parameter estimation procedure, that is, values in the right order of magnitude. As explained above, one can try to get such approximate values from the literature and/or by a hand-fitting of the parameters until the solutions of the ODE system are at least in the neighborhood of the data. Regarding k, we can use the literature value $k = 10^{-4}$ mentioned above. Since we do not have any information on N_0, the hand-fitting procedure is used for this parameter. A hand-fit of N_0 (setting `NonlinRegress=FALSE` and `PlotStateData=TRUE` in the "program control parameters" Section of `Fermentation.r`) leads us, for example, to Figure 3.19, which is obtained for $k = 10^{-4}$ and $N_0 = 0.1$.

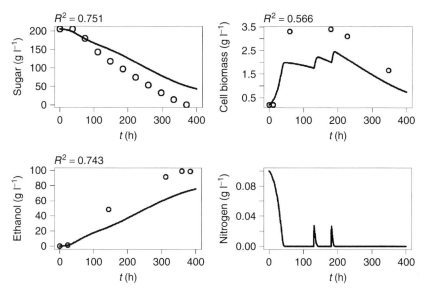

Fig. 3.19 Solution of Equations 3.295–3.298 using $k = 10^{-4}$, $N_0 = 0.1$ and the parameters in Table 3.4.

3.10.2.6 Parameter Estimation

The fine tuning of the parameters to be estimated can then be performed based on R's nls function as discussed in Section 3.9, using $k = 10^{-4}$ and the hand-fitted $N_0 = 0.1$ as starting values of k and N_0. Setting NonlinRegress=TRUE in the "program control parameters" section of Fermentation.r, the result shown in Figure 3.20 is obtained. As can be seen, a very good fit between the solution of the ODE system and the data is obtained, with all R^2 values above 0.97. This is a very good result since this is a fit between three nonlinear curves and data that was obtained by a tuning of only two parameters. If you have achieved a result like this, the next step is to test the predictive power of your model, applying it to data not used for the parameter estimation. This has been done by Blank in [122]. Blank shows that this model performs very well on unknown data in several cases, although there are also datasets where the results are unsatisfactory – which is not really surprising considering the aforementioned complexity of the wine fermentation process. From the point of view of mathematical modeling, the latter datasets producing the unsatisfactory results are the interesting ones, because they give us hints for further improvements of the model.

Note that the relative and absolute tolerances rtolDef and atolDef have both been set to 10^{-5} in Fermentation.r. This is based on the "rule of thumb" explained in Section 3.8.3.5: Considering the fact that the computations involve nitrogen additions of 0.03 gl^{-1} (see t_{add}^i and N_{add}^i in Table 3.4), it is clear that atolDef must be smaller than 10^{-2}. For example, we can set it to 10^{-3}. Regarding rtolDef, Figure 3.18 shows that the sugar and ethanol data have at least three

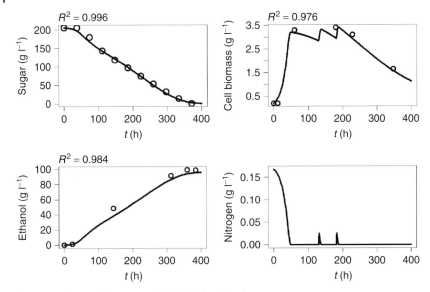

Fig. 3.20 Solution of Equations 3.295–3.298 using the estimated coefficients $k = 6.637202 \times 10^{-5}$ and $N_0 = 1.678096 \times 10^{-1}$, and the parameters in Table 3.4.

significant digits, which means we should set rtolDef also, for example, to 10^{-3}. In Fermentation.r, both tolerances were set to 10^{-5}, that is, two orders of magnitude smaller, which was based on the heuristical procedure for the choice of rtolDef and atolDef explained in Section 3.8.3.5 (a further decrease in the tolerances below 10^{-5} did not affect the result anymore).

3.10.2.7 Problems with Nonautonomous Models

Finally, let us look through Fermentation.r, focusing on new aspects in this code. As it was said above, Fermentation.r was derived from ODEFitEx1.r which was discussed in Section 3.9.1. Within Fermentation.r, the ODE system Equations 3.295–3.298 is defined in a function dfn as follows:

```
 1:  dfn <-
 2:    function(t, y, p)
 3:      {
 4:        X=y[1]
 5:        N=y[2]
 6:        E=y[3]
 7:        S=y[4]
 8:        k=p[1]
 9:        dXdt=mu(T(t),N)*X-k*E*X
10:        dNdt= r(t)-mu(T(t),N)*X/Yxn
11:        dEdt=beta(T(t),S)*X
```
(3.299)

```
12:        dSdt=-beta(T(t),S)*X/Yes
13:        list(c(dXdt,dNdt,dEdt,dSdt))
14:    }
```

The general structure of the dfn function has been discussed in Section 3.8.3. The correspondence between Equations 3.295–3.298 with lines 9–12 of program 3.299 is obvious. You may be irritated by the fact that the time dependence of r(t) and T(t) is written explicitly in the code, while this is not done for the state variables X, N, E and S. This is related with the fact that Equations 3.295–3.298 are a *nonautonomous* system of ODEs. As it was discussed in Section 3.5.3, this means that at least one of the right-hand sides of these equations depends on t not only (implicitly) via the state variables, but also via some explicitly given functions of t. Looking at the equations, you see that all right-hand sides of the system depend explicitly on the temperature $T(t)$ or on the nitrogen supply rate $r(t)$. When writing down the dfn function for nonautonomous ODEs, you must take care that only the state variables are written without an explicit "(t)", while all other time-dependent function must be written similar to T(t), r(t) and so on.

But even if you have done this correctly, you might run into another problem characteristic of nonautonomous ODEs when you try to compute the numerical solution. Remember our definition of the nitrogen supply rate in Equation 3.294. In the above example, nitrogen is supplied at two times only, $t^1_{add} = 130$ h and $t^2_{add} = 181$ h. Since the overall time interval in the example is from 0 to 400 h, this means that the function $r(t)$ defined in Equation 3.294 is almost everywhere zero, except for two small 1-h intervals around $t^1_{add} = 130$ h and $t^2_{add} = 181$ h. Remember also the discussion of lsoda in Section 3.8: lsoda determines its stepsize h automatically, choosing the stepsize as small as it is required by the error tolerances, but not smaller. Now if lsoda sets the stepsize relatively large around $t^1_{add} = 130$ h and $t^2_{add} = 181$ h, it may happen that it ignores the nitrogen supply rate $r(t)$ in the sense that $r(t)$ is never evaluated at one of its nonzero points. Then, although you prescribe a substantial nitrogen supply in two small 1-h intervals around $t^1_{add} = 130$ h and $t^2_{add} = 181$ h, you will not see this in the solution, that is, the problem is solved by lsoda in a way as if there would be no nitrogen supply at all ($r(t) = 0$). Implementing the fermentation model yourself by an appropriate editing of ODEFitEx1.r, you will see that you get exactly this problem when you do not apply appropriate measures to avoid this.

Note 3.10.2 (Problem with short external events) Numerical solutions of nonautonomous ODE systems may ignore short external events (such as nitrogen addition in the wine fermentation model).

The first and simplest thing that one can do is to prescribe a maximum stepsize in lsoda, which can be done using an argument of lsoda called hmax (see [108, 109] and R's hep pages). Choosing hmax sufficiently small, one can make sure that $r(t)$ is used in the computation as desired. For example,, based on hmax=1/10, $r(t)$

will be evaluated 10 times during each of those 2 h of nitrogen supply. Similar to the heuristics for an optimal choice of the stepsize described above (Note 3.8.1), one can then optimize the size of hmax, making it small enough such that a further decrease of hmax does not affect your results, but not smaller since this would further increase your computation times, roundoff errors and so on as discussed in Section 3.8.

A second option that can avoid this unwanted "blindness" of lsoda with respect to $r(t)$ lies in an appropriate choice of the time vector supplied in the call of lsoda. As it was mentioned in the above discussion of ODEEx1.r, the time vector in the call of lsoda (line 2 in program 3.249) defines the times at which lsoda is expected to generate numerical approximations of the state variables. Now if we want lsoda to "see" those 2 h where the nitrogen supply rate $r(t)$ is different from zero, we can define that time vector such that it involves several times within those 2 h. Exactly this has been done in Fermentation.r, using a time vector supplied to lsoda called tInt. Check the definition of tInt in Fermentation.r to see that it really does what has been described above.

3.10.2.8 Converting Data into a Function
The temperature $T(t)$ in Equations 3.295–3.298 needs some special treatment since it is not known as a *mathematical function*, but rather in the form of experimental data (columns 1–2 in fermentation.csv in the book software). The conversion of such experimental data into a form that can be treated by software such as lsoda is a standard problem in the numerical treatment of ODEs, and it is usually solved by a simple linear interpolation of the data. Denoting with (t_i, T_i) $(i = 1, \ldots, n)$ the experimental data, this means that $T(t)$ is defined in each of the intervals $[t_{i-1}, t_i]$ $(i = 2, \ldots, n)$ as follows:

$$T(t) = T_{i-1} + \frac{t - t_{i-1}}{t_i - t_{i-1}} \cdot (T_i - T_{i-1}) \tag{3.300}$$

3.10.2.9 Using Weighting Factors
The wine fermentation model is an example where it makes sense to use the weighting factors w_{ij} in the residual sum of squares, Equation 3.260 (Section 3.9.2). As it was discussed there, weighting factors should be used in situations where the measurement data cover several orders of magnitude. As Figure 3.18 shows, the sugar and cell biomass data differ by about two orders of magnitude, and the sugar and ethanol data cover two orders of magnitude. In [41] it is suggested that a weighting $w_{ij} = 1/\sigma_i^2$ $j = 1, \ldots, m_i$ should be used in this situation, where σ_i is the standard deviation of the measurement data referring to state variable i. [122] reports the following values for the state variables of the wine fermentation model: $\sigma_X = 0.1$, $\sigma_S = 0.7$, $\sigma_E = 0.7$. This gives the weighting factors

$$w_X = \frac{1}{\sigma_X^2} = 100 \tag{3.301}$$

$$w_S = \frac{1}{\sigma_S^2} \approx 2 \tag{3.302}$$

$$w_E = \frac{1}{\sigma_E^2} \approx 2 \tag{3.303}$$

Here, w_X refers to all w_{ij} in the residual sum of squares, Equation 3.260, where i refers to X (similar for w_S and w_E). In Fermentation.r, this weighting is prescribed in the "program control parameters" section as follows:

$$\texttt{WhichData=c(1,1,50)} \tag{3.304}$$

which means that X (third entry in WhichData) receives 50 times more weighting compared to S and E (first and second entries in WhichData). Looking into the call of nls further below in Fermentation.r, you can see how WhichData is used to define an option of nls called weights.

3.10.3
Pharmacokinetics

Most therapeutic drugs are ingested orally and then enter the blood stream via gastrointestinal (GI) absorption. For a therapeutic drug to be effective, it is of course important to know how fast the drug enters the blood stream, and how the drug concentration changes depending on time. In (probably) the simplest possible approach, we would just consider two state variables: $G(t)$, the concentration of the drug in the GI tract, and $B(t)$, the concentration of the drug in the blood (we assume units of micrograms per milliliter). If the application rate $D(t)$ (micrograms per milliliter per hour) expresses the drug dosage regime as seen by the GI tract, the dynamics of the system can be described by the following ODE system [116, 129]:

$$G'(t) = -aG(t) + D(t) \tag{3.305}$$

$$B'(t) = aG(t) - bB(t) \tag{3.306}$$

Equation 3.306 says that the blood concentration $B(t)$ grows proportionally to the concentration in the GI tract, $G(t)$, and it decays proportionally to $B(t)$, where a and b (per hour) are the constants of proportionality. The term $aG(t)$ that describes the increase of the blood concentration of the drug by absorption appears with an inverse sign in Equation 3.305, expressing the mass balance of the drug in the sense that precisely the amount of drug that enters the blood by absorption is lost in the GI tract. Of course, the dosage $D(t)$ causes an increase of $G(t)$ and thus appears with a "+"-sign in Equation 3.305.

To show the qualitative behavior of the solution, we use the parameter settings suggested in [116]: $G(0) = B(0) = 0$ (i.e. no drug has been applied before $t = 0$), $a = \ln(2)/2$, $b = \ln(2)/5$, and

$$D(t) = 2 \cdot \sum_{n=0}^{10} \left(H(t - 6n) - H\left(t - \left(6n + \frac{1}{2} \right) \right) \right) \tag{3.307}$$

where $H(x)$ is the Heaviside step function:

$$H(x) = \begin{cases} 0 & \text{if } x < 0 \\ 1 & \text{if } x \geq 0 \end{cases} \tag{3.308}$$

Equation 3.307 yields the drug dosage regime shown in Figure 3.21, which can be thought of as representing a situation where the drug is applied every 6 h, for example, at 6 a.m., 12 a.m., and 6 p.m. As can be seen in the figure (and as it follows from Equation 3.307), it is assumed here that the drug is supplied at a continuous rate of $2\,\mu g\,ml^{-1}\,h^{-1}$, and that this rate is held constant for $1/2$ h after drug ingestion. Figure 3.22 shows the resulting patterns of $B(t)$ and $G(t)$ that are obtained by a solution of the model, Equations 3.305 and 3.306. To solve the mathematical problem, the code ODEEx3.r has been used. This code is a part of the book software (see Appendix A), and it has been obtained by an editing of ODEEx2.r, a similar code that has been discussed in Section 3.8.3 above.

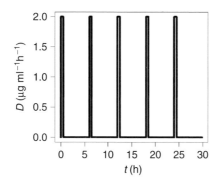

Fig. 3.21 Assumed pattern of drug dosage $D(t)$ corresponding to Equation 3.307.

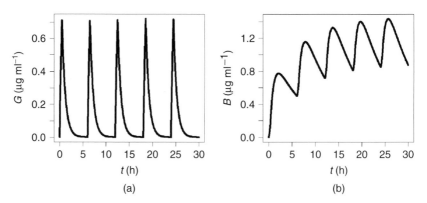

Fig. 3.22 Drug concentration (a) in the gastrointestinal tract and (b) in the blood.

Figure 3.22a shows the drug concentration in the GI tract that is implied by the drug dosage regime in Figure 3.21. As can be seen, the five peaks of $G(t)$ correspond with the five peaks of $D(t)$. While $D(t)$ goes abruptly to zero after 1/2 h according to the dosage regime assumed above, $G(t)$ decreases much slower, reflecting the gradual absorption of the drug from the GI tract into the blood stream. The drug concentration in the blood in Figure 3.22b exhibits a periodical pattern which is also showing five peaks corresponding to the five peaks of the dosage regime in Figure 3.21. The peaks of $B(t)$, however, are somewhat delayed compared with the $D(t)$ peaks, which again reflects the gradual absorption of the drug from the GI tract into the blood stream. If the dosage regime is continued in the same way for $t > 30$, $B(t)$ becomes a periodical function that oscillates between 0.8 and $1.4\,\mu\mathrm{g}\,\mathrm{ml}^{-1}$, that is, the simulation shows that the above dosage regime guarantees a blood concentration between 0.8 and $1.4\,\mu\mathrm{g}\,\mathrm{ml}^{-1}$. Simulations of this kind can thus be used to optimize drug dosage regimes in a way that guarantees certain limit blood concentrations of the drug that are required for medical reasons. Of course, this model needs validation before it can be applied in this way (see [129, 130] for a comparison of this and similar models with data). Using phase plane plots similar as in Section 3.10.1, it can be shown that the above model exhibits an interesting dynamical behavior involving *limit cycles* which you will find further discussed in [114] (see also the remarks on the theory of dynamical systems in Section 3.10.1.2).

The above drug model and many other pharmacokinetic models are examples of a concept called *compartment models* [130]. Compartment models are mathematical models where each of the state variables expresses a specific property of some part of a system, and these parts are referred to as the compartments of the model. Usually, this reflects the assumption that the property of the compartment expressed by the state variable is homogeneously distributed within the compartment. In this sense, the above drug model is a two-compartment model. It involves two state variables: $G(t)$ expresses a property of the GI tract (which is compartment 1), and $B(t)$ expresses a property of the blood volume (which is compartment 2). The properties expressed by $G(t)$ and $B(t)$ – the drug concentration in the GI tract and in the blood volume, respectively – are assumed to be independent of space, that is, it is assumed that the drug is homogeneously distributed within the GI tract and within the blood volume (like in a well-stirred container). Compartment models thus are examples of what we have called *lumped models* in Section 1.6.3. Typically, compartment models are visualized similar to Figure 3.23, that is, the compartments are represented, for example, as rectangles, and arrows are used to

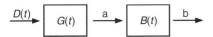

Fig. 3.23 Drug model as a two-compartment model.

indicate the mutual exchange of mass, energy etc. between the compartments as well as the mass and energy flows between the compartments and the outside world.

3.10.4
Plant Growth

A great number of mathematical models of plant growth has been developed, both from the scientific perspective (for example: *"how* do plants grow?") and from the engineering perspective (for example: "how can crop *yield* be maximized?"), see e.g. the books of Richter/Söndgerath [41] and Overman/Scholtz [131], which emphasize the scientific and engineering perspectives, respectively. In its simplest form (which applies to plants and to other growing organisms in general), plant growth models can be written in the form [41]

$$B'(t) = r \cdot B(t) \cdot \Phi(B(t)) \tag{3.309}$$

where $B(t)$ denotes the overall biomass of the plant at time t (e.g. in kg ha^{-1}), r (e.g. in day^{-1}) is the growth rate of the plant and $\Phi(x)$ is some (dimensionless) nonlinear real function. As usual, an initial condition is needed before this equation can be solved, that is, we need to specify the value of the biomass at some time, for example, in the form $B(0) = B_0$. Depending on the choice of the function $\Phi(B)$, several plant growth models can be derived from Equation 3.309. For example , setting $\Phi(B) = 1$ generates the *exponential growth model*:

$$B'(t) = r \cdot B(t) \tag{3.310}$$

while $\Phi(B) = 1 - B/K$ (for $K \in \mathbb{R}$) gives the *logistic growth model*:

$$B'(t) = r \cdot B(t) \cdot \left(1 - \frac{B(t)}{K}\right) \tag{3.311}$$

Here, K (kg ha^{-1}) can be interpreted as the (genetically fixed) maximum possible biomass of the plant. Figure 3.24 shows the typical patterns of the exponential and logistic growth models. As can be seen, the exponential plant growth model yields an exponential increase of the biomass, which holds true at the early stages of plant growth. As the plant approaches its maximum possible biomass, the growth curve will asymptotically slow down, which can be expressed using the logistic growth model similar to Figure 3.24b.

Figure 3.24a,b has been produced using the codes `Plant1.r` and `Plant2.r` in the book software (see Appendix A), which were obtained by an appropriate editing of `ODEEx1.r`, and which are based on a numerical solution of the ODEs Equations 3.310 and 3.311 (see Section 3.8.3). Of course, closed form solutions of Equations 3.310 and 3.311 can also be easily obtained using the methods described

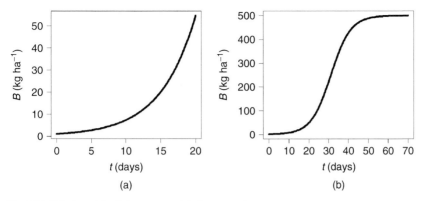

Fig. 3.24 Solutions (a) of the exponential plant growth model, Equation 3.310 and (b) of the logistic plant growth model, Equation 3.311 using: $B(0) = 1$, $r = 0.2$, $K = 500$ (plots generated using Plant1.r and Plant2.r).

in Section 3.7. Assuming $B(0) = B_0$ as the initial condition, the closed form solution of Equation 3.310 is

$$B(t) = B_0 e^{rt} \tag{3.312}$$

and the closed form solution of Equation 3.311 is

$$B(t) = \frac{K}{1 + e^{\beta - rt}} \quad \text{where } \beta = \ln\left(\frac{K}{B_0} - 1\right) \tag{3.313}$$

A great number of plant growth models are obtained by appropriate modifications of this approach, depending on the data that are analyzed and depending on the question that one is asking. For example, *multiorgan plant growth models* can be formulated basically by adding a growth model of the above kind for each of the plant organs such as roots, stems, leaves and fruits, and by balancing the mass flows between these "compartments" following the idea of the compartmental approach described in Section 3.10.3 above [41, 132, 133]. The above equations can also be supplemented, for example, by equations describing the kinetics of enzymes that are involved in photosynthesis if the focus of the investigation is on air pollutants that affect these enzymes during plant growth [134, 135].

Different plant growth models may apply to one and the same plant depending on the growth conditions. While the logistic growth model works well for the asparagus biomass data in [136], it was inapplicable to the staircase-like asparagus biomass data shown in Figure 3.25. As discussed in [137], the staircase-like structure of the data in the figure reflects the fact that several groups of asparagus spears had been growing successively, that is, first group 1 began to grow and stopped growing after some time, then group 2 began to grow and stopped growing after some time and so on. The data in Figure 3.25 look like several logistic growth functions stacked on top of each other, and they can indeed be described by a system of ODEs that

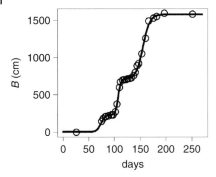

Fig. 3.25 Asparagus spear biomass data asparagus.csv compared with model Equations 3.314 and 3.315, where $r_1 = 0.23$, $r_2 = 0.42$, $r_3 = 0.15$, $K_1 = 230$, $K_2 = 480$, $K_3 = 870$, $t_1 = 50$, $t_2 = 93$, $t_3 = 110$, $B_1(0) = B_2(0) = B_3(0) = 1$. Data from [137] (note: biomass in "centimeter" based on a constant assumed spear diameter and density). Plot generated using Plant3.r.

involves three logistic growth equations (one for three groups of spears in the above sense) as follows:

$$B'_i(t) = r_i \cdot H(t - t_i) \cdot \left(1 - \frac{B_i(t)}{K_i}\right) \cdot B_i(t) \text{ for } i = 1, 2, 3 \tag{3.314}$$

$$B(t) = \sum_{i=1}^{3} B_i(t) \tag{3.315}$$

where $H(x)$ is the Heaviside step function that was introduced in Section 3.10.3. Basically, the Heaviside function is used here to successively "switch on" the growth functions for the three groups of asparagus spears. Figure 3.25 shows an almost perfect coincidence of this model with the data. The figure has been plotted using the code Plant3.r in the book software, which is based on the numerical solution of Equation 3.314 using ODEEx1.r again (as before, the closed form solution Equation 3.313 could also have been used to get the same result). Note that the parameters shown in Figure 3.25 have been obtained from a (quick) manual tuning of the parameters, although the (slightly more tedious) automated procedure described in Section 3.9 could also have been used.

4

Mechanistic Models II: PDEs

4.1
Introduction

4.1.1
Limitations of ODE Models

Ordinary differential equation (ODE) models are restricted in the sense that they involve derivatives with respect to one variable only, which means that they describe the dynamical behavior of the quantity of interest with respect to this one variable only. In the wine fermentation model, for example, the quantities of interest have been the sugar, ethanol, nitrogen, and yeast cell biomass concentrations, and all these quantities were considered as a function of time only (Section 3.10.2). Looking at the examples in Section 3, you will note that time was the independent variable in most examples, although, of course, any kind of variable can be used in principal (e.g. a space coordinate was used as the independent variable in the metal rod example, see Section 3.5.5)

Now, obviously, we live in a world where everything depends on many variables simultaneously. Using ODE models, therefore, usually means we are referring to special situations where it can be assumed that the independent variable used in the ODE model is the most important factor affecting our quantity of interest, while the influence of other factors with a possible impact on our quantity of interest can be assumed to be negligible. In the wine fermentation model, for example, it is obvious that quantities such as the yeast biomass concentration will depend not only on time but also on the space coordinates x, y, and z. Assuming the yeast biomass concentration would not depend on the spatial coordinates would require that quantity to be exactly the same at every particular spatial coordinate within the fermenter. You do not need to be a wine fermentation specialist to understand that this is an unrealistic assumption, and that it may of course happen that you have variable yeast biomass concentrations in the fermenter; for example, gravitation may increase the yeast biomass at the bottom of the fermenter as discussed in [124].

The fact that the wine fermentation model – as well as the other ODE models discussed in Section 3 – can be applied successfully in some situations, thus, does not "prove" the spatial homogeneity of the state variables, or the absence

Mathematical Modeling and Simulation: Introduction for Scientists and Engineers. Kai Velten
Copyright © 2009 WILEY-VCH Verlag GmbH & Co. KGaA, Weinheim
ISBN: 978-3-527-40758-8

of any other variables affecting the state variables. It just means that spatial dishomogeneities, if they exist, and any other variables have a negligible effect on your state variables in those particular situations where you apply the model successfully. And you must always keep in mind that you are making a strong assumption when you are neglecting all those other possible influences on your state variables. This is particularly important when you observe deviations between your model and data. In the wine fermentation model, substantial deviations from data might indicate that you are, for example, in a situation where the dishomogeneity of the yeast biomass concentration is so high that it can no longer be neglected. Then a possible solution would be to use partial differential equations (PDEs), which describe the dynamics of the yeast biomass concentration in time *and* space.

> **Note 4.1.1 (Limitations of ODE models)**   Deviations between an ODE model and data may indicate that its state variables depend on more than one variable (e.g. on time *and* space variables). Then, it may be appropriate to use PDE models instead.

4.1.2
Overview: Strange Animals, Sounds, and Smells

In contrast to ODEs, PDE models involve derivatives with respect to at least two independent variables, and hence they can be used to describe the dynamics of your quantities of interest with respect to several variables at the same time. A great number of the classical laws of nature can be formulated as PDEs, such as the laws of planetary motion, thermodynamics, electrodynamics, fluid flow, elasticity, and so on. As a whole, PDEs are a really big topic. In particular, their structure is much more variable compared to ODEs since they involve several variables and derivatives. There are many different subtypes of PDEs, which need specifically tailored numerical procedures for their solution. Many volumes could be filled with a thorough discussion of all those subtypes and their appropriate treatment, and it is hence obvious that we need to confine ourselves here to a first introduction into the topic, with the aim of introducing the reader to some of the main ideas and procedures that are applied when people formulate and solve PDE models. If you imagine the PDE topic as a dense and big jungle, then the intention of this chapter can be described as cutting a small machete path, which you can follow to get first sensual impressions of those strange animals, sounds, and smells within the jungle – so do not mistake yourself for a PDE expert after reading the following pages. To know more about PDEs, readers are referred to an abundant literature on the topic, for example, books such as [101, 138–142].

As a guide and compass for our machete path we will take the heat equation that was already discussed in Section 3.5.5. This equation will serve as our main example in the following introduction into PDEs and their numerical procedures. The heat equation provides a way to compute temperature distributions, and since so many processes in science and engineering are affected by temperature, it is important in all fields of science and engineering. This is why the two heat equation problems posed in Section 4.1.3 really are "problems you should be able to solve".

Sections 4.2 and 4.3 provide some theoretical background on the heat equation and on PDEs, in general. Sections 4.4–4.7 are devoted to the solution of PDEs in closed form or based on numerical procedures, respectively. In Sections 4.8 and 4.9, software is discussed that solves PDEs based on the finite-element method, which is one of the most important numerical procedures that can be used to solve PDEs. Section 4.9 provides a sample session using the *Salome-Meca* software, an open-source finite-element software with a general workflow similar to commercial finite-element software (Appendix A). Then, Section 4.10 introduces you to some of the most important PDE models "beyond the heat equation". This includes, for example, computational fluid dynamics (CFD) and structural mechanics models and appropriate open-source software to solve these models in 3D. Finally, Section 4.11 ends the chapter with some examples of mechanistic modeling approaches beyond differential equations.

4.1.3
Two Problems You Should Be Able to Solve

We begin with the formulation of two problems that are solved below using PDEs, and that will be used to motivate and illustrate the material in subsequent sections. These problems are concerned with the computation of temperature distributions in the geometries shown in Figure 4.1:

Problem 1:
Consider the cylinder in Figure 4.1a. Assuming
- a perfect insulation of the cylinder surface in $0 < x < 1$,
- constant temperatures in the y and z directions at time $t = 0$, that is, no temperature variations across transverse sections,
- a known initial temperature distribution $T_i(x)$ at time $t = 0$,
- and constant temperatures T_0 and T_1 at the left and right ends of the cylinder for all times $t > 0$,

what is the temperature $T(x, t)$ for $x \in (0, 1)$ and $t \in (0, T]$?

Problem 2:
Referring to the configuration in Figure 4.1b and assuming
- a constant temperature T_c at the top surface of the cube ($z = 1$),
- a constant temperature T_s at the sphere surface,
- and a perfect insulation of all other surfaces of the cube,

what is the stationary temperature distribution $T(x, y, z)$ within the cube (i.e. in the domain $[0, 1]^3 \setminus S$ if S is the sphere)?

Note the *practical relevance* that problems of this kind have *in both science and engineering*. As was mentioned above, PDEs are a very broad topic and involve a great deal of really sophisticated mathematics. Taking one of the more mathematical oriented books on PDEs, and then gazing at endless formulas, many of us will be tempted to ask: "How can *this* be useful for me?" The above two problems give the answer: PDEs are useful for everybody in science and engineering since they provide the only way to solve absolutely fundamental and elementary problems such as the computation of temperature distributions. No one can seriously doubt the fundamental importance of temperature in science and engineering – just remember, for example, the role of temperature in the wine fermentation model discussed in Section 3.10.2. If you do not know how to solve this kind of problems, then you lack to know one of the really fundamental methods in science and engineering. Without too much exaggeration, one can say, it is a bit like not being able to compute the surface area of a circle. Fortunately, you will be able to learn about some PDE basics in this chapter from a very practical and software-oriented point of view.

To make **Problem 1** a little bit more concrete, you may imagine a situation like this: the cylinder in Figure 4.1a is a metallic cylinder with a constant initial temperature of $100\,^\circ$C. At time $t = 0$, you keep the ends of the cylinder in ice water $(0\,^\circ$C) and maintain this situation unchanged for all times $t > 0$. Then, you know that the temperature of the cylinder will be approximately $0\,^\circ$C after "some time". The problem is to make this precise, and to be able to predict temperatures at any particular time $t > 0$ and at any particular location $x \in (0, 1)$ within the cylinder. Exactly this problem is solved in Section 4.6.

Regarding **Problem 2**, remember from the discussion of the metallic rod problem in Section 3.5.5 the meaning of stationarity: the stationary temperature distribution is what you obtain for $t \to \infty$ if you do not change the environment. Referring to Figure 4.1b, "unchanged environment" means that the temperatures imposed at the top of the cube and at the sphere surface remain unchanged for all times $t > 0$, the other sides of the cube remain perfectly insulated, and so on. Stationarity

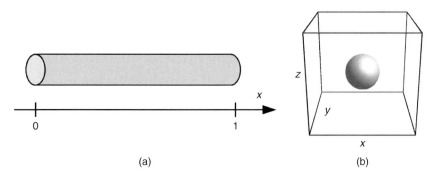

(a)   (b)

Fig. 4.1 (a) Cylinder used in *Problem 1*. (b) Cube $[0, 1]^3$, containing a cylinder with radius 0.1 centered at $(0.5, 0.5, 0.5)$, as used in *Problem 2*.

can also be explained referring to the above discussion of *Problem 1*, where it was said that the temperature of the cylinder will be $0\,°$C after "some time", which can be phrased as follows: given the conditions in the above discussion of *Problem 1*, $T(x) = 0\,°$C is the stationary solution, which is approached for $t \to \infty$ (this will be formally shown in Section 4.4).

Again, let us spend a few thoughts on what the situation in *Problem 2* could mean in a practical situation. As was said above, temperatures are of great importance in all kinds of applications in science and engineering. Many of the devices used in science and engineering contain certain parts specially designed to control the temperature. The question then is whether the design of these temperature-controlling parts is good enough in the sense that the resulting temperature distribution in the device satisfies your needs. In *Problem 2* above, the small sphere inside the cube can be viewed as a temperature-controlling part. Once we are able to solve *Problem 2*, we can, for example, compare the resulting temperature distribution with the distributions obtained if we use a small cube, tetrahedron, and so on, instead of the sphere, and then try to optimize the design of the temperature-controlling part in this way.

This procedure has been used, for example, in [143, 144] to optimize cultivation measures affecting the temperature distribution in *asparagus dams*. In the *wine fermentation example* discussed above, temperature is one of the variables that is used to control the process (Section 3.10.2). To be able to adjust temperatures during fermentation, various cooling devices are used inside fermenters, and the question then is where these cooling devices should be placed and how they should be controlled such that the temperature distribution inside the fermenter meets the requirements. In this example, the computation of temperature distributions is further complicated by the presence of convectional flows, which transport heat through the fermenter. To compute temperature distributions in a situation like this, we would have to solve a coupled problem, which would also involve the computation of the fluid flow pattern in the fermenter. This underlines again that *Problem 2* can be seen as a first approximation to a class of problems of great practical importance.

4.2
The Heat Equation

To solve the problems posed in Section 4.1.3, we need an equation that describes the dynamics of temperature as a function of space and time: the heat equation. The heat equation is used here in the following form:

$$\frac{\partial T}{\partial t} = \frac{K}{C\rho}\left(\frac{\partial^2 T}{\partial x^2} + \frac{\partial^2 T}{\partial y^2} + \frac{\partial^2 T}{\partial z^2}\right) \tag{4.1}$$

where K ($\text{WK}^{-1}\text{m}^{-1}$) is the thermal conductivity, C ($\text{J kg}^{-1}\,\text{K}^{-1}$) is the specific heat capacity, and ρ (kg m^{-3}) is the density.

Using the so-called Laplace operator

$$\Delta = \frac{\partial^2}{\partial x^2} + \frac{\partial^2}{\partial y^2} + \frac{\partial^2}{\partial z^2} \tag{4.2}$$

the heat equation is also frequently written as

$$\frac{\partial T}{\partial t} = \frac{K}{C\rho} \Delta T \tag{4.3}$$

We could end our discussion of the heat equation here (before it actually begins) and work with this equation as it is. That is, we could immediately turn to the mathematical question of how this equation can be solved, and how it can be applied to the problems in Section 4.1.3. Indeed, the reader could skip the rest of this section and continue with Section 4.3 if he or she is interested in the technical aspects of solving this kind of problems. It is, nevertheless, recommended to read the rest of this section, not only because it is always good to have an idea about the background of the equations that one is using. The following discussion introduces you into an important way how this and many other PDEs can be derived from balance considerations, and beyond this you will understand why most PDEs in the applications are of second order.

4.2.1
Fourier's Law

The *specific heat capacity* C in the above equations is a measure of the amount of heat energy required to increase the temperature of 1 kg of the material under consideration by $1\,^\circ$C. Values of C can be found in the literature, as well as values for the *thermal conductivity* K, which is the proportionality constant in an empirical relation called *Fouriers's law*:

$$\begin{pmatrix} q_x \\ q_y \\ q_z \end{pmatrix} = -K \cdot \begin{pmatrix} \dfrac{\partial T}{\partial x} \\ \dfrac{\partial T}{\partial y} \\ \dfrac{\partial T}{\partial z} \end{pmatrix} \tag{4.4}$$

Using $\mathbf{q} = (q_x, q_y, q_z)^t$ and the *nabla operator*

$$\nabla = \begin{pmatrix} \dfrac{\partial}{\partial x} \\ \dfrac{\partial}{\partial y} \\ \dfrac{\partial}{\partial z} \end{pmatrix} \tag{4.5}$$

Fourier's law can also be written as

$$q = -K \cdot \nabla T \tag{4.6}$$

In these equations, q is the *heat flow rate* (W m^{-2}). In a situation where you do not have temperature gradients in the y and z directions (such as *Problem 1* in Section 4.1.3), Fourier's law attains the one-dimensional form

$$q_x = -K \cdot \frac{dT}{dx} \tag{4.7}$$

So, you see that Fourier's law expresses a very simple proportionality: the heat flow in the positive x direction is negatively proportional to the temperature gradient. This means that a temperature which is going down in the positive x direction (corresponding to a negative value of dT/dx) generates a heat flow in the positive x direction. This expresses our everyday experience that heat flows are always directed from higher to lower temperatures (e.g. from your hot coffee cup toward its surroundings).

4.2.2
Conservation of Energy

Let us now try to understand the heat equation. Remember that in the ODE models discussed in Chapter 3, the left-hand sides of the ODEs usually expressed rates of changes of the state variables, which were then expressed in the right-hand sides of the equations. Take the body temperature model as an example: as was explained in Section 3.4.1, Equation 3.28 is in an immediate and obvious correspondence with a statement describing the temperature adaption such as *the sensor temperature changes proportionally to the difference between body temperature and actual sensor temperature*. Regarding PDEs, the correspondence with the process under consideration is usually not so obvious. If you are concerned with PDEs for some time, you will of course learn to "read" and understand PDEs and their correspondence with the processes under consideration to some extent, but understanding a PDE still needs one or two thoughts more as compared to ODEs. Equation 4.1, for example, involves a combination of first- and second-order derivatives with respect to different variables which expresses more than just a rate of change of a quantity as was the case with ODEs. The problem is that PDEs simultaneously express rates of changes of several quantities.

Fourier's law provides us with a good starting point for the derivation of Equation 4.1. It is in fact one of the two main ingredients in the heat equation. What could the other essential ingredient possibly be? From the ODE point of view, we could say that Fourier's law expresses the consequences of rates of changes of the temperature with respect to the spatial variables x, y, and z. So, it seems obvious that we have to look for equations expressing the rate of change of temperature with respect to the remaining variable, which is time t. Such an equation can be derived from the *conservation of energy* principle [145].

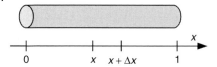

Fig. 4.2 Interval $[x, x + \Delta x]$ corresponding to a small part of a one-dimensional body.

> **Note 4.2.1 (Deriving PDEs using conservation principles)** Many of the important PDEs of mathematical physics can be derived from conservation principles such as *conservation of energy*, *conservation of mass*, or *conservation of momentum* (see the examples in Section 4.10).

The most intuitive and least technical way to apply energy conservation is the consideration of a small interval $[x, x + \Delta x]$ corresponding to a small part of a one-dimensional body (Figure 4.2).

Remember from your physics courses that the energy conservation principle states that energy cannot be created or destroyed, which means that the total amount of energy in any closed system remains constant. Applied to $[x, x + \Delta x]$ (you should always think of the part of the physical body corresponding to $[x, x + \Delta x]$ when we write down this interval in the following) this means that any change in the energy content of $[x, x + \Delta x]$ must equal the amount of heat that flows into $[x, x + \Delta x]$ through its ends at x and $x + \Delta x$. Note that there can be no flow in the y and z directions since we assume a one-dimensional body here. For example, imagine the outer surface of the cylinder in Figure 4.2 to be perfectly insulated against heat flow (see Section 4.3.3 for more on dimensionality considerations). The heat balance of $[x, x + \Delta x]$ can be written as

$$
\left\{ \begin{array}{c} \text{Rate of change} \\ \text{of energy content} \\ \text{in } [x, x + \Delta x] \end{array} \right\} = \left\{ \begin{array}{c} \text{Net heat inflow} \\ \text{through } x \text{ and } x + \Delta x \end{array} \right\} \tag{4.8}
$$

4.2.3
Heat Equation = Fourier's Law + Energy Conservation

Basically, the heat equation can now be derived by putting together the results of the last two sections. To this end, consider some small time interval $[t, t + \Delta t]$. Within this time interval, the temperature at x will change from $T(x, t)$ to $T(x, t + \Delta t)$. This corresponds to a change in the energy content of $[x, x + \Delta x]$ that we are going to estimate now. First of all, note that for sufficiently small Δx, we have

$$
T(\tilde{x}, t) \approx T(x, t) \quad \forall \tilde{x} \in [x, x + \Delta x] \tag{4.9}
$$

and

$$T(\tilde{x}, t + \Delta t) \approx T(x, t + \Delta t) \quad \forall \tilde{x} \in [x, x + \Delta x] \tag{4.10}$$

The last two equations say that we may take $T(x, t)$ to $T(x, t + \Delta t)$ as the "representative temperatures" in $[x, x + \Delta x]$ at times t and $t + \Delta t$, respectively. This means that we are now approximately in the following situation: we have a physical body corresponding to $[x, x + \Delta x]$, which changes its temperature from $T(x, t)$ to $T(x, t + \Delta t)$. Using the above explanation of C and denoting the mass of $[x, x + \Delta x]$ with m, we, thus, have

$$\left\{ \begin{array}{c} \text{Rate of change} \\ \text{of energy content} \\ \text{in } [x, x + \Delta x] \end{array} \right\} = C \cdot m \cdot (T(x, t + \Delta t) - T(x, t)) \tag{4.11}$$

Denoting the cross-sectional area of the body in Figure 4.2 with A, we have $m = A\Delta x\rho$, which turns Equation (4.11) into

$$\left\{ \begin{array}{c} \text{Rate of change} \\ \text{of energy content} \\ \text{in } [x, x + \Delta x] \end{array} \right\} = C \cdot A\Delta x\rho \cdot (T(x, t + \Delta t) - T(x, t)) \tag{4.12}$$

This gives us the left-hand side of Equation 4.8. The right-hand side of this equation can be expressed as

$$\left\{ \begin{array}{c} \text{Net heat inflow} \\ \text{through } x \text{ and } x + \Delta x \end{array} \right\} = A\Delta t(q_x(x) - q_x(x + \Delta x)) \tag{4.13}$$

using the heat flow rate q_x introduced above. Note that the multiplication with $A\Delta t$ is necessary in the last formula since q_x expresses heat flow per units of time and surface area: its unit is $(\text{W m}^{-2}) = (\text{J s}^{-1}\text{m}^{-2})$. Note also that the signs used in the difference $q_x(x) - q_x(x + \Delta x)$ are chosen such that we get the net heat *inflow* as required. Using Equations 4.8, 4.12, and 4.13, we obtain the following:

$$\frac{T(x, t + \Delta t) - T(x, t)}{\Delta t} = \frac{1}{C\rho} \frac{q_x(x) - q_x(x + \Delta x)}{\Delta x} \tag{4.14}$$

Using Fourier's law, Equation 4.7 gives

$$\frac{T(x, t + \Delta t) - T(x, t)}{\Delta t} = \frac{K}{C\rho} \frac{\frac{\partial T(x + \Delta x, t, t)}{\partial x} - \frac{\partial T(x, t)}{\partial x}}{\Delta x} \tag{4.15}$$

From this, the heat equation (4.1) is obtained by taking the limit for $\Delta t \to 0$ and $\Delta x \to 0$.

4.2.4
Heat Equation in Multidimensions

The above derivation of the heat equation can be easily generalized to multidimensions by expressing energy conservation in volumes such as $[x, x + \Delta x] \times [y, y + \Delta y]$. Another option is to use an integral formulation of energy conservation over generally shaped volumes, which leads to the heat equation by an application of standard integral theorems [101]. In any case, the "small volumes" that are used to consider balances of conserved quantities such as energy, mass, momentum, and so on, similar to above are usually called *control volumes*. Beyond this, you should note that the parameters C, ρ, and K may depend on the space variables. Just check that the above derivation works without problems in the case where C and ρ depend on x. If K depends on x, the above derivation gives

$$\frac{\partial T(x, t)}{\partial t} = \frac{1}{C\rho} \frac{\partial}{\partial x} \left(K(x) \cdot \frac{\partial T(x, t)}{\partial x} \right) \tag{4.16}$$

instead of Equation 4.1 and

$$\frac{\partial T(\mathbf{x}, t)}{\partial t} = \frac{1}{C\rho} \left(\frac{\partial}{\partial x_1} \left(K(\mathbf{x}) \cdot \frac{\partial T(\mathbf{x}, t)}{\partial x_1} \right) + \frac{\partial}{\partial x_2} \left(K(\mathbf{x}) \cdot \frac{\partial T(\mathbf{x}, t)}{\partial x_2} \right) \right. $$
$$\left. + \frac{\partial}{\partial x_3} \left(K(\mathbf{x}) \cdot \frac{\partial T(\mathbf{x}, t)}{\partial x_3} \right) \right) \tag{4.17}$$

in the case where K depends on the space coordinates $\mathbf{x} = (x_1, x_2, x_3)^t$. Using the nabla operator ∇ (Equation 4.5), the last equation can be written more compactly as

$$\frac{\partial T(\mathbf{x}, t)}{\partial t} = \frac{1}{C\rho} \nabla \left(K(\mathbf{x}) \cdot \nabla T(\mathbf{x}, t) \right) \tag{4.18}$$

4.2.5
Anisotropic Case

Until now, the thermal conductivity was assumed to be independent of the direction of measurement. Consider a situation with identical temperature gradients in the three main space directions, that is,

$$\frac{\partial T}{x_1} = \frac{\partial T}{x_2} = \frac{\partial T}{x_3} \tag{4.19}$$

Then, Fourier's law (Equation 4.6) says that

$$q_1 = K \cdot \frac{\partial T}{x_1} = q_2 = K \cdot \frac{\partial T}{x_2} = q_3 = K \cdot \frac{\partial T}{x_3} \tag{4.20}$$

that is, the heat flux generated by the above temperature gradients is independent of the direction in space (note that this argument can be generalized to arbitrary directions). A material with a thermal conductivity that is direction independent in this sense is called *isotropic*. The term "isotropy" is used for other material properties such as fluid flow permeability or electrical conductivity in a similar way. Materials that do not satisfy the isotropy condition of directional independence are called *anisotropic*. For example, an anisotropic material may have a high thermal conductivity in the x_1 direction and smaller conductivities in the other space directions. A single number K obviously does not suffice to describe the thermal conductivity in such a situation, which means you need a multi- or matrix-valued thermal conductivity **K** such as

$$\mathbf{K} = \begin{pmatrix} K_{11} & K_{12} & K_{13} \\ K_{21} & K_{22} & K_{23} \\ K_{31} & K_{32} & K_{33} \end{pmatrix} \tag{4.21}$$

Many other anisotropic material properties are described using matrices or tensors in a similar way, while isotropic material properties are described using a single scalar quantity such as the scalar thermal conductivity in Fourier's law (Equation 4.6) above. Thanks to matrix algebra, Equation 4.6 and the heat equation (4.18) remain almost unchanged when we use **K** from Equation 4.21:

$$\mathbf{q}(\mathbf{x}, t) = -\mathbf{K} \cdot \nabla T(\mathbf{x}, t) \tag{4.22}$$

$$\frac{\partial T(\mathbf{x}, t)}{\partial t} = \frac{1}{C\rho} \nabla \left(\mathbf{K}(\mathbf{x}) \cdot \nabla T(\mathbf{x}, t) \right) \tag{4.23}$$

The diagonal entries of **K** describe the thermal conductivities in the main space directions, that is, K_{11} is the thermal conductivity in x_1 direction, K_{22} in x_2 direction, and so on.

4.2.6
Understanding Off-diagonal Conductivities

To understand the off-diagonal entries, consider the following special case:

$$\mathbf{K} = \begin{pmatrix} K_{11} & K_{12} & 0 \\ 0 & K_{22} & 0 \\ 0 & 0 & K_{33} \end{pmatrix} \tag{4.24}$$

Applying this in Equation 4.22, the heat flow in the x_1 direction is obtained as follows:

$$q_1(\mathbf{x}, t) = K_{11} \cdot \frac{\partial T(\mathbf{x}, t)}{\partial x_1} + K_{12} \cdot \frac{\partial T(\mathbf{x}, t)}{\partial x_2} \tag{4.25}$$

So, you see that $K_{12} \neq 0$ means that the heat flow in the x_1 direction depends not only on the temperature gradient in that direction, $\partial T / \partial x_1$, but also on the temperature gradient in the x_2 direction, $\partial T / \partial x_2$. You may wonder how this is

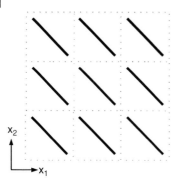

Fig. 4.3 Periodicity cells of a medium with an anisotropic effective thermal conductivity.

possible, so let us consider an example. Figure 4.3 shows a "microscopic picture" of a periodic medium in the x_1/x_2 plane. Here, "periodic medium" means that the medium is assumed to consist of a great number of periodicity cells such as those shown in the figure. "Microscopic picture" is to say that the figure just shows a few of the periodicity cells, which we assume to be very small parts of the overall geometry of the medium in the x_1/x_2 plane (we do not need the overall geometry for our argument). Now assume that the black bars in the medium consist of a material having zero thermal conductivity, while the matrix surrounding the bars has some finite (scalar) thermal conductivity. If we now apply a temperature gradient in the x_2 direction, this will initiate a heat flow in the same direction:

$$q_2(\mathbf{x}, t) = K_M \cdot \frac{\partial T(\mathbf{x}, t)}{\partial x_2} \tag{4.26}$$

Here, K_M denotes the (isotropic) thermal conductivity of the matrix surrounding the black bars in Figure 4.3. But due to the heat-impermeable bars in the medium, this heat flow cannot go straight through the medium in the x_2 direction. Rather, it will be partially deflected into the x_1 direction by the action of the bars. In this way, a temperature gradient in the x_2 direction can indeed initiate a heat flow in the x_1 direction. If one then assumes the periodicity cells in Figure 4.3 to be very small in relation to the overall size of the medium and derives "effective", averaged thermal conductivities for the medium, one arrives at thermal conductivity matrices with off-diagonal entries similar to Equation 4.24. Such effective ("homogenized") thermal conductivities of periodic media can be derived, for example, using the methods of *homogenization theory* [146,147]. Note that the effective thermal conductivity matrix of a medium such as the one shown in Figure 4.3 will be symmetric. Denoting this matrix as

$$\tilde{\mathbf{K}} = \begin{pmatrix} \tilde{K}_{11} & \tilde{K}_{12} \\ \tilde{K}_{21} & \tilde{K}_{22} \end{pmatrix} \tag{4.27}$$

means that we would have $\tilde{K}_{12} = \tilde{K}_{21}$ for the medium in Figure 4.3.

Note that the heat equation can be further generalized beyond Equation 4.23 to include effects of heat sources, convective heat transfer, and so on, [97, 101].

4.3
Some Theory You Should Know

This section explains some theoretical background of PDEs. You can skip this in a first reading if you just want to gain a quick understanding of how the problems posed in Section 4.1.3 can be solved. In that case, go on with Section 4.4.

4.3.1
Partial Differential Equations

The last section showed that the problems posed in Section 4.1.3 lead us to the heat equation (4.23). In this equation, the temperature $T(x, t)$ (or $T(x, y, t)$, or $T(\mathbf{x}, t)$, depending on the dimensionality of the problem) serves as the unknown, and the equation involves partial derivatives with respect to at least two variables. Hence, the heat equation is a PDE, which can be generally defined as follows

> **Definition 4.3.1 (Partial differential equation)**   A *PDE* is an equation that satisfies the following conditions:
> - A function $u : \mathbb{R}^n \supset \Omega \to \mathbb{R}$ serves as the unknown of the equation.
> - The equation involves partial derivatives of u with respect to at least two independent variables.

This definition can be generalized to several unknowns, vector-valued unknowns, and systems of PDEs, [101,142]. The *fundamental importance of PDEs*, which can hardly be overestimated, arises from the fact that there is an abundant number of problems in science and engineering which lead to PDEs, just as *Problem 1* and *Problem 2* led us to the heat equation in Section 4.2. One may find this surprising, but remember the qualitative argument for the distinguished role of differential equations (ODEs or PDEs) that has been given in Section 3.1 (Note 3.1.1): scientists or engineers usually are interested in rates of changes of their quantities of interest, and since rates of changes are derivatives in mathematical terms, writing down equations involving rates of changes thus means writing down differential equations in many cases. The *order of a PDE* is the degree of the highest derivative appearing in the PDE, which means that the heat equation discussed in Section 4.2 is a second-order PDE.

> **Note 4.3.1 (Distinguished role of up to second-order PDEs)**   Most PDEs used in science and engineering applications are first- or second-order equations.

The derivation of the heat equation in Section 4.2 gives us an idea why this is so. We have seen there that the heat equation arises from a combined application of Fourier's law and the energy conservation principle. Analyzing the formulas in Section 4.2, you will see that the application of the conservation principle basically amounted to balancing the conserved quantity (energy in this case) over some "test volume", which was $[x, x + \Delta x]$ in Section 4.2. This balance resulted in one of the two orders of the derivatives in the PDE, the other order was a result of Fourier's law, a simple empirical rate of change law similar to the "rate of change-based" ODEs considered in Section 3. Roughly speaking, one can, thus, say that you can expect conservation arguments to imply one order of your derivatives, and "rate of change arguments" to imply another derivative order. A great number of the PDEs used in the applications is based on similar arguments. This is also reflected in the PDE literature, which has its main focus on first- and second-order equations. It is, therefore, not a big restriction if we confine ourselves to up to second-order PDEs in the following. Note also that many of the formulas below will refer to the 2D case (two independent variables x and y) just to keep the notation simple, although everything can be generalized to multidimensions (unless otherwise stated).

4.3.1.1 First-order PDEs
The general form of a first-order PDE in two dimensions is [142]

$$F(x, y, u, u_x, u_y) = 0 \tag{4.28}$$

Here, $u = u(x, y)$ is the unknown function, x and y are the independent variables, $u_x = u_x(x, y)$ and $u_y = u_y(x, y)$ are the partial derivatives of u with respect to x, and y, respectively, and F is some real function. Since we are not going to develop any kind of PDE theory here, there is no need to go into a potentially confusing discussion of domains of definitions, differentiability properties, and so on, of the various functions involved in this and the following equations. The reader should note that our discussion of equations such as Equation 4.28 in this section is purely formal, the aim just being a little sightseeing tour through the "zoo of PDEs", showing the reader some of its strange animals and giving an idea about their classification. Readers with a more theoretical interest in PDEs are referred to specialized literature such as [101, 142]. Note that Equation 4.28 can also beinterpreted as a vector-valued equation, that is, as a compact vector notation of a *first-order PDE system* such as

$$
\begin{aligned}
F_1(x, y, u, u_x, u_y) &= 0 \\
F_2(x, y, u, u_x, u_y) &= 0 \\
&\ \vdots \\
F_n(x, y, u, u_x, u_y) &= 0
\end{aligned}
\tag{4.29}
$$

In a PDE system like this, u will also typically be a vector-valued function such as $u = (u_1, \ldots, u_n)$. The *shock wave equation*

$$u_x + u \cdot u_y = 0 \tag{4.30}$$

is an example of a PDE having the form of Equation 4.28. Comparing Equations 4.28 and 4.30, you see that the particular form of F in this example is

$$F(x, y, u, u_x, u_y) = u_x + u \cdot u_y \tag{4.31}$$

Equations like 4.30 are used to describe abrupt, nearly discontinuous changes of quantities such as the pressure or density of air, which appear, for example, in supersonic flows.

4.3.1.2 Second-order PDEs

Writing down the general form of a second-order PDE just amounts to adding second-order derivatives to the expression in Equation 4.28:

$$F(x, y, u, u_x, u_y, u_{xx}, u_{xy}, u_{yy}) = 0 \tag{4.32}$$

An example is the one-dimensional heat equation, that is, the heat equation in a situation where T depends only on x and t. Then, the partial derivatives with respect to y and z in Equation 4.1 vanish, which leads to

$$\frac{\partial T}{\partial t} = \frac{K}{C\rho} \frac{\partial^2 T}{\partial x^2} \tag{4.33}$$

Using the index notation for partial derivatives similar to Equation 4.32, this gives

$$F(t, x, T, T_t, T_x, T_{tt}, T_{tx}, T_{xx}) = T_t - \frac{K}{C\rho} T_{xx} = 0 \tag{4.34}$$

Note that this is exactly analogous to Equation 4.32, except for the fact that different names are used for the independent variables and for the unknown function.

4.3.1.3 Linear versus Nonlinear

As it holds true for other types of mathematical equations, linearity is a property of PDEs which is of particular importance for an appropriate choice of the solution method. As usual, it is easier to solve linear PDE's. Nonlinear PDEs, on the other hand, are harder to solve, but they are also often more interesting in the sense that they express a richer and more multifaceted dynamical behavior of their state variables. Equation 4.34 is a *linear PDE* since it involves a sum of the unknown function and its derivatives where the unknown function and its derivatives are multiplied by coefficients independent of the unknown function or its derivatives. Equation 4.30, on the other hand, is an example of a *nonlinear PDE*, since this equation involves a product of the unknown function u with its derivative u_y.

The general *strategy to solve nonlinear equations* in all fields of mathematics is linearization, that is, the consideration of linear equations that approximate a given nonlinear equation. The Newton method for the solution of nonlinear equations

$f(x) = 0$ is an example, which is based on the local approximation of the function $f(x)$ by linear equations $ax + b$ describing tangents of $f(x)$ [148]. Linearization methods such as the Newton method usually imply the use of iteration procedures, which means that a sequence of linearized solutions $x_1, x_2, \ldots$ is generated, which converges to the solution of the nonlinear equation. Such iteration procedures (frequently based on appropriate generalizations of the Newton method) are also used to solve nonlinear PDEs [141]. Of course, the theoretical understanding of linear PDEs is of great importance as a basis of such methods, and there is in fact a well-developed theory of linear PDEs, which is described in detail in specialized literature such as [101, 142]. We will now just sketch some of the main ideas of the linear theory that are relevant within the scope of this book.

4.3.1.4 Elliptic, Parabolic, and Hyperbolic Equations

The general form of a linear second-order PDE in two dimensions is

$$A u_{xx} + B u_{xy} + C u_{yy} + D u_x + E u_y + F = 0 \tag{4.35}$$

Here, the coefficients $A, \ldots, F$ are real numbers, which may depend on the independent variables x and y. Depending on the sign of the *discriminant* $d = AC - B^2$, linear second-order PDEs are called

- *elliptic* if $d > 0$,
- *parabolic* if $d = 0$, and
- *hyperbolic* if $d < 0$

Since we have allowed the coefficients of Equation 4.35 to be x and y dependent, the type of an equation in the sense of this classification may also depend on x and y. An example is the *Euler–Tricomi* equation:

$$u_{xx} - x \cdot u_{yy} = 0 \tag{4.36}$$

This equation is used in models of *transonic flow*, that is, in models referring to velocities close to the speed of sound [149]. The change of type in this equation occurs due to the nonconstant coefficient of u_{yy}. The discriminant of Equation 4.36 is

$$d = AC - B^2 = x \begin{cases} >0 & \text{if } x > 0 \\ <0 & \text{if } x < 0 \end{cases} \tag{4.37}$$

which means that the Euler–Tricomi equation is an elliptic equation in the positive half plane $x > 0$ and a hyperbolic equation in the negative half plane $x < 0$.

The above classification is justified by the fact that Equation 4.35 can be brought into one of three standard forms by a linear transformation of the independent variables, where the standard form corresponding to any particular equation depends on the discriminant d [142]. Using "$\cdots$" to denote terms that do not involve second-order derivatives, the *standard forms* of elliptic, parabolic, and hyperbolic PDEs are (in this order):

$$u_{xx} + u_{yy} + \cdots = 0 \tag{4.38}$$

$$u_{xx} + \ldots = 0 \tag{4.39}$$

$$u_{xx} - u_{yy} + \cdots = 0 \tag{4.40}$$

Referring to the standard form, one can, thus, say that elliptic PDEs are characterized by the fact that they contain second-order derivatives with respect to all independent variables, which all have the same sign when they are written on one side of the equation. Parabolic PDEs involve one second-order derivative and at least one first-order derivative. Note that Definition 4.1 requires that there must be at least one first-order derivative in the "$\cdots$" of Equation 4.39 (otherwise, that equation would be an ODE). Finally, hyperbolic equations can be described similar to elliptic equations except for the fact that the second-order derivatives have opposite signs when brought on one side of the equation.

Comparing Equations 4.34 and 4.39, you see that the heat equation is an example of a parabolic PDE. The stationary case of the two-dimensional heat equation (4.1) is an example of an elliptic PDE. As was mentioned before, a *stationary solution* of a PDE is a solution referring to the case where the time derivatives in the PDE vanish. Solutions of this kind can usually be interpreted as expressing the state of the system which is attained in a constant environment after a "very long" time (mathematically, the state that is approached for $t \to \infty$). In this sense, stationary solutions of the heat equation express the temperature distribution attained by a system in a constant environment after a "very long" time. Using the index notation for partial derivatives in Equation 4.1 and assuming $T_t = 0$ in that equation, it turns out that the two-dimensional stationary heat equation is

$$T_{xx} + T_{yy} = 0 \tag{4.41}$$

which corresponds to Equation 4.38, so we see that the stationary heat equation is elliptic. Hyperbolic equations (Equation 4.40) are used to describe all kinds of wave phenomena such as sound waves, light waves, or water waves [142].

Using appropriate methods of matrix algebra, the above classification of PDEs into elliptic, parabolic, and hyperbolic PDEs can be generalized to multidimensions (i.e. PDEs depending on $x_1, \ldots, x_n$) [142]. This classification is of particular importance in the numerical treatment of PDEs and will be discussed in Sections 4.6 and 4.8.

4.3.2
Initial and Boundary Conditions

Above we have seen that ODEs are usually solved by an entire family of solutions unless initial or boundary conditions are imposed, which select one particular solution among those many solutions (Section 3.7.1.1). For the same reason, initial or boundary conditions are used together with PDEs. From the mathematical point of view, initial or boundary conditions are needed to make the mathematical problem uniquely solvable. From the applications point of view, they are a necessary part of the description of the system that is investigated. Considering *Problem 1* in Section 4.1.3, for example, it is obvious from the applications point of view, i.e. without any mathematical considerations, that this problem cannot be solved unless we know the temperatures at the left and right ends of the cylinder (T_0 and T_1) and the initial temperature distribution, $T_i(x)$. To solve *Problem 1*, we will use the one-dimensional form of the heat equation (4.33) (Section 4.6). Above we have seen that this is a parabolic equation, and so we see here that this *parabolic equation* requires a boundary condition referring to the spatial variable and an initial condition referring to the time variable.

4.3.2.1 Well Posedness

Generally, the appropriate choice of initial and boundary conditions for PDEs can be a subtle matter as discussed in [142]. It is related to the mathematical concept of a *well-posed problem*, which was originally introduced by the French mathematician Hadamard. A well-posed differential equation problem satisfies the following conditions [142]:

- *existence*: a solution exists;
- *uniqueness*: the solution is unique;
- *stability*: the solution depends continuously on the data of the problem.

Let us explain the last point referring to *Problem 1* again. Assume a small change in the temperature imposed at the left or right end of the cylinder, or small changes in the initial temperature distribution. Then stability in the above sense means that the change in the temperature distribution $T(x, t)$ implied by your change in the problem data should go to zero if you let the size of this change in the data go to zero. This expresses our physical intuition, but you should know that the discussion of stability matters and well-posedness matters in general can be subtle (e.g. the appropriate definition of what is meant by "continuity"). From a mathematical point of view, we can thus say that initial or boundary conditions have to be chosen in a way that makes the overall problem well posed in the above sense.

4.3.2.2 A Rule of Thumb

The following "rule of thumb" for the proper selection of initial or boundary conditions applies to many equations [138]:

Note 4.3.2 (Rule of thumb)
- Elliptic equation: add a boundary condition
- Parabolic equation: add a boundary condition for the space variables and an initial condition at $t = 0$

- Hyperbolic equation: add a boundary condition and two initial conditions at $t = 0$

Note that this is consistent with our above discussion of the initial and boundary conditions of *Problem 1*, which was motivated by physical intuition. Generally, one can say that *physical intuition* applied as above to *Problem 1* frequently is a good guideline for the proper choice of initial or boundary conditions.

Problem 2 from Section 4.1.3 gives us an example of boundary conditions for an *elliptic equation*. In Section 4.9, this problem is solved using the stationary heat equation. We have seen above that this is an elliptic equation in the two-dimensional case (Section 4.3.1.3). Using appropriate methods of matrix algebra, the same can be shown for the three-dimensional version of the stationary heat equation that is used in the solution of *Problem 2*. Now looking at the formulation of *Problem 2* in Section 4.1.3, you see that this problem involves conditions for the space variables at the boundary of the computational domain only, which is consistent with the above "rule of thumb". Again, this is also consistent with physical intuition since any change in the initial temperature distribution within the cube will certainly affect the temperatures inside the cube for some time, but in the long run physical intuition tells us that everything is determined by the constant conditions applied at the boundaries of the cube and at its internal boundaries on the sphere surface of Figure 4.1. For the hyperbolic case in the above "rule of thumb", appropriate wave equation examples may be found in [138].

4.3.2.3 Dirichlet and Neumann Conditions

Note that two different kinds of boundary conditions are used in *Problem 2*. On the top boundary of the cube and on the sphere surface, the value of the temperature is prescribed. Boundary conditions that prescribe the values of the unknown function of a PDE in this way are called *Dirichlet boundary conditions*. The remaining boundaries of the cube are assumed to be "perfectly insulated" in *Problem 2*. Perfect insulation means that there is no heat flow across these boundaries. Using vector notation, this can be expressed as follows: Let S denote one of the cube boundaries, and let $\mathbf{q}(\mathbf{x})$ (W m^{-2}) be the heat flow rate introduced in Section 4.2. Then, if $\mathbf{n}(\mathbf{x})$ denotes the outward facing normal vector on S, $\mathbf{q}(\mathbf{x}) \cdot \mathbf{n}(\mathbf{x})$ (W m^{-2}) is the (outward going) heat flow through S, which is zero since S is insulated:

$$\mathbf{q}(\mathbf{x}) \cdot \mathbf{n}(\mathbf{x}) = 0 \quad \mathbf{x} \in S \tag{4.42}$$

Using Fourier's law (Equation 4.6), this turns into

$$\frac{\partial T}{\partial n} = \nabla T \cdot \mathbf{n}(\mathbf{x}) = 0 \quad \mathbf{x} \in S \tag{4.43}$$

where $\partial T / \partial n$ denotes the normal derivative of T with respect to $\mathbf{n}$, which expresses T's rate of change in the direction of $\mathbf{n}$. Boundary conditions that prescribe the normal derivative of the unknown function such as Equation 4.43 are called *Neumann boundary conditions*. If u is the unknown function in a PDE defined on a domain $\Omega \times I \subset \mathbb{R}^n \times \mathbb{R}$ and if $S \subset \partial\Omega$ is a subset of Ω's boundary, the general form of Dirichlet and Neumann boundary conditions in S can be written as [142]

$$\text{Dirichlet:} \quad u(\mathbf{x}, t) = f_1(\mathbf{x}, t) \qquad\qquad (\mathbf{x}, t) \in S \times I \qquad (4.44)$$

$$\text{Neumann:} \quad \frac{\partial u(\mathbf{x}, t)}{\partial n} = f_2(\mathbf{x}, t) \qquad\qquad (\mathbf{x}, t) \in S \times I \qquad (4.45)$$

where f_1 and f_2 are given real functions. Another frequently used boundary condition is the *Robin boundary condition*, which specifies a linear combination of u and $\partial u/\partial n$ as follows:

$$\text{Robin:} \quad a(\mathbf{x}, t)u(\mathbf{x}, t) + b(\mathbf{x}, t)\frac{\partial u(\mathbf{x}, t)}{\partial n} = f_3(\mathbf{x}, t) \quad (\mathbf{x}, t) \in S \times I \qquad (4.46)$$

Any of the above three boundary conditions is called *homogeneous* if its right-hand side vanishes (e.g. $f_1(\mathbf{x}, t) = 0$), otherwise *inhomogeneous*. A homogeneous Neumann condition ($f_2(\mathbf{x}, t) = 0$ in Equation 4.45) is also known as a *no-flow condition*. Equation 4.43 is an example of such a homogeneous Neumann condition, which forbids any heat flow through S as discussed above. Similar interpretations apply in many other situations, and this is why the term "no-flow condition" is used. Note that no-flow conditions can sometimes also be interpreted as expressing the symmetry of a problem, and hence the term *symmetry boundary condition* is also used. An example of this along with further explanations is given in Section 4.3.3.

Regarding *Problem 1* and *Problem 2* in Section 4.1.3, the above discussion can be summarized as follows:

- *Problem 1* imposes an initial condition in the domain $0 < x < 1$ and two Dirichlet conditions at $x = 0$ and $x = 1$. The Dirichlet condition at $x = 0$ (or $x = 1$) is inhomogeneous if a nonzero temperature is imposed there, homogeneous otherwise.
- *Problem 2* imposes Dirichlet conditions at the top surface of the cube and at the sphere surface (homogeneous or inhomogeneous Dirichlet conditions depending on the actual values of T_c and T_s), and a homogeneous Neumann condition (no-flow condition) at the remaining sides of the cube.

4.3.3
Symmetry and Dimensionality

In practice, most PDEs cannot be solved in closed form, which means that they are solved using numerical algorithms in most cases. The application of such numerical algorithms can be extremely expensive in terms of computation time and machine requirements such as memory or processor speed requirements [150].

As has been already mentioned in Section 3.6.1, it may take several hours, days, or even longer to solve complex coupled multidimensional PDE problems, even if you are using supercomputers or large computer clusters (i.e. a great number of coupled computers that work together to solve your problem). The continuous increase of processor speeds and available memory does not change this situation since the increase in computer power is accompanied by a similarly increasing complexity of the problems that are solved. As faster computers become available, people begin to solve problems that were beyond the scope of the old computer generation. Therefore, the reduction of computation time and machine requirements is an important issue in the solution of PDEs. As is explained below, a lot can be achieved by the use of fast and efficient numerical algorithms, by the intelligent use of these algorithms, and by using appropriate fast and efficient software. But there is one thing you can do at an early stage before the application of numerical algorithms: you can analyze the symmetry and dimensionality of your problem.

> **Note 4.3.3 (Symmetry/dimensionality should be exploited)**   Based on a wise consideration of symmetry and dimensionality, the complexity of a problem and its computation time and machine requirements can often be substantially reduced.

4.3.3.1   1D Example
To understand the point, consider the following *Problem 3* (a modification of *Problem 2* discussed in Section 4.1.3):

> **Problem 3:**
> Referring to the configuration in Figure 4.4a and assuming
> - a constant temperature T_t at the top surface of the cube ($z = 1$),
> - a constant temperature T_b at the bottom surface of the cube
>   ($z = 0$),
> - and a perfect thermal insulation of all other surfaces of the cube,
>
> what is the stationary temperature distribution $T(x, y, z)$ within the cube?

It is simple to see that the solution of this problem will depend on z only, that is, the resulting stationary temperature distribution will be of the form $T(z)$ (see the discussion of *Problem 2* in Section 4.1.3 to understand what is meant by the term "stationary temperature distribution"). If there would be any temperature differences in some given plane $z = a$ inside the cube ($0 \leq a \leq 1$), then there would be a point (x, y, a) in that plane where we would have either $\frac{\partial T(x,y,a)}{\partial x} \neq 0$ or $\frac{\partial T(x,y,a)}{\partial y} \neq 0$. This would initiate a heat flow according to Fourier's law (Equation 4.6), which would then tend to flatten out the temperature gradient at that point until $\frac{\partial T(x,y,a)}{\partial x} = \frac{\partial T(x,y,a)}{\partial y} = 0$ would be achieved in the stationary limit. Similar to the

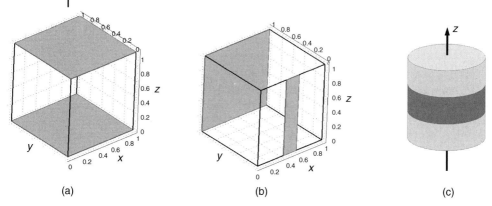

(a)   (b)   (c)

Fig. 4.4 (a) Cube $[0, 1]^3$ with highlighted boundaries $z = 0$ (bottom surface) and $z = 1$ (top surface) used in *Problem 3*. (b) Cube $[0, 1]^3$ with highlighted boundaries $y = 1$ (back surface) and $y = 0$, $0.4 \leq x \leq 0.6$ (strip on the front surface) used in *Problem 4*. (c) Cylinder used in *Problem 5*.

derivation of Equation 3.70 in Section 3.5.5, the stationary temperature distribution that solves *Problem 3* can be derived from the heat equation as follows:

$$T(x, y, z) = T_b + (T_t - T_b) \cdot z \tag{4.47}$$

Since $T(x, y, z)$ depends on z only, this can also be written as

$$T(z) = T_b + (T_t - T_b) \cdot z \tag{4.48}$$

Such a temperature distribution that depends on one space coordinate only is called a *one-dimensional* temperature distribution, and the corresponding physical problem from which it is derived (*Problem 4* in this case) is called a *one-dimensional* problem. Correspondingly, the solution of a *two-dimensional* (*three-dimensional*) problem depends on two (three) space coordinates. Intuitively, it is clear that it is much less effort to compute a one-dimensional temperature distribution $T(x)$ compared to higher dimensional distributions such as $T(x, y)$ or $T(x, y, z)$. The discussion of the PDE solving numerical algorithms in Section 4.5 indeed shows that the number of unknowns in the resulting systems of equations – and hence the overall computational effort – depends dramatically on the dimension.

Note 4.3.4 (Dimension of a PDE problem) The solution of *one-dimensional* (*two-dimensional, three-dimensional*) PDE problems depends on one (two, three) independent variables. To reduce the computational effort, PDEs should always be solved using the lowest possible dimension.

4.3.3.2 2D Example
Consider the following problem:

Problem 4:
Referring to the configuration in Figure 4.4b and assuming
- a constant temperature T_b at the back surface of the cube ($y = 1$),
- a constant temperature T_f at the strip $y = 0, 0.4 \leq x \leq 0.6$ at the front surface of the cube,
- and a perfect thermal insulation of all other surfaces of the cube,

what is the stationary temperature distribution $T(x, y, z)$ within the cube?

In this case, it is obvious that the stationary temperature distribution will depend on x and y. Assuming $T_f > T_b$, for example, it is clear that the stationary temperature will go down toward T_b as we move in the y direction toward the (colder) back surface, and it is likewise clear that the stationary temperature will increase as we move in the x direction toward $x = 0.5$, the point on the x axis closest to the (warm) strip on the front surface (see also the discussion of Figure 4.5b further below). On the other hand, there will be no gradients of the stationary temperature in the z direction. To see this, a similar "flattening out" argument could be used as in our above discussion of *Problem 3*. Alternatively, you could observe that the same stationary temperature distribution would be obtained if the cube would extend infinitely in the positive and negative z direction, which expresses the *symmetry*

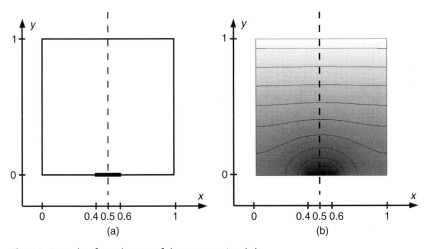

Fig. 4.5 Example of a reduction of the computational domain due to symmetry: (a) Geometry. (b) Solution of *Problem 4* for $T_b = 20\,^\circ$C and $T_f = 0\,^\circ$C (computed using *Salome-Meca*). Colors range from white (20 $^\circ$C) to black (0 $^\circ$C), lines in the domain are temperature isolines.

of this problem in the z direction. More precisely, this expresses the *translational symmetry* of this problem in the z direction, as you could translate the cube of *Problem 4* along the "infinitely extended cube" arbitrarily in the z direction without changing the stationary temperature distribution. This and other kinds of spatial symmetries can often be used to reduce the spatial dimensionality of a problem, and hence to reduce the computation time and machine requirements of a PDE problem [151].

4.3.3.3 3D Example

Problem 2 in Section 4.1.3 is an example of a three-dimensional problem. In this case, there are no symmetries that could be used to reduce the spatial dimensionality of the problem. The stationary temperature in any particular point (x, y, z) depends in a complex three-dimensional manner on the point's relative position toward the sphere and the top side of the cube in Figure 4.1b. There is, however, a kind of symmetry in *Problem 2* that can at least be used to reduce the computational effort (although the problem remains three dimensional). Indeed, it is sufficient to compute the temperature distribution in a quarter of the cube of Figure 4.1b, which corresponds to $(x, y) \in (0, 0.5) \times (0, 0.5)$. The temperature distribution in this quarter of the cube can then be extended to the other three quarters using the *mirror symmetry* of the temperature distribution inside the cube with respect to the planes $y = 0.5$ and $x = 0.5$. See the discussion of Figure 4.5 for another example of a mirror symmetry.

In the discussion of the finite-element method (Section 4.7), it will become clear that the number of unknowns is reduced by about 75% if the computational domain is reduced to a quarter of the original domain, so you see that the computational effort is substantially reduced if *Problem 2* is solved using the mirror symmetry. However, note that we will *not* use the mirror symmetry when *Problem 2* is solved using *Salome-Meca* in Section 4.9, simply because the computational effort for this problem is negligible even without the mirror symmetry (and we save the effort that would be necessary to extend the solution from the quarter of the cube into the whole cube in the postprocessing step, see Section 4.9.4).

4.3.3.4 Rotational Symmetry

As another example of how the spatial dimension of a problem can be reduced due to symmetry, consider

Problem 5:
Referring to the configuration in Figure 4.4c and assuming
- a constant temperature T_t at the top surface of the cylinder,
- a constant temperature T_s at the dark strip around the cylinder,
- and a perfect thermal insulation of all other surfaces of the cylinder,

what is the stationary temperature distribution $T(x, y, z)$ within the cylinder?

In this case, it is obvious that the problem exhibits *rotational symmetry* in the sense that the stationary temperature distribution is identical in any vertical section through the cylinder that includes the z axis (a reasoning similar to the "flattening out" argument used in Section 4.3.3.1 would apply to temperature distributions deviating from this pattern). Using cylindrical coordinates (r, ϕ, z), this means that we will get identical stationary temperature distributions on any plane $\phi = $ const. Hence, the stationary temperature distribution depends on the two spatial coordinates r and z only, and *Problem 5*, thus, is a two-dimensional problem. Note that in order to solve this problem in two dimensions, the heat equation must be expressed in cylindrical coordinates (in particular, it is important to choose the appropriate model involving cylindrical coordinates if you are using software).

4.3.3.5 Mirror Symmetry

Consider Figure 4.5a, which shows the geometrical configuration of the two-dimensional problem corresponding to *Problem 4*. As in *Problem 4*, we assume a constant temperature T_b for $y = 1$ (the top end of the square in Figure 4.5a), a constant temperature T_f in the strip $y = 0$, $0.4 \leq x \leq 0.6$ (the thick line at the bottom end of the square in Figure 4.5a), and a perfect thermal insulation at all other boundary lines of the square. Again, we ask for the stationary temperature within the square. As discussed above, the solution of this problem will then also solve *Problem 4* due to the translational symmetry of *Problem 4* in the z direction.

Note that the situation in Figure 4.5a is mirror symmetric with respect to the dashed line in the figure (similar to the mirror symmetry discussed in Section 4.3.3.3). The boundary conditions imposed on each of the two sides of the dashed line are mirror symmetric with respect to that line, and hence the resulting stationary temperature distribution is also mirror symmetric with respect to that dashed line. This can be seen in Figure 4.5b, which shows the solution of *Problem 4* computed using *Salome-Meca* (Section 4.9). So we see here that it is sufficient to compute the solution of *Problem 4* in one half of the square only, for example, for $x < 0.5$, and then to extend the solution into the other half of the square using mirror symmetry. In this way, the size of the computational domain is reduced by one half. This reduces the number of unknowns by about 50% and leads to a substantial reduction of the computational effort necessary to solve the PDE problem (see Section 4.5 for details).

4.3.3.6 Symmetry and Periodic Boundary Conditions

In all problems considered above, we have used "thermal insulation" as a boundary condition. As discussed in Section 4.3.2.3 above, this boundary condition is classified as a Neumann boundary condition, and it is also called a *no-flow* condition since it forbids any heat flow across the boundary. Referring to *Problem 4* and Figure 4.4b, the thermal insulation condition serves as a "no-flow" condition in this sense, for example, at the cube's side surfaces $x = 0$ and $x = 1$ and at the cube's front surface besides the strip. At the top and bottom surfaces of the cube, the thermal insulation condition can be interpreted in the same way, but it can also be interpreted there as expressing the symmetry of the problem in the z direction. As was explained in Section 4.3.3.2, *Problem 4* can be interpreted as describing a situation where the cube extends infinitely into the positive and negative z directions. In this case, the "no-flow" condition at the top and bottom surfaces of the cube in Figure 4.4b would not be a consequence of a thermal insulation at these surface – rather, it would be a consequence of the symmetry of the problem which implies that there can be no temperature gradients in the z direction. This is why no-flow conditions are often also referred to as *symmetry boundary conditions*. Finally, we remark that substantial reductions of the complexity of a PDE problem can also be achieved in the case of periodic media. For periodic media such as the medium shown in Figure 4.3, it will usually be sufficient to compute the solution of a PDE on a single periodicity cell along with appropriate *periodic boundary conditions* at the boundaries of the periodicity cell. An example of this and some further explanations is given in Section 4.10.2.1.

4.4
Closed Form Solutions

In Section 3.6, we have seen that ODEs can be solved either in closed form or numerically. As was discussed there, closed form solutions can be expressed in terms of well-known functions such as the exponential function and the sine function, while the numerical approach is based on the approximate solution of the equations on the computer. All these hold for PDEs as well, including the fact that closed form solutions cannot be obtained in most cases – they are like "dust particles in the ODE/PDE universe" as discussed in Section 3.7.4. Since PDEs involve derivatives with respect to several variables and, thus, can express a much more complex dynamical behavior compared to ODEs, it is not surprising that one can generally say that it is even harder to find closed form solutions of PDEs compared to ODEs.

Nevertheless, as was explained in Section 3.6, it is always a good idea to look for closed form solutions of differential equations since they may provide valuable information about the dependence of the solution on the parameters of the system under investigation. A great number of techniques to derive closed form solutions of differential equations – PDEs as well as ODEs – has been developed, but an exhaustive treatment of this topic is beyond the scope of a first introduction into mathematical modeling. Since closed form solutions are unavailable in most cases, we will confine ourselves here to an example derivation of a closed form solution for the one-dimensional heat equation. Readers who want to know more on closed form solution techniques are referred to appropriate literature such as [152].

4.4.1
Problem 1

So let us reconsider the one-dimensional heat equation now (Equation 4.1 in Section 4.2):

$$\frac{\partial T}{\partial t} = \frac{K}{C\rho} \cdot \frac{\partial^2 T}{\partial x^2} \qquad (4.49)$$

As discussed in Section 4.3.2, this equation needs boundary conditions for the space variables and an initial condition at $t = 0$. Assuming $(0, L)$ as the spatial domain in which Equation 4.49 is solved (that is, $x \in (0, L)$), let us consider the following initial and boundary conditions:

$$T(0, t) = 0 \quad \forall t \geq 0 \qquad (4.50)$$

$$T(L, t) = 0 \quad \forall t \geq 0 \qquad (4.51)$$

$$T(x, 0) = T_0 \quad \forall x \in (0, L) \qquad (4.52)$$

where $T_0 \in \mathbb{R}$ is a constant initial temperature. Note that this corresponds to *Problem 1* in Section 4.1.3 with $T_i(x) = T_0$ and $T_0 = T_1 = 0$. As was explained there, you can imagine, for example, a cylindrical body as in Figure 4.1a with an initial, constant temperature T_0, the ends of this body being in contact with ice water after $t = 0$.

4.4.2
Separation of Variables

The above problem can now be solved using a separation of variables technique [111]. Note that a similar technique has been described for ODEs in Section 3.7.2. This method assumes a solution of the form

$$T(x, t) = a(x) \cdot b(t) \qquad (4.53)$$

where the variables x and t involve separate functions a and b. Substituting Equation 4.53 in Equation 4.49 you get

$$\frac{db(t)/dt}{\frac{K}{C\rho} \cdot b(t)} = \frac{d^2 a(x)/dx^2}{a(x)} \qquad (4.54)$$

Both sides of the last equation must equal some constant $-k^2$ with $k \neq 0$ (it can be shown that T would be identically zero otherwise). This leads to a system of two uncoupled ODEs:

$$\frac{db(t)}{dt} + k^2 \frac{K}{C\rho} b(t) = 0 \qquad (4.55)$$

$$\frac{d^2 a(x)}{dx^2} + k^2 a(x) = 0 \tag{4.56}$$

Equation 4.55 can be solved using the methods in Section 3.7, which leads to

$$b(t) = A e^{-k^2 K / C \rho t} \tag{4.57}$$

where A is some real constant. The general solution of Equation 4.56 is

$$a(x) = B_1 \cdot \sin(kx) + B_2 \cdot \cos(kx) \tag{4.58}$$

where B_1 and B_2 are real constants. Equation 4.50 implies

$$B_2 = 0 \tag{4.59}$$

and Equation 4.51 gives

$$kL = n\pi \tag{4.60}$$

or

$$k = k_n = \frac{n\pi}{L} \tag{4.61}$$

for some $n \in \mathbb{N}$. Thus for any $n \in \mathbb{N}$, we get a solution of Equation 4.56:

$$a(t) = B_1 \cdot \sin(k_n x) \tag{4.62}$$

Using Equations 4.53, 4.57, and 4.62 gives a particular solution of Equations 4.49–4.51 for any $n \in \mathbb{N}$ as follows:

$$T_n(x, t) = A_n \cdot \sin(k_n x) \cdot e^{-k_n^2 \frac{K}{C\rho} t} \tag{4.63}$$

Here, A_n is a real constant. The general solution of Equations 4.49–4.51 is then obtained as a superposition of the particular solutions in Equation 4.63:

$$T(x, t) = \sum_{n=1}^{\infty} T_n(x, t) = \sum_{n=1}^{\infty} A_n \cdot \sin(k_n x) \cdot e^{-k_n^2 \frac{K}{C\rho} t} \tag{4.64}$$

The coefficients A_n in the last equation must then be determined in a way such that the initial condition in Equation 4.52 is satisfied:

$$T_0 = T(x, 0) = \sum_{n=1}^{\infty} A_n \cdot \sin(k_n x) \tag{4.65}$$

This can be solved for the A_n using Fourier analysis as described in [111], which finally gives the solution of Equations 4.49–4.52 as follows:

$$T(x, t) = \sum_{n=1,3,5,\ldots}^{\infty} \frac{4T_0}{n\pi} \cdot \sin(k_n x) \cdot e^{-k_n^2 \frac{K}{C\rho} t} \tag{4.66}$$

The practical use of this expression is, of course, limited since it involves an infinite sum, which can only be evaluated approximately. An expression such as Equation 4.66 is at what might be called the *borderline* between closed from solutions and numerical solutions, which will be discussed in the subsequent sections. The fact that we are getting an infinite series solution for a problem such as Equations 4.49–4.52 – which certainly is among the most elementary problems imaginable for the heat equation – confirms our statement that numerical solutions are even more important in the case of PDEs compared to ODEs.

4.4.3
A Particular Solution for Validation

Still, closed form solutions are of great use, for example, as a test of the correctness of the numerical procedures. Remember that a comparison with closed form solutions of ODEs was used in Section 3.8 to demonstrate the correctness of the numerical procedures that were discussed there (e.g. Figure 3.9 in Section 3.8.2). In the same way, we will use the particular solution $T_1(x, t)$ from Equation 4.63 as a means to validate the finite difference method that is described in Section 4.6. Assuming $A_1 = 1$ in Equation 4.63, T_1 turns into

$$T^*(x, t) = \sin\left(\frac{\pi}{L}x\right) \cdot e^{-\frac{\pi^2}{L^2}\frac{K}{C\rho}t} \tag{4.67}$$

which is the closed form solution of the following problem:

$$\frac{\partial T}{\partial t} = \frac{K}{C\rho} \cdot \frac{\partial^2 T}{\partial x^2} \tag{4.68}$$

$$T(0, t) = 0 \quad \forall t \geq 0 \tag{4.69}$$

$$T(L, t) = 0 \quad \forall t \geq 0 \tag{4.70}$$

$$T(x, 0) = T^*(x, 0) = \sin\left(\frac{\pi}{L}x\right) \quad \forall x \in (0, L) \tag{4.71}$$

4.5
Numerical Solution of PDE's

Remember the discussion of the Euler method in Section 3.8.1.1. There, we wanted to solve an initial-value problem for a general ODE of the form

$$y'(x) = F(x, y(x)) \tag{4.72}$$

$$y(0) = a \tag{4.73}$$

Analyzing the formulas discussed in Section 3.8.1, you will find that the main idea used in the Euler method is the approximation of the derivative in Equation 4.72

by the difference expression

$$y'(x) \approx \frac{y(x+h) - y(x)}{h} \tag{4.74}$$

Using similar difference expressions to approximate the derivatives, the same idea can be used to solve PDEs, and this leads to a class of numerical methods called *finite difference methods* (often abbreviated as *FD methods*) [140]. An example application of the FD method to the one-dimensional heat equation is given in Section 4.6. The application of the FD method is often the easiest and most natural thing to do in situations where the computational domain is geometrically simple, for example, where you solve an equation on a rectangle (such as the domain $[0, L] \times [0, T]$ used in Section 4.6).

If the computational domain is geometrically complex, however, the formulation of FD methods can be difficult or even impossible since FD methods always need regular computational grids to be mapped onto the computational domain. In such situations, it is better to apply e.g. the *finite-element method* (often abbreviated as the *FE method*) described in Section 4.7. In the FE method, the computational domain is covered by a grid of approximation points that do not need to be arranged in a regular way as it is required by the FD method. These grids are often made up of triangles or tetrahedra, and it is seen later that they can be used to describe even very complex geometries.

Similar to the numerical methods solving ODEs discussed in Section 3.8.1, numerical methods for PDEs are *discretization methods* in the sense that they provide a discrete reformulation of the original, continuous PDE problem. While the discussion in this book will be confined to FD and FE methods, you should note that there is a number of other discretization approaches, most of them specifically designed for certain classes of PDEs. For example, PDEs that are formulated as an initial-value problem in one of its variables can be treated by the *method of lines*, which basically amounts to a reformulation of the PDE in terms of a system of ODEs or differential-algebraic equations [153]. *Spectral methods* are a variant of the FE method involving nonlocal basis functions such as sinusoids or Chebyshev polynomials, and they work best for PDEs having very smooth solutions as described in [154]. *Finite volume methods* are often applied to PDEs based on conservation laws, particularly in the field of CFD [155, 156].

4.6
The Finite Difference Method

4.6.1
Replacing Derivatives with Finite Differences

The FD method will be introduced now referring to the one-dimensional heat equation 4.49 (see Section 4.2):

$$\frac{\partial T}{\partial t} = \frac{K}{C\rho} \cdot \frac{\partial^2 T}{\partial x^2} \tag{4.75}$$

As was already mentioned above, the idea of the FD method is similar to the idea of the Euler method described in Section 3.8.1.1: a replacement of the derivatives in the PDE by appropriate difference expression. In Equation 4.75, you see that we need numerical approximations of a first-order time derivative and of a second-order space derivative. Similar to the above derivation of the Euler method, the time derivative can be approximated as

$$\frac{\partial T(x,t)}{\partial t} \approx \frac{T(x,t+\Delta t) - T(x,t)}{\Delta t} \tag{4.76}$$

if Δt is a sufficiently small *time step* (which corresponds to the stepsize h of the Euler method). To derive an approximation of the second-order space derivative, let Δx be a sufficiently small *space step*. Then, second-order Taylor expansions of $T(x,t)$ give

$$T(x+\Delta x,t) = T(x,t) + \frac{\partial T(x,t)}{\partial x}\Delta x + \frac{1}{2}\frac{\partial^2 T(x,t)}{\partial x^2}(\Delta x)^2 + \cdots \tag{4.77}$$

$$T(x-\Delta x,t) = T(x,t) - \frac{\partial T(x,t)}{\partial x}\Delta x + \frac{1}{2}\frac{\partial^2 T(x,t)}{\partial x^2}(\Delta x)^2 - \cdots \tag{4.78}$$

Adding these two equations, a so-called *central difference approximation* of the second-order space derivative is obtained:

$$\frac{T(x,t+\Delta t) - T(x,t)}{\Delta t} \approx \frac{T(x+\Delta x,t) + T(x-\Delta x,t) - 2T(x,t)}{(\Delta x)^2} \tag{4.79}$$

The right-hand sides of Equations 4.76 and 4.79 are called *finite difference approximations* of the derivatives. As discussed in [157], a great number of finite difference approximations can be used for any particular equation. An important criterion for the selection of finite difference approximations is their *order of accuracy*. For example, it can be shown that the approximation in Equation 4.79 is fourth order accurate in Δx, which means that the difference between the left- and right-hand side in Equation 4.79 can be expressed as $c_1 \cdot (\Delta x)^4 + c_2 \cdot (\Delta x)^6 \cdots$ where $c_1, c_2, \ldots$ are constants [157]. Hence, the approximation error made in Equation 4.79 decreases quickly as $\Delta x \to 0$. Note that the high accuracy of this approximation is related to the fact that all odd terms cancel out when we add the two Taylor expansions (4.77) and (4.78). Substituting Equations 4.76 and 4.79 in Equation 4.75 gives

$$\frac{T(x,t+\Delta t) - T(x,t)}{\Delta t} \approx \frac{K}{C\rho} \cdot \frac{T(x+\Delta x,t) + T(x-\Delta x,t) - 2T(x,t)}{(\Delta x)^2} \tag{4.80}$$

4.6.2
Formulating an Algorithm

Assume now we want to solve Equation 4.75 for $t \in (0, T]$ and $x \in (0, L)$ where

$$T = N_t \cdot \Delta t \tag{4.81}$$

$$L = N_x \cdot \Delta x \tag{4.82}$$

for $N_t, N_x \in \mathbb{N}$. Let us define

$$x_i = i \cdot \Delta x, \quad i = 0, \ldots, N_x \tag{4.83}$$

$$t_j = j \cdot \Delta t, \quad j = 0, \ldots, N_t \tag{4.84}$$

$$T_{i,j} = T(x_i, t_j), \quad i = 0, \ldots, N_x, j = 0, \ldots, N_t \tag{4.85}$$

Then, Equation 4.80 suggests the following approximation for $i = 1, \ldots, N_x - 1$ and $j = 0, \ldots, N_t - 1$:

$$T_{i,j+1} = T_{i,j} + \eta \left(T_{i+1,j} + T_{i-1,j} - 2T_{i,j} \right) \tag{4.86}$$

where

$$\eta = \frac{K \Delta t}{C \rho \Delta x^2} \tag{4.87}$$

We will now use Equation 4.86 to solve Equations 4.68–4.71, since the closed form solution of these equations is the known expression $T^*(x, t)$ in Equation 4.67, which can then be used to assess the accuracy of the numerical procedure. Using the notation of this section, Equations 4.69–4.71 can be written as follows:

$$T_{0,j} = 0, \quad j = 0, \ldots, N_t \tag{4.88}$$

$$T_{N_x,j} = 0, \quad j = 0, \ldots, N_t \tag{4.89}$$

$$T_{i,0} = \sin \left(\frac{\pi}{L} i \Delta x \right), \quad i = 0, \ldots, N_x \tag{4.90}$$

Let us now write down Equation 4.86 for $j = 0$:

$$T_{i,1} = T_{i,0} + \eta \left(T_{i+1,0} + T_{i-1,0} - 2T_{i,0} \right) \tag{4.91}$$

On the right-hand side of the last equation, everything is known from the initial condition, Equation 4.90. Hence, Equation 4.91 can be used to compute $T_{i,1}$ for $i = 1, \ldots, N_x - 1$ (note that $T_{0,1}$ and $T_{N_x,1}$ are known from Equations 4.88 and 4.89). Then, all $T_{i,j}$ with $j = 0$ or $j = 1$ are known and we can proceed with Equation 4.86 for $j = 1$:

$$T_{i,2} = T_{i,1} + \eta \left(T_{i+1,1} + T_{i-1,1} - 2T_{i,1} \right) \tag{4.92}$$

Again, everything on the right-hand side is known, so we get $T_{i,2}$ for $i = 1, \ldots, N_x - 1$ from the last equation. Proceeding in this way for $j = 2, 3, \ldots$, all $T_{i,j}$ for $i = 0, \ldots, N_x$ and $j = 0, \ldots, N_t$ are obtained successively.

4.6.3
Implementation in R

Obviously, the procedure just described consists of the application of two nested iterations, or, in the language of programming languages: it consists of two nested *loops*. The first (outer) loop advances j from 0 to $N_t - 1$, and within this loop there is a second loop, which advances i from 1 to $N_x - 1$. The entire algorithm is realized in the *R*-program HeatClos.r which is a part of the book software. The essential part of this code consists of the two loops just mentioned:

```
1: for (j in 0:Nt){
2: for (i in 2:Nx)
3: T[i,2]=T[i,1]+eta*(T[i+1,1]+T[i-1,1]-2*T[i,1])
4: for (i in 2:Nx)
5: T[i,1]=T[i,2]
6: ...
7: }
```
(4.93)

The outer loop over j is defined in line 1 using *Rs* for command, which works similar to *Maximas* for command that was discussed in Section 3.8.1.2. This commands iterates anything enclosed in brackets "{...}" in lines 1–7, using successive values $j = 0, j = 1, \ldots, j = $ Nt. The inner loop over i is in lines 2–3. The range of both the loops is slightly different from the ranges discussed above for purely technical reasons (*R* does not allow arrays having index "0", and hence all indices in HeatClos.r begin with "1" instead of "0"). The temperatures are computed in line 3, which is in exact correspondence with Equation 4.86, except for the fact that the index "2" is used here instead of "$j + 1$" and index "1" instead of j. This means that HeatClos.r stores the temperatures at two times only, the "new" temperatures in T[i,2] and the "old" temperatures in T[i,1]. This is done to save memory since one usually wants to have the temperature at a few times only as a result of the computation. In HeatClos.r, a vector to defines these few points in time and the 6 of the code (which we have left out here for brevity). After each time step, the next time step is prepared in lines 4–5 where the new "old" temperature is defined as the old "new" temperature.

Figure 4.6 shows the result of the computation. Obviously, there is a perfect coincidence between the numerical solution (lines) and the closed form solution, Equation 4.67. As was mentioned above, you can imagine Figure 4.6 to express the temperature distributions within a one-dimensional physical body such as the cylinder shown in Figure 4.1a. At time 0, the top curve in the figure shows a sinusoidal temperature distribution within that body, with temperatures close to $100\,°C$ in the middle of that body and temperatures around $0\,°C$ at its ends. The ends of the body are kept at a constant temperature of $0\,°C$ according to Equations 4.69 and 4.70. Physically, you can imagine the ends of that body to be in contact with ice water. In such a situation, physical intuition tells us that a continuous heat flow will leave that body through its ends, accompanied by a continuous decrease of the temperature within the body until $T(x) = 0$ ($x \in [0, 1]$) is achieved. Exactly this can be seen in the figure, and so you see that this solution of the heat equation corresponds with our physical intuition. Note that the material parameters used in this example (Figure 4.6 and HeatClos.r) correspond to a body made up of aluminum.

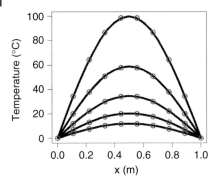

Fig. 4.6 Solution of Equations 4.68–4.71 for $K = 210$ (cal/$^\circ$C/g), $C = 900$ (cal $^\circ$C g s), $\rho = 2700$ (g/cm^3), and $L = 1$. Lines: numerical solution computed using HeatClos.r based on the finite difference scheme, Equation 4.86, at times $t = 0, 625, 1250, 1875, 2500$ s, which correspond to the lines from top to bottom. Circles: closed form solution, Equation 4.67.

4.6.4
Error and Stability Issues

In the general case, you will not have a closed form solution, which could be used as in Figure 4.6 to assess the accuracy of the numerical solution. To control the error within FD algorithms, you may look for error controlling algorithms that apply to your specific FD scheme in the literature [3, 138, 150, 157]. Alternatively, you may use the heuristic procedure described in Note 3.8.1, which means in this case that you compute the solution first using Δx and Δt, and then the second time using $\Delta x/2$ and $\Delta t/2$. As explained in Section 3.8.2.1, you may assume a sufficiently precise result if the differences between these two solutions are negligibly small. Any newly written FD program, however, should first be checked using a closed form solution similar to above.

Looking into HeatClos.r, you will note that the time step Δt – which is denoted as Dt within HeatClos.r – is chosen depending on the space step Δx as follows:

$$\Delta t = \frac{C\rho\Delta x^2}{8K} \tag{4.94}$$

If you increase Δt above this value, you will note that beyond a certain limit the solution produced by HeatClos.r becomes numerically unstable, that is, it exhibits wild unphysical oscillations similar to those that we have seen in the discussion of stiff ODEs in Section 3.8.1.4. The stability of the solution, hence, is an issue for all kinds of differential equation including PDEs. If you apply finite difference schemes such as Equation 4.86, you should always consult appropriate literature such as [140] to make sure that your computations are performed within the stability limits of the particular method you are using. In the case of the FD scheme Equation 4.86, a *von Neumann stability analysis* shows that stability requires

$$\eta = \frac{K\Delta t}{C\rho\Delta x^2} < \frac{1}{4} \tag{4.95}$$

to be satisfied [111]. This criterion is satisfied by choosing Δt according to Equation 4.94 in `HeatClos.r`.

4.6.5
Explicit and Implicit Schemes

The FD scheme Equation 4.86 is also known as the *leap frog* scheme since it advances the solution successively from time level $j = 1$ to $j = 2, 3, \ldots$, similar to the children's game *leapfrog* [111]. Since the solution at a new time level $j + 1$ can be computed explicitly based on the given right-hand side of Equation 4.86, this FD scheme is called an *explicit FD method*.

An *implicit* variant of the FD scheme Equation 4.86 is obtained if Equation 4.80 is replaced by

$$\frac{T(x, t + \Delta t) - T(x, t)}{\Delta t}$$
$$\approx \frac{K}{C\rho} \cdot \frac{T(x + \Delta x, t + \Delta t) + T(x - \Delta x, t + \Delta t) - 2T(x, t + \Delta t)}{(\Delta x)^2} \tag{4.96}$$

which leads to the approximation

$$\frac{T_{i,j+1} - T_{i,j}}{\Delta t} = \frac{K}{C\rho} \cdot \frac{T_{i+1,j+1} + T_{i-1,j+1} - 2T_{i,j+1}}{\Delta x^2} \tag{4.97}$$

for $i = 1, \ldots, N_x - 1$ and $j = 0, \ldots, N_t - 1$. The last equation cannot be solved for $T_{i,j+1}$ as before. $T_{i,j+1}$ is expressed *implicitly* here since the equation depends on other unknown temperatures at the time level $j + 1$: $T_{i+1,j+1}$ and $T_{i-1,j+1}$. Given some time level j, this means that the equations (4.97) for $i = 1, \ldots, N_x - 1$ are coupled and must be solved as a system of linear equations. Methods of this kind are called *implicit FD methods*. Implicit FD methods are usually more stable compared to explicit methods at the expense of higher computation time and memory requirements (since a system of linear equations must be solved in any time step for the implicit scheme Equation 4.97, while the temperatures in the new time step could be obtained directly from the explicit scheme Equation 4.86). The implicit method Equation 4.97, for example, can be shown to be *unconditionally stable*, which means that no condition such as Equation 4.95 must be observed to ensure stable results without those unphysical oscillations discussed above [140].

The FD method Equation 4.86 is also known as the *FTCS method* [158]. To understand this, consider Equation 4.80 on which Equation 4.86 is based: as was mentioned above, the right-hand side of Equation 4.80 is a *central difference approximation* of the second space derivative, which gives the letters "CS" (Central difference approximation in *Space*) of "FTCS". The left-hand side of Equation 4.80, on the other hand, refers to the time level t used for the approximation of the space derivative and to the time level $t + \Delta t$, that is, to a time level in the forward direction. This gives the letters "FT" (*Forward* difference approximation in *Time*) of "FTCS". The implicit scheme (Equation 4.97) can be described in a similar way, except for the fact that in this case the approximation of the time derivative goes into the backward direction, and this is why Equation 4.97 is also known as the *BTCS method*.

Note that there is a great number of other FD methods beyond the methods discussed so far, most of them specifically tailored for special types of PDEs [138, 150, 157, 158].

4.6.6
Computing Electrostatic Potentials

After solving the heat equation, let us consider a problem from classical electrodynamics: the computation of a two-dimensional electrostatic potential $U(x, y)$ based on a given distribution of electrical charges, $\rho(x, y)$. The electrostatic potential is known to satisfy the following PDE [111]:

$$\frac{\partial^2 U(x, y)}{\partial x^2} + \frac{\partial^2 U(x, y)}{\partial y^2} = -4\pi\rho(x, y) \tag{4.98}$$

According to the classification in Section 4.3.1.3, this is an elliptic PDE which is known as *Poisson's equation*. Assume we want to solve Equation 4.98 on a square $(0, L) \times (0, L)$. To formulate an FD equation, let us assume the same small space step $\Delta = L/N$ in the x and y directions ($N \in \mathbb{N}$). Then, denoting $U_{i,j} = U(i\Delta, j\Delta)$, $\rho_{i,j} = \rho(i\Delta, j\Delta)$ and using the central difference approximation for the derivatives similar as above, the following FD equation is obtained for $i = 1, \ldots, N-1$ and $j = 1, \ldots, N-1$:

$$U_{i,j} = \frac{1}{4}\left(U_{i+1,j} + U_{i-1,j} + U_{i,j+1} + U_{i,j-1}\right) + \pi\rho_{i,j}\Delta^2 \tag{4.99}$$

Note that all U terms in this equation are unknown, and hence this is again a coupled system of linear equations, similar to the implicit BTCS scheme discussed above. Equations of this kind are best solved using the iterative methods that are explained in the next section.

4.6.7
Iterative Methods for the Linear Equations

Systems of coupled linear equations such as Equation 4.99 arise quite generally in the numerical treatment of PDEs, regardless of whether you are using FD methods, FE methods (Section 4.7), or any of the other methods mentioned in

Section 4.5. Principally, problems of this kind can be solved as matrix equations using the standard methods of linear algebra such as Gaussian elimination and LU decomposition techniques [159]. This kind of solution methods are called *direct solution methods* since they solve the general matrix problem $Ax = b$ in a single step, basically by inverting the matrix A. Direct methods, however, are usually expensive in terms of computer memory since you need to keep the matrix (at least the nonzero elements) in the memory, and also in terms of their computational requirements, that is, in terms of the number of arithmetical operations needed until the solution is obtained [160]. This is a problem even though the matrices arising in the numerical treatment of PDEs are usually *sparse matrices* in the sense that most matrix entries are zero, which reduces the memory requirements.

Thus, linear equations derived from PDEs are usually solved based on another class of methods called *iterative methods*. These methods start with an initial guess for the unknowns, which is then successively improved until a sufficiently precise approximation of the solution of the PDE is achieved. In the above example, we could start with an initial guess $U_{i,j}^{(0)}$ $(i,j = 0, \ldots, N)$ and then use Equation 4.99 to compute $U_{i,j}^{(1)}$, $U_{i,j}^{(2)}, \ldots$ from

$$U_{i,j}^{(k+1)} = \frac{1}{4} \left(U_{i+1,j}^{(k)} + U_{i-1,j}^{(k)} + U_{i,j+1}^{(k)} + U_{i,j-1}^{(k)} \right) + \pi \rho_{i,j} \Delta^2 \qquad (4.100)$$

This particular iterative method is called the *Jacobi method*, but there are more effective methods such as the *Gauss–Seidel method* and *relaxation methods*, which are discussed in [111, 160]. In [111] you may find an example application of Equation 4.99, along with a program solving this equation using a relaxation method.

> **Note 4.6.1 (Nested iterations)** The numerical solution of nonlinear PDEs may involve several nested iterations (which often can be observed in the output of PDE solving software): for example, outer iterations resolving the time scale (for instationary equations) and/or the nonlinearity (Section 4.3.1.3), and an inner iteration solving the resulting system of linear equations.

4.6.8
Billions of Unknowns

As was just mentioned, iterative methods are used mainly due to the large size of the linear equation systems derived from PDEs, which may increase the memory and computation time requirements of direct methods beyond all limits. Referring to the FD approach just discussed, it can be demonstrated why the linear equation systems derived from PDEs can be so large. Assume we want to use the FD method to compute some unknown quantity U in the cube $[0, 1]^3$. Let us assume that we use the same spatial step $\Delta = 1/N$ $(N \in \mathbb{N})$ in all three space directions. Then, an application of the FD method similar to above would generate a linear equation system with about N^3 unknowns $U_{i,j,k}$ (the real number of unknowns may be slightly different due to the application of the appropriate boundary conditions.) The value of N will depend on your application. $N = 10$ may be sufficient in some cases, which would give about $N^3 = 1000$ unknowns. If, however, your application requires a very fine resolution, you might need $N = 100$ and then come up with an overall number of unknowns in the order of millions.

Now you should note that this example refers to the computation of only *one* state variable U. Many applications require the solution of systems of PDEs involving the computation of several state variables on your computational grid, for example, the pressure and several velocity components in CFD applications. In the $N = 1000$ case, you would get a billion unknowns for each of your state variables. Then, you may want to observe the development of your state variables in time, that is, you may want to solve your problem as an instationary problem, which means you have to solve problems involving billions of unknowns in every single time step of your numerical algorithm. In applications of the FE method (Section 4.7), your geometry may involve tiny structures that need to be resolved by a very fine computational grid, and again every point of that fine grid will increase the number of the unknowns. In this way, your PDE application may easily generate far more than a billion unknowns, and hence it is not surprising that the solution of PDEs can be very expensive in terms of memory requirements and computation time.

Note 4.6.2 (Need for computational efficiency) Computational efficiency is an important issue in the numerical treatment of PDEs. It involves a proper analysis of the dimensionality and symmetry of the problem (Section 4.3.3) as well as the use of fast and efficient iterative methods to solve the equations.

Note that $N = 100$ would generate only about $10\,000$ unknowns (instead of a million) if you can solve the problem in 2D (instead of 3D), or about $500\,000$ unknowns would suffice if you can do your computation in half of the original domain due to symmetry considerations.

4.7
The Finite-Element Method

As was discussed in Section 4.5, a main disadvantage of the FD method is its lack of geometrical flexibility. It always needs regular computational grids to be mapped onto the computational domain, which restricts the complexity of the geometries that can be treated. The finite-element method (FE method), on the other hand, is particularly well suited for complex geometries. It uses grids made up of simple geometrical forms such as triangles in 2D or tetrahedra in 3D. These simple geometrical forms are called *finite elements*, and by putting together a great number of these finite elements virtually any imaginable geometrical object can be

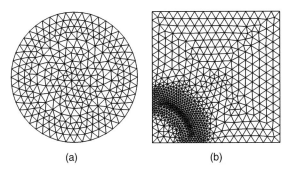

(a)   (b)

Fig. 4.7 (a) Example triangulation of a circle. (b) Example of a locally refined mesh.

constructed. As an example, Figure 4.7a shows the construction of a circular domain using triangles as finite elements. A construction as in Figure 4.7a is called a *mesh* or *grid*, and the generation of a mesh covering the geometry is an important step of the finite-element method (see the discussion of mesh generation using software in Section 4.8). If a 2D grid is constructed using triangles similar to Figure 4.7a, the resulting mesh is also called a *triangulation* of the original geometry.

Note that the geometry shown in Figure 4.7a cannot be covered by the *structured grids* that are used in the FD method (Section 4.6). The grid shown in Figure 4.7a is called an *unstructured grid*, since the crossing points of the grid – which are also called the *knots* of the grid – do not have any kind of regular structure (e.g. in contrast to the FD grids, they do not imply any natural ordering of these crossing points). Similar to the FD method, the crossing points of an FE mesh are the points where the numerical algorithm produces an approximation of the state variables (more precisely, weights of basis functions are determined at the crossing points, see below). Depending on your application, you might need a higher or lower resolution of the numerical solution – and hence a higher or lower density of those crossing points – in different regions of the computational domain. This can be realized fairly easily using unstructured FE meshes. Figure 4.7b shows an example where a much finer grid is used in a circular strip. Such meshes are called *locally refined meshes*. Locally refined meshes can not only be used to achieve a desired resolution of the numerical solution but may also help to circumvent inherent problems with the numerical algorithms [161–163].

4.7.1
Weak Formulation of PDEs

The difference between the computational approaches of the FD and FE methods can be phrased as follows:

Note 4.7.1 The FD method approximates the equation and the FE method approximates the solution.

To make this precise, consider the Poisson equation in 2D (Section 4.6.6):

$$\frac{\partial^2 U(x, y)}{\partial x^2} + \frac{\partial^2 U(x, y)}{\partial y^2} = f(x, y) \tag{4.101}$$

where $(x, y) \in \Omega \subset \mathbb{R}^2$ and f is some real function on Ω. Now following the idea of the FD approach and using the central difference approximation described in Section 4.6.1, this equation turns into the following difference equation:

$$\frac{U(x + \Delta x, y) + U(x - \Delta x, y) - 2U(x, y)}{\Delta x^2}$$
$$+ \frac{U(x, y + \Delta y) + U(x, y - \Delta y) - 2U(x, y)}{\Delta y^2} = f(x, y) \tag{4.102}$$

Equation 4.102 approximates Equation 4.101, and in this sense it is valid to say that "the FD method approximates the equation". Now to understand what it means that the "FE method approximates the solution", let us consider a 1D version of Equation 4.101:

$$u''(x) = f(x) \quad x \in (0, 1) \tag{4.103}$$
$$u(0) = u(1) = 0 \tag{4.104}$$

Of course, this is now an ODE, but this problem is, nevertheless, very well suited to explain the idea of the FE method in simple terms. Let v be a smooth function which satisfies $v(0) = v(1) = 0$. Here, "smoothness" means that v is assumed to be infinitely often differentiable with compact support, which is usually denoted as $v \in C_0^\infty(0, 1)$, see [138] (we do not need to go into more details here since we are not going to develop a theory of PDEs). Equation 4.103 implies

$$\int_0^1 u''(x)v(x)\, dx = \int_0^1 f(x)v(x)\, dx \tag{4.105}$$

Using $v(0) = v(1) = 0$, an integration by parts of the left-hand side gives

$$-\int_0^1 u'(x)v'(x)\, dx = \int_0^1 f(x)v(x)\, dx \tag{4.106}$$

Let $H_0^1(0, 1)$ be a suitable set of real functions defined on $(0, 1)$. $H_0^1(0, 1)$ is a so-called *Sobolev space*, which is an important concept in the theory of PDEs (see the precise definitions in [138]). Defining

$$\phi(u, v) = \int_0^1 u'(x)v'(x)\, dx \tag{4.107}$$

for $u, v \in H_0^1(0, 1)$, it can be shown that the following problem is uniquely solvable [80]:

Find $u \in H_0^1(0, 1)$ such that
$$\forall v \in H_0^1(0, 1): -\phi(u, v) = \int_0^1 f(x)v(x)\, dx \tag{4.108}$$

Problem (4.108) is called the *variational formulation* or *weak formulation* of Equations 4.103 and 4.104. Here, the term *weak* is motivated by the fact that some of the solutions of problem (4.108) may not solve Equations 4.103 and 4.104 since they do not have a second derivative in the usual sense, which is needed in Equation 4.103. Instead, solutions of problem (4.108) are said to be *weakly differentiable*, and these solutions themselves are called *weak solutions* of the PDE in its original formulation, that is, of Equations 4.103 and 4.104. Although we confine ourselves to a discussion of Equations 4.103 and 4.104 here, you should note that similar weak formulations can be derived for linear second-order PDEs in general, and these weak formulations can then also be approximated in a similar way using finite elements as will be explained below [138].

4.7.2
Approximation of the Weak Formulation

The idea of the finite-element method is to replace the infinite-dimensional space $H_0^1(0, 1)$ by a finite-dimensional subspace $V \subset H_0^1(0, 1)$, which turns Equation 4.108 into the following problem:

Find $u \in V$ such that
$$\forall v \in V : -\phi(u, v) = \int_0^1 f(x)v(x)\, dx \tag{4.109}$$

Remember from your linear algebra courses that the *dimension* of a vector space V is $n \in \mathbb{N}$ if there are linear independent *basis vectors* $v_1, \ldots, v_n \in V$ that can be used to express any $w \in V$ as a linear combination $w = a_1 v_1 + \cdots + a_n v_n$ ($a_1, \ldots, a_n \in \mathbb{R}$). For the space $H_0^1(0, 1)$ used in problem (4.108), no such basis can be found which means that this is an infinite-dimensional space [138]. Such a space is unsuitable for a numerical algorithm that can perform only a finite number of steps, and this is why problem (4.108) is replaced by problem (4.109) based on the finite-dimensional subspace V. If $v_1, \ldots, v_n$ is a set of basis functions of V, any function $u \in V$ can be written as

$$u = \sum_{j=1}^{n} u_j v_j \tag{4.110}$$

where $u_1, \ldots, u_n \in \mathbb{R}$. Problem (4.109) can now be written as

Find $u_1, \ldots, u_n \in \mathbb{R}$ such that for $i = 1, \ldots, n$
$$-\sum_{j=1}^{n} \phi(v_i, v_j) \cdot u_j = \int_0^1 v_i(x)f(x)\, dx \tag{4.111}$$

Using the definitions

$$\mathbf{A} = (\phi(v_i, v_j))_{i=1\ldots n, j=1\ldots n} \tag{4.112}$$

$$\mathbf{u} = (u_j)_{j=1\ldots n} \tag{4.113}$$

$$\mathbf{f} = \left(\int_0^1 v_i(x)f(x)\, dx \right)_{i=1\ldots n} \tag{4.114}$$

problem (4.111) can be written in matrix form as follows:

$$
\begin{aligned}
&\text{Find } \mathbf{u} \in \mathbb{R}^n \text{ such that} \\
&\mathbf{A} \cdot \mathbf{u} = \mathbf{f}
\end{aligned}
\tag{4.115}
$$

The FE method thus transforms the original PDE Equations 4.103 and 4.104 into a *system of linear equations*, as was the case with the FD method discussed above.

Remember that it was said above that the FD method approximates the equation expressing the PDE, while the FE method approximates the solution of the PDE (Note 4.7.1). This can now be made precise in terms of the last equations. Note that the derivation of Equation 4.115 did not use any discrete approximations of derivatives, which would lead to an approximation of the equation expressing the PDE similar to the FD method. Instead, Equation 4.110 was used to approximate the solution of the PDE – which lies in the infinite-dimensional Sobolev space $H_0^1(0, 1)$ as explained above – in terms of the finite-dimensional subspace $V \subset H_0^1(0, 1)$.

4.7.3
Appropriate Choice of the Basis Functions

To simplify the solution of Equation 4.115, the basis $v_1, \ldots, v_n$ of V is chosen in a way that turns the matrix A into a *sparse matrix*, that is, into a matrix with most of its entries being zero. Based on a decomposition of $[0, 1]$ into $0 = x_0 < x_1 < x_2 \cdots < x_{n+1} = 1$, one can, for example, use the *piecewise linear functions*

$$
v_k(x) =
\begin{cases}
\dfrac{x - x_{k-1}}{x_k - x_{k-1}} & \text{if } x \in [x_{k-1}, x_k], \\
\dfrac{x_{k+1} - x}{x_{k+1} - x_k} & \text{if } x \in [x_k, x_{k+1}], \\
0 & \text{otherwise}
\end{cases}
\tag{4.116}
$$

Then, the functions v_k for $k = 1, \ldots, n$ span up an n-dimensional vector space V that can be used as an approximation of $H_0^1(0, 1)$ as discussed above. More precisely, these functions span up the vector space of all functions being piecewise linear on the subintervals $[x_{k-1}, x_k]$ ($k = 1, \ldots, n + 1$). In terms of our above discussion, $0 = x_0 < x_1 < x_2 \cdots < x_{n+1} = 1$ defines a decomposition of $[0, 1]$ into finite elements where the subintervals $[x_{k-1}, x_k]$ are the finite elements, and the mesh (or grid) consists of the points $x_0, \ldots, x_{n+1}$.

Figure 4.8 shows an example for $n = 4$. The figure shows four "hat functions" centered at x_1, x_2, x_3, x_4 corresponding to the basis functions v_1, v_2, v_3, v_4, and a function being piecewise linear on the subintervals $[x_{k-1}, x_k]$ ($k = 1, \ldots, 5$), which is some linear combination of the v_k such as

$$
u(x) = \sum_{k=1}^{4} a_k v_k(x)
\tag{4.117}
$$

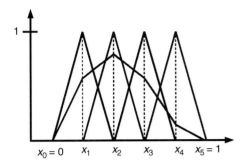

Fig. 4.8 Example one-dimensional "hat" basis functions, and a piecewise linear function generated by a linear combination of these basis functions.

where $a_1, a_2, a_3, a_4 \in \mathbb{R}$. Since v_k is nonzero only on the interval $[x_{k-1}, x_{k+1}]$, the entries of the matrix $\mathbf{A}$ in Equation 4.115

$$\phi(v_i, v_j) = \int_0^1 v_i'(x)v_j'(x)\,dx \tag{4.118}$$

are zero whenever $|i - j| > 1$, which means that these basis functions, indeed, generate a sparse matrix $\mathbf{A}$ as discussed above. Obviously, the sparsity of $\mathbf{A}$ is a consequence of the fact that the basis functions v_k are zero almost everywhere. Since the set of points where a function gives nonzero values is called the *support* of that function, we can say that a basic trick of the FE method is to use basis functions having a small support.

4.7.4
Generalization to Multidimensions

The same procedure can be used for higher-dimensional PDE problems. For example, to solve the Poisson equation (4.101), the first step would be to find an appropriate *weak formulation* of Equation 4.101 similar to Equation 4.108. This weak formulation would involve an appropriate *Sobolev Space* of functions, which would then again be approximated by some finite-dimensional vector space V. As a basis of V, piecewise linear functions could be used as before (although there are many alternatives [138]). A decomposition of $[0, 1]$ into $0 = x_0 < x_1 < x_2 \cdots < x_{n+1} = 1$ was used above to define appropriate piecewise linear functions. In 2D, the corresponding step is a decomposition of the two-dimensional domain into finite elements, similar to the decomposition of a circular domain into triangles in Figure 4.7a. In analogy with the above discussion, one would then use piecewise linear basis functions v_k, which yield 1 on one particular crossing point of the finite-element grid and 0 on all other crossing points. Using linear combinations of these basis functions, arbitrary piecewise linear functions can then be generated. In 3D, the same procedure can be used, for example, based on a decomposition of the computational domain in tetrahedra (Section 4.9.2 and Figure 4.17).

The FE method as described above is an example of a more general approach called the *Galerkin method* [161, 162], which applies also, for example, to the *boundary element method* for solving integral equations [164, 165] or to *Krylov subspace methods* for the iterative solution of linear equation systems [160].

4.7.5
Summary of the Main Steps

The general procedure of the FE method can be summarized as follows:

Main steps of the FE method

1. *Geometry definition:* definition of the geometry of the domain in which the PDE is to be solved.
2. *Mesh generation:* decomposition of the geometry into geometric primitives called *finite elements* (e.g. intervals in 1D, triangles in 2D, and tetrahedra in 3D).
3. *Weak problem formulation:* formulation of the PDE in weak form. Finite-dimensional approximation of the weak problem using the mesh.
4. *Solution:* solution of a linear equation system derived from the weak problem or of an iterated sequence of linear equation systems in the case of nonlinear and/or instationary PDEs (Note 4.6.1).
5. *Postprocessing:* generation of plots, output files, and so on.

The *geometry definition step* corresponds to the definition of the interval [0, 1] in the above discussion of Equations 4.103 and 4.104. If we want to solve, for example, the two-dimensional Poisson equation (4.101) in a circular domain, this step would involve the definition of that circular domain. Although these examples are simple, geometry definition can be quite a complex task in general. For example, if we want to use the PDEs of CFD to study the air flow within the engine of a tractor, the complex geometry of that engine's surface area needs to be defined. In practice, this is done using appropriate CAD software tools. FE software (such as *Salome-Meca*, Section 4.8) usually offers at least some simple CAD tools for geometry construction as well as options for the import of files generated by external CAD software.

The *mesh generation step* involves the decomposition of the geometry into finite elements as discussed above. Although triangles (in 2D) and tetrahedra (in 3D) are used in most practical applications, all kinds of geometrical primitives can be used here in principle. For example, some applications use rectangles or curvilinear shapes in 2D or hexahedra, prisms, or pyramids in 3D. Of course, mesh generation is an integral part of any FE software. Efficient algorithms for this task have been developed. An important issue is the *mesh quality* generated by these algorithms, that is, the compatibility of the meshes with the numerical solution procedures. For example, too small angles in the triangles of a FE mesh can obstruct the numerical solution procedures, and this can be avoided, for example, by the use of *Delaunay triangulations* [166]. Another important aspect of mesh generation is *mesh refinement*. As was discussed above, one might want to use locally refined meshes such as the one shown in Figure 4.7b to achieve the desired resolution of the numerical solution or to avoid problems with the numerical algorithms. FE software such as *Salome-Meca* offers a number of options to define locally refined meshes as required (Section 4.9.2). The mesh generation step may also be coupled with the solution of the FE problem in various ways. For example, some applications require "moving meshes", such as coupled fluid–structure problems where a flowing fluid interacts with a deforming solid structure [167]. Some algorithms use *adaptive mesh refinement* strategies where the mesh is automatically refined or coarsened depending on *a posteriori error estimates* computed from the numerical solution [139].

The *weak problem formulation step* is the most technical issue in the above scheme. This step involves, for example, the selection of basis functions of the finite-dimensional subspace V in which the FE method is looking for the solution. In the above discussion, we used piecewise linear basis functions, but all kinds of other basis functions such as piecewise quadratic or general piecewise polynomial basis functions can also be used [166]. Note that some authors use the term *finite element* as a name for the basis functions, rather than for the geometrical primitives of the mesh. This means that if you read about "quadratic elements", the FE method is used with second-order polynomials as basis functions. If the PDE is nonlinear, the weak problem formulation must be coupled with appropriate linearization strategies [139]. Modern FE software such as *Salome-Meca* (Section 4.8) can be used without knowing too much about the details of the weak problem formulation step. Typically, the software will use reasonable standard settings depending on the PDE type specified by the user, and as a beginner in the FE method it is usually a good idea to leave these standard settings unchanged.

The *solution step* of the FE method basically involves the solution of linear equation systems involving sparse matrices as discussed above. As was mentioned in Section 4.6.7, large sparse linear equation systems are most efficiently solved using appropriate iterative methods. In the case of instationary PDEs, the FE method can be combined with a treatment of the time derivative similar as was done above for the heat equation (Section 4.6). Basically, this means that a sequence of linear equation systems must be solved as we move along the time axis [166]. In the case of nonlinear PDEs, the FE method must be combined with linearization methods such as Newton's method [139], which again leads to the solution of an iterated sequence of linear equation systems. Again, all this as well as the final *postprocessing step* of the FE method is supported by modern FE software such as *Salome-Meca* (Section 4.8).

Note 4.7.2 (Comparison of the FD and FE methods) The FD method is relatively easy to implement (Section 4.6), but is restricted to simple geometries. The FE method can treat complex geometries, but its implementation is a tedious task and so it is usually efficient to use existing software tools such as *Salome-Meca*.

4.8
Finite-element Software

As was mentioned above, the software implementation of the FE method is a demanding task, and this is why most people do not write their own FE programs. A great number of both open-source and commercial FE software packages are available, see Table 4.1 for a list of examples which is by no means exhaustive. Generally speaking, there is a relatively broad gap between open-source and commercial FE software. It is relatively easy to do your everyday office work, for example, using the open-source *OpenOffice* software suite instead of the commercial *Microsoft Office* package, or to perform a statistical analysis using the open-source *R* package instead of commercial products such as *SPSS*, but it is much less easy to work with open-source FE software if you are used to commercial products such as *Fluent* or *Comsol Multiphysics*. Given a particular FE problem, it is highly likely that you will find open-source software that can solve your problem, but you will need time to understand that software before you can use it. Having solved that particular problem, your next problem may be beyond the scope of that software, so you may have to find and understand another suitable open-source FE package. The advantages of commercial FE software can be summarized as follows:

- *Range of application:* commercial FE software usually provides a great number of models that can be used.
- *User-friendliness:* commercial FE software usually provides sophisticated graphical user interfaces, a full integration of all steps of the FE analysis from the geometry definition to the postprocessing step (Section 4.7.5), a user friendly workflow involving, for example, the extraction of parameters from material libraries, and so on.
- *Quality:* commercial FE software usually has been thoroughly tested.
- *Maintenance and support:* commercial FE software usually provides a continuous maintenance and support.

Of course, these points describe the general advantages of commercial software not only in the field of finite-element analysis. All these have particular relevance here due to the complexity of this kind of analysis: an enormous amount of resources is needed to set up general FE software packages such as *Fluent* or *Comsol Multiphysics*.

Table 4.1 Examples of finite-element and finite-volume software packages.

Program	Open source?	Comment
Salome-Meca	Yes	Professional design. Workflow for a limited number of FE models similar to the commercial products. Part of CAELinux, see Appendix A, Section 4.9 and *www.caelinux.com*.
Code_Aster	Yes	Part of the *Salome-Meca* package. ISO 9001 certified FE solver. Focus on structural mechanics problems. Part of CAELinux, see Appendix A, and *www.code-aster.org*.
Code_Saturne	Yes	Computational fluid dynamics (CFD) software based on finite volumes. Part of CAELinux, see Appendix A, Section 4.10.3, and *www.code-saturne.org*
OpenFoam	Yes	Finite-element software. Limited user-friendliness, but large range of applications ranging from complex fluid flows involving chemical reactions, turbulence, and heat transfer to solid dynamics, electromagnetics, and the pricing of financial options. Part of CAELinux, see *www.openfoam.org*.
Many other open-source programs	Yes	Search for "finite element" or "finite volume" at *www.sourceforge.net*.
ABAQUS	No	General FE software, particularly good in structural mechanics applications. See *www.simulia.com*.
Comsol Multiphysics	No	General FE software, particularly suitable for coupled phenomena such as fluid–structure interactions. See *www.comsol.com*.
Fluent	No	General FE software, particularly good in computational fluid dynamics (CFD) applications. See *www.fluent.com*.
LS-DYNA	No	General FE software, particularly good in structural mechanics applications. See *www2.lstc.com*.

Does this mean that one should use commercial software in the field of FE analysis? Yes – and no. Yes: for example, if your daily work involves a large range of applications that is not covered by any particular open-source FE software, if you need fast results, if your time for software training is limited, if you do not have time or enough competences to assess the numerical quality of the results, and if you need software support. No: for example, if you find commercial software too expensive. Looking at the prices of commercial FE software, you will find that the complexity of FE software corresponds to the fact that you have to pay a really substantial amount of money for these packages.

> **Note 4.8.1 (Open-source versus commercial FE software)** Every user of FE software has to trade off the advantages of commercial software mentioned above against the disadvantage of paying a lot of money for this, and a general answer to the question whether one should use open-source or commercial FE software, thus, cannot be given.

For a beginner in FE analysis, however, who just wants to get a first impression of FE software and its general procedures, an answer can be given as follows: try the open-source *Salome-Meca*. It is based on the ISO 9001 certified FE solver *Code_Aster*, and it is currently one of the best available approximations of commercial FE software, for example, in terms of its general workflow that features a full integration of all steps of the analysis beginning with the geometry definition and ending with the postprocessing step as described in Section 4.7.5. If you understand and reproduce the *Salome-Meca* sample session described in the next section, you will have a fairly good idea of the way in which an FE analysis is performed using software.

4.9
A Sample Session Using *Salome-Meca*

The general steps of an FE analysis have been described in Section 4.7.5: geometry definition, mesh generation, weak problem formulation, solution, and postprocessing. These steps define the general structure of a typical software-based FE analysis, and we will go through each of these steps now in the corresponding subsections. We will use *Salome-Meca* – which is available in *CAELinux*, see Appendix A – to solve *Problem 2* from Section 4.1.3 with $T_c = 20\,^\circ$C and $T_s = 0\,^\circ$C:

> **Problem 6:**
> Referring to the configuration in Figure 4.1b and assuming
> - a constant temperature $20\,^\circ$C at the top surface of the cube $(z = 1)$,
> - a constant temperature $0\,^\circ$C at the sphere surface,
> - and a perfect insulation of all other surfaces of the cube,
>
> what is the stationary temperature distribution $T(x, y, z)$ within the cube (i.e. in the domain $[0, 1]^3 \setminus S$ if S is the sphere)?

4.9.1
Geometry Definition Step

4.9.1.1 Organization of the GUI

To start *Salome-Meca* in *CAELinux*, select `CAE-Software/Salome_Meca` under the `PC` button of the desktop (see Appendix A for more details on *CAELinux*). Choose the geometry module using the drop-down list as shown in Figure 4.9a. The remaining steps of the FE analysis discussed in the next sections (except for the problem definition step) will also begin with the selection of an appropriate module from that same drop-down list. Figure 4.9a shows the general structure of the *Salome-Meca* window: the left subwindow shows an *object browser*, which is used to select and explore various items that are generated during the analysis procedure, such as geometrical forms, meshes, solutions, and plots. The right subwindow is used to visualize these items. In the geometry module, an *OCC viewer* is used for visualization. Here, "OCC" refers to the fact that this viewer was developed using *Open CasCade*, a development platform for a variety of 3D graphical applications (*www.opencascade.org*).

Standard *Salome-Meca* will also display a third subwindow called *Python Console*, which we do not need here. If you want to get the same picture as shown in Figure 4.9a, deactivate the Python Console using the menu entry `View/Windows`. *Python* is a general programming language, which can be used in connection with *Salome-Meca* to perform the analysis in *batch mode*, that is, independent of the GUI [168]. This can be effective, for example, in situations where you perform a complex analysis several times, changing just a few parameters of the analysis between subsequent runs. Using the GUI, you would have to do almost the same work (except for those few parameters) again and again. Using the batch mode, on the other hand, you do your work just one time using the GUI, and then save your work in a *Python* script. Editing that script, you can then impose any desired parameter changes and rerun the analysis without using the GUI again. Similar batch mode facilities are provided by most FE software packages.

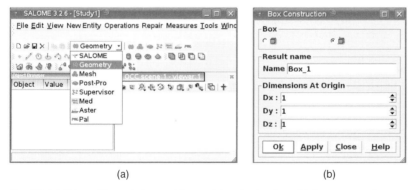

(a) (b)

Fig. 4.9 (a) *Salome-Meca* window and its module selection box. (b) Box construction window.

4.9.1.2 Constructing the Geometrical Primitives

In *Problem* 2, the geometry involves two main ingredients: the cube $[0, 1]^3$ and a sphere with radius 0.1 in the center of the box. The usual procedure now is to construct each of these geometrical primitives one after the other, and then to put them together as required. So let us begin with the construction of the cube, which can be done using the menu option New Entity/Primitives/Box. (Alternatively, you could use the appropriate symbol on *Salome-Meca's* toolbars, but this is omitted here and in the following for the sake of simplicity.) Figure 4.9b shows the *box construction window* that opens up when you click on New Entity/Primitives/Box. In this window, the cube data are entered in a self-explanatory way. After confirming the box construction window, the cube receives a name (Box_1 in this case), which is then used to display and access that geometrical primitive in the object browser window as it is shown in Figure 4.10a. In the right *OCC viewer* subwindow of Figure 4.10a, you can see a 3D picture of Box_1. Note that if you use *Salome-Meca* with its standard settings, this 3D picture will look different, that is, you will have another background color, wireframe graphics, and so on. All this can be changed via the menu File/Preferences or by using the context menu in the *OCC viewer* (right-click of your mouse).

To construct the sphere, we can start similar to above using the menu option New Entity/Primitives/Sphere. This activates the *sphere construction window* shown in Figure 4.11a. In this window, you cannot enter the sphere center in terms of coordinates. Rather, you have to enter the center of the sphere by reference to an existing point object in the object browser window, which is a general principle used in *Salome-Meca*. So, the right procedure is to begin with the definition of the sphere center using New Entity/Basic/Point, which opens up the *point construction window* shown in Figure 4.10b. There you can enter the point coordinates and supply a name for that point. Calling that point SphereCenter, it will appear under that name in the object browser window as shown in Figure 4.11b. Now you can open the sphere construction window using New Entity/Primitives/Sphere and define the sphere center by a mouse click on SphereCenter in the object browser

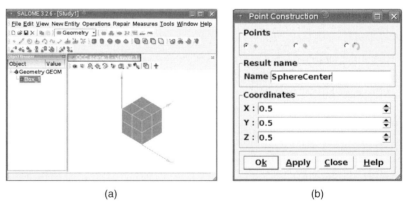

(a)   (b)

Fig. 4.10 (a) *Salome-Meca* showing Box_1. (b) Point construction window.

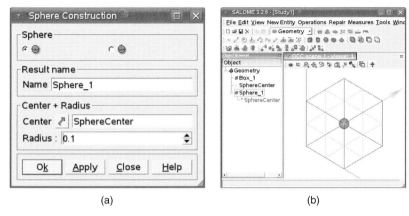

(a)   (b)

Fig. 4.11 (a) Sphere construction window. (b) *Salome-Meca* showing Box_1 in "wireframe" mode and the sphere Sphere_1 inside the box in "shading" mode.

and a subsequent click on the arrow button in the sphere construction window (Figure 4.11a). The same principle is used generally in *Salome-Meca* to import geometrical information into subwindows.

After confirming the sphere construction window, the sphere appears as the new entity Sphere_1 in the object browser. To see the box and the sphere inside the box at the same time, you can use the context menu in the *OCC viewer* to display the box in *wireframe mode* and the sphere in *shading mode*, as it is shown in Figure 4.11b. Other options would have been to display the box in a transparent shading mode, using the *transparency* option in the context menu of the OCC viewer, or to use the *clipping plane* option of the OCC viewer, which allows you to look into the geometry along an arbitrarily oriented plane. See Figure 4.13b for an example application of a clipping plane.

4.9.1.3 Excising the Sphere

So far the box and the sphere have been constructed as separate geometrical items. They are displayed together in Figure 4.11b, but they virtually do not "know" of their mutual existence. For the physical problem expressed in *Problem* 2, it is, of course, important that the cube "knows" that there is a cold sphere in its center. In terms of the discussion in Section 4.3.2, we can say that what we need here is a boundary condition inside the cube that imposes $0\,^{\circ}\text{C}$ at the sphere surface. To be able to impose such a boundary condition, the sphere surface must appear as a boundary of the computational domain, and this means that the computational domain must be what could be phrased in rough terms as "cube minus sphere".

To get this kind of domain, the sphere needs to be *excised* from the cube, and this is again a standard operation that is used quite generally when a finite-element analysis is performed using software. In *Salome-Meca*, this is done in two steps. First, you choose the menu option Operations/Partition, which gives the

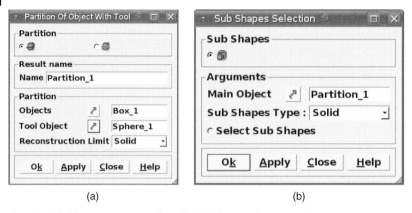

Fig. 4.12 (a) Object partition window. (b) Subshapes selection window.

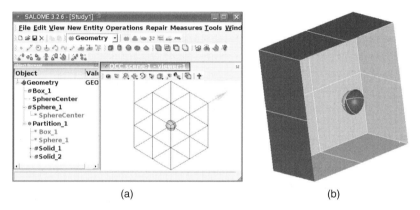

Fig. 4.13 (a) *Salome-Meca* after the definition of Solid_1 and Solid_2. (b) Solid_1 displayed using the clipping plane $x = 0.5$.

object partition window shown in Figure 4.12a. Within this window, there is an object named Partition_1, which involves Box_1 and Sphere_1 as subobjects (verify this using the object browser). These two objects need now to be separated from each other as explained above, which can be done using the menu option New Entity/Explode (a really meaningful name, is it not. . .). This gives the subshapes selection window in Figure 4.12b, along with two new geometrical entities Solid_1 and Solid_2 (see the object browser in Figure 4.13a). Using *Salome_Meca*'s OCC viewer, you can verify that Solid_1 is what we have called *cube minus sphere*, while Solid_2 is the sphere. To achieve this, you can, for example, use the *OCC viewer*'s clipping plane tool mentioned above. After defining a clipping plane at $x = 0.5$, it can be seen that Solid_2 indeed consists of the cube with the sphere being excised from its center (Figure 4.13b).

4.9.1.4 Defining the Boundaries

After this, names need to be assigned to those boundaries of the geometry where boundary conditions are applied, which can then be used later to access these boundaries in a simple way when the boundary conditions are defined (Section 4.9.3). This is done using the menu option New Entity/Group/Create, which brings up the "Create Group" window displayed in Figure 4.14. Using Solid_1 as the main shape and the "select subshapes" button, select the top side of the cube in the *OCC viewer* where a temperature of 20 °C is required (*Problem 6*). This side of the cube can then be given a name in the name field of the "Create Group" window. Let us denote it as CubeTop in the following text. In the same way (referring to Solid_2 as the main shape), a name can be assigned to the sphere surface where a temperature of 0 °C is required in *Problem 2*. This boundary will be denoted as SphereSurf in the following text.

4.9.2
Mesh Generation Step

The next step is mesh generation. Choose the mesh module using the drop-down list shown in Figure 4.9a as before. As the mesh module is started, a second subwindow opens up at the right-hand side of the *Salome-Meca* window on top of the *OCC viewer*: the Visualization Toolkit *VTK 3D viewer*, which is *Salome-Meca*'s default viewer for mesh visualization (Figure 4.16a). The *VTK 3D viewer* is based on an open-source graphics application called *Visualization Toolkit (VTK)*, see *http://www.vtk.org/*. To construct the mesh, choose Mesh/Create Mesh, which opens up the *create mesh window* shown in Figure 4.15a. Within this window, the geometry object needs to be specified for which a mesh is to be created, which is Partition_1 in our case. A number of algorithms can then be used to generate the mesh using the drop-down list supplied in the create mesh window. Within the scope of this book, we are unable to go into a detailed discussion of the various meshing algorithms – the interested reader is referred to specialized literature such as [163]. As a default choice we recommend the NETGEN algorithm developed by Schöberl [169], which is a reliable 3D tetrahedral mesh generator.

Fig. 4.14 Group creation window.

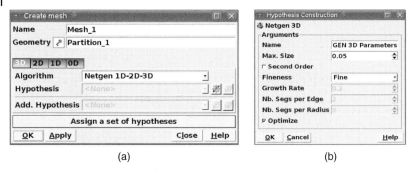

Fig. 4.15 (a) Create mesh window. (b) Hypothesis construction window.

After choosing the algorithm, various parameters can be set in the hypothesis field of the "create mesh" window.

A click on the hypothesis button of the "create mesh" window opens up the *hypothesis construction* window shown in Figure 4.15b. In this window, you can limit the maximal size of the tetrahedra within the mesh, adjust the general fineness of the mesh, and so on. Since the computation time increases substantially with the fineness of the mesh particularly in 3D applications (see the discussion of computational efficiency in Section 4.6.8), these parameters should be chosen with care. Generally, it is a good idea to start with a relatively coarse mesh and then to increase the mesh fineness in small steps as required. For example, a finer mesh may be necessary since a finer resolution of the result is required in your application, or to control the computational error as described in Section 4.6.4).

Choosing "ok" in the create mesh window generates an entity called Mesh_1 in *Salome-Meca*'s object browser (Figure 4.16a). A right mouse click on Mesh_1 brings up the context menu shown in the figure. Choose "Compute" in this menu to start the *mesh computation*. After mesh computation is finished, a mesh statistic is

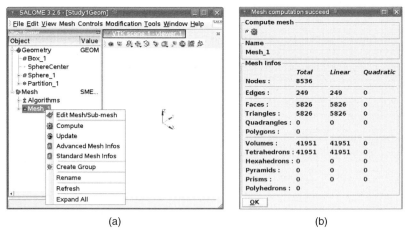

Fig. 4.16 (a) Drop-down list for mesh computation. (b) Mesh statistics.

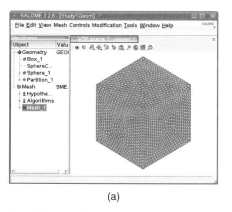
(a)

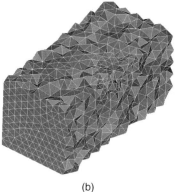
(b)

Fig. 4.17 (a) VTK viewer showing the mesh. (b) Internal structure of the mesh visualized using the clipping planes $z = 0.5$ and $x = 0.5$.

displayed, as shown in Figure 4.16b, if all parameters are chosen as in Figure 4.15. In the mesh statistics, you see, for example, that the mesh consists of a total of 8536 nodes, 41 951 tetrahedrons, and so on. This information can, for example, be used to assess the size of the mesh and to get an idea about the computation time and memory requirements. The mesh is displayed in *Salome-Meca*'s VTK viewer (Figure 4.17a). The VTK viewer can be used for a detailed investigation of the mesh. For example, clipping planes can be used as described above to see the internal structure of the mesh. Figure 4.17b shows an example that has been generated using the clipping planes $z = 0.5$ and $x = 0.5$ in the VTK viewer. The figure shows that the NETGEN algorithm generated a uniform tetrahedral layout of the mesh as required.

4.9.3
Problem Definition and Solution Step

After geometry definition and mesh generation, the next steps in the FE procedure explained in Section 4.7.5 are the *weak problem formulation step* and the *solution step*. In *Salome-Meca*, these two steps are performed within the *Aster module*, which is selected as before using the drop-down list shown in Figure 4.9a. This module is based on the ISO 9001 certified FE solver Code_Aster (Section 4.8). To solve *Problem* 2, choose the menu option `Code_Aster Wizards/Linear Thermal`. This fixes the PDE that is to be solved (the stationary heat equation). The weak problem formulation of this PDE is then done by the software, invisible for a standard user. Experienced users may apply a number of options provided by *Salome-Meca* which affect the weak problem formulation. For example, standard *Salome-Meca* uses piecewise linear shape functions (Section 4.7.3), which can be replaced, for example, by quadratic shape functions using the appropriate options in the "Create Mesh" window (Section 4.9.2). The menu option `Code_Aster Wizards/Linear Thermal` starts a wizard that prompts the user for all information required to set up

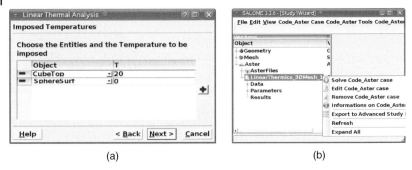

(a)   (b)

Fig. 4.18 (a) Imposing temperature boundary conditions in the linear thermal analysis wizard. (b) Context menu to start the solution engine.

the weak problem. To get the result shown in the next section, enter the following information:

- *Model:* 3D
- *Salome Object:* Geometry
- *Main Shape:* Partition_1
- *Thermal Conductivity:* any value (see below)
- *Imposed Temperatures:* T = 20 for Object = CubeTop and T = 0 for Object = SphereSurf (use the "+"-key of the wizard, see Figure 4.18a)
- *Imposed Flux:* remove everything using the "-"-key of the wizard
- *Applied Sources:* remove everything using the "-"-key of the wizard

Note that any value can be used here for the thermal conductivity since these settings are based on the isotropic stationary heat equation, which can be written as

$$\Delta T = 0 \qquad (4.119)$$

since we have $\partial T / \partial t = 0$ in the stationary case (see Equation 4.3 in Section 4.2). The settings in the "imposed temperatures" option of the wizard are based on the definition of the boundary surfaces CubeTop and SphereSurf in Section 4.9.1.4. The "imposed flux" option of the wizard can be used to impose boundary conditions involving a heat flow across the boundary, while the "applied sources" option refers to situations where sinks or sources of heat are present, that is, where heat is generated or removed somewhere inside the domain [97].

After the wizard is finished, an entity called Aster appears in *Salome-Meca's* object browser (Figure 4.18b). Opening this entity as shown in the figure, the context menu on the item "LinearThermics_3DMesh_1" allows the user to solve the problem via *Solve Code_Aster case.*

4.9.4
Postprocessing Step

The solution procedure described in the last section generates an entity called
Post-Pro in *Salome-Meca*'s object browser (Figure 4.19a). To use this entity for
postprocessing, activate *Salome-Meca*'s *Post-Pro module* using the drop-down list
shown in Figure 4.9a as before. Then, opening the "Post-Pro" entity in the object
browser as shown in Figure 4.19a, the context menu on "0, INCONNUE" allows
you to display the solution in various ways. (The french word "INCONNUE"
reminds us of the fact that *Code_Aster*, the software behind *Salome-Meca*'s Aster
module, is a French development, see Section 4.8.) Figure 4.19b shows how the
plots (an isosurface plot in this case) are displayed in *Salome-Meca*'s VTK 3D viewer.
A number of options can be used to affect the way in which the plots are displayed,
such as arbitrary 3D rotations using the mouse.

Figure 4.20a and b shows the solution displayed using *cut plane plots*, that is,
plots made up of planes inside the domain, which are colored corresponding to
the values of the unknown. Although Figures 4.20a and b show screenshots in
black and white, you can see that the darkest colors – which correspond to the
lowest temperatures – are concentrated around the cold sphere inside the cube,
compare Figure 4.1b and the description of *Problem* 6 above. The various kinds of
plots offered in the postprocessing context menu (Figure 4.19a) can be combined
arbitrarily as required. For example, Figure 4.20c shows a combination of the
isosurface plot from Figure 4.19b with a cut plane plot. Depending on the PDE that
is solved, *Salome-Meca*'s postprocessing module also provides a number of other
classical plots that can be used to visualize PDE solutions, such as arrow plots and
streamline plots.

Since we are restricted to black and white plots here, the solution of *Problem* 6 is
best discussed using isosurface plots. Note that an *isosurface plot* shows a surface
made up of points where some quantity of interest (temperature in this case) attains
some given, fixed value. Figure 4.21 shows a number of isosurface plots where the

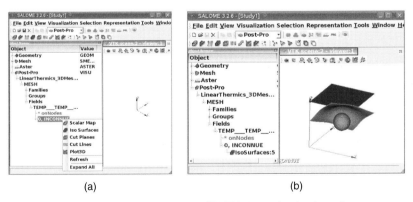

(a) (b)

Fig. 4.19 (a) Postprocessing context menu. (b) VTK viewer showing isosurfaces.

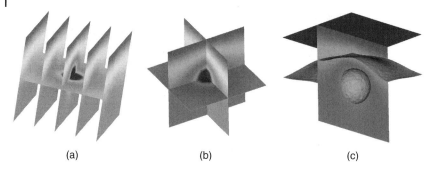

(a) (b) (c)

Fig. 4.20 (a) and (b) Solution of *Problem* 6 displayed using cut planes. (c) Combination of a cut plane with isosurfaces.

temperature of the isosurfaces has been varied in the various subplots, such that we can virtually "move through the solution" along these subplots. Figure 4.21a shows three isosurfaces corresponding to 20, 15, and 10 °C (from top to bottom). Since 20 °C was prescribed on the top surface of the cube in *Problem* 6, the 20 °C isosurface corresponds exactly to this top surface of the cube, and hence this is a flat surface as it can be seen in the figure. The 10 °C isosurface, on the other hand, is an approximately spherical surface, which surrounds the 0 °C sphere in Figure 4.1b like a wrapping. This suggests that the isosurfaces referring to temperatures below 10 °C constitute a sequence of nested (approximate) spheres, that is, the radius of these isosurfaces approaches the radius of the 0 °C sphere in Figure 4.1b as the temperature approaches 0 °C. You can easily verify this using *Salome-Meca*'s Post-Pro module as described above. Finally, the isosurface corresponding to 15 °C in Figure 4.21a is located inside the cube somewhere between the 10 and 20 °C isosurfaces, and its shape can also be described as lying somewhere between the other two isosurfaces. Observing the shape of this intermediate isosurface through the plots in Figure 4.21a–e, you can see how its shape gradually deforms toward a spherical form. Similar to the cut planes discussed above, plots of this kind are very well suited to understand the results of 3D FE computations.

4.10
A Look Beyond the Heat Equation

As was explained above, differential equations arise naturally in science and engineering in many cases where the processes under consideration involve rates of changes of the quantities of interest (Note 3.1.1). Not surprisingly, hence, PDEs are applied in a great number of ways in science and engineering. To give you an idea of PDEs "beyond the heat equation", a few classical applications of PDEs are treated in this section: diffusion and convection processes (Section 4.10.1), porous media flow (Section 4.10.2), computational fluid dynamics (CFD) (Section 4.10.3), and structural mechanics (Section 4.10.4). Among these applications, CFD is perhaps the "most classical" application of PDEs in the sense that it involves

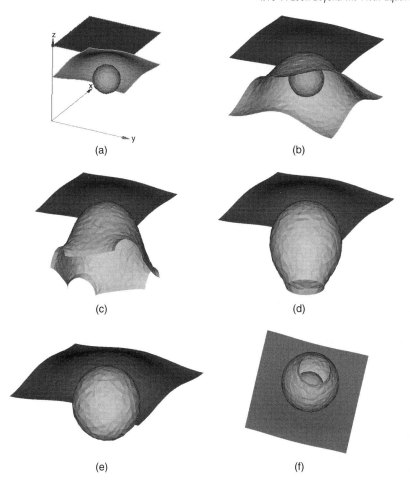

(a)

(b)

(c)

(d)

(e)

(f)

Fig. 4.21 Solution of Problem 6: (a) Iso-planes corresponding to 20, 15, and 10 °C (from top to bottom). (b)–(e) Gradual decrease of the isoplane temperatures (top isoplane temperatures in (b)–(e) are 16, 15, 14, and 13 °C, respectively). (f) A look into (d) from the bottom, showing the spherical isosurface that can be seen in (a) and (b) inside the pear-shaped isosurface.

many applications that have gained attention in the public, such as meteorological flow simulations on the earth's surface, or the simulation of air flow around cars, airplanes and space shuttles. You should note, however, that there is also a great number of PDE applications in fields where you probably would not expect it, which includes applications in economics such as the famous *Black–Scholes equation* that is used as an option pricing model, and which can be transformed into the form of the heat equation [170, 171]. Another aspect that cannot be treated here in sufficient detail is the coupling of differential equations, that is, the coupling of PDEs with PDEs, PDEs with ODEs, or differential equations with other types of mathematical equations. This is an important issue for the simple reason that real systems usually involve a coupling of several phenomena. An example is *fluid–structure interaction* [167], that is, the coupling of the PDEs of CFD with the PDEs of structural mechanics (see sections 4.10.3 and 4.10.4, and the example in [5]) or the coupling of the PDEs describing porous media flow with the "free flow" described by the Navier–Stokes equation (see Sections 4.10.2 and 4.10.3 and the examples in [172, 173]).

4.10.1
Diffusion and Convection

In a sense, a look beyond the heat equation is a look *at* the heat equation: A number of different phenomena can be treated by this equation, which includes general diffusion phenomena as well as porous media flow models (Section 4.10.2). To derive the general diffusion equation, let us write Fourier's law (Equation 4.22), in a different notation as follows:

$$\mathbf{q}(\mathbf{x}, t) = -\mathbf{D}(\mathbf{x}) \cdot \nabla N(\mathbf{x}, t) \tag{4.120}$$

This equation is known as *Fick's first law* [170], where $\mathbf{q} \in \mathbb{R}^3$ is the diffusion flux, for example, in $(\mathrm{g\,m^{-2}\,s^{-1}})$ or $(\mathrm{mol\,m^{-2}\,s^{-1}})$, $\mathbf{D} \in \mathbb{R}^{3 \times 3}$ is the diffusion coefficient matrix, e.g. in $(\mathrm{m^2\,s^{-1}})$, and N is the concentration of a substance, e.g. in $(\mathrm{g\,m^{-3}})$ or $(\mathrm{mol\,m^{-3}})$.

Fick's first law pertains e.g. to a situation where there is a tank filled with a fluid in which some substance is dissolved. Equation 4.120 says that the diffusion flux of the substance is proportional to the gradients of its concentration, and that this flux will always be directed toward regions of low concentration. Note that Fourier's law describes the diffusion of heat in a similar way. In Section 4.2, the heat equation was derived from Fourier's law by an application of the energy conservation principle. Since Fick's first law refers to the diffusion of the mass of a substance rather than to the diffusion of energy, mass conservation must be applied instead of energy conservation here. This can be done in a similar way as was done in Section 4.2.2, that is, by balancing the mass flows in a small control volume, which leads to the *diffusion equation* [174]

$$\frac{\partial N(\mathbf{x}, t)}{\partial t} = \nabla \left(\mathbf{D}(\mathbf{x}) \cdot \nabla N(\mathbf{x}, t) \right) \tag{4.121}$$

Except for different notation and interpretation, this equation is identical with the heat equation 4.23. It can be solved, for example, using *Salome-Meca* and the procedure described in Section 4.9.

Equation 4.121 holds in situations where the fluid is at rest. In many practical cases, however, the fluid will move during the diffusion process, which means that the concentration of the fluid changes due to the combined effect of diffusion and convection. Examples include fermentation processes, such as the wine fermentation process discussed in Section 3.10.2. As it was discussed there, temperature control is an important issue in fermentation processes. Usually, this is achieved by cooling down certain parts inside the tank, and a standard question is which particular configuration of these cooling parts should be used in order to achieve a desired temperature distribution at minimal costs. To answer this question, we need to compute the temperature distribution that results from a given configuration of those cooling parts. To do this, it would *not* be sufficient if we would just apply the heat equation 4.23, since there are substantial movements of the fluid inside the fermenter, which are caused by density gradients as well as by the natural evolution of carbon dioxide bubbles during the fermentation process [175].

A convection flow field $\mathbf{v}(\mathbf{x}) = (v_x(\mathbf{x}), v_y(\mathbf{x}), v_z(\mathbf{x}))$ $(\mathrm{m\,s^{-1}})$ generates a convective flux $\mathbf{J}(\mathbf{x})$ (e.g. in $(\mathrm{g\,m^{-2}\,s^{-1}})$) of the substance, which can be written as [139]

$$\mathbf{J}(\mathbf{x}, t) = \mathbf{v}(\mathbf{x}) N(\mathbf{x}, t) \tag{4.122}$$

Combining Equations 4.120 and 4.122, the overall flux of the substance becomes

$$\mathbf{q}(\mathbf{x}, t) + \mathbf{J}(\mathbf{x}, t) = -\mathbf{D}(\mathbf{x}) \cdot \nabla N(\mathbf{x}, t) + \mathbf{v}(\mathbf{x}) N(\mathbf{x}, t) \tag{4.123}$$

Applying mass balance to this flux as before, the *convection–diffusion equation* is obtained:

$$\frac{\partial N(\mathbf{x}, t)}{\partial t} = \nabla \left(\mathbf{D}(\mathbf{x}) \cdot \nabla N(\mathbf{x}, t) - \mathbf{v}(\mathbf{x}) N(\mathbf{x}, t) \right) \tag{4.124}$$

This equation can be further generalized to include source terms that can be used to describe situations where the substance is supplied or destroyed in certain parts of the tank [139]. To compute the temperature distribution in a fermenter for a given convective flow field $\mathbf{v}(\mathbf{x})$, we would use this equation in the form

$$\frac{\partial T(\mathbf{x}, t)}{\partial t} = \frac{1}{C\rho} \nabla \left(\mathbf{K}(\mathbf{x}) \cdot \nabla T(\mathbf{x}, t) - \mathbf{v}(\mathbf{x}) T(\mathbf{x}, t) \right) \tag{4.125}$$

that is, we would just reinterpret Equation 4.124 in terms of the heat equation 4.23.

To get acquainted with the convection–diffusion equation, you can try R's SoPhy package, a package contributed by Schlather [176]. SoPhy solves the one-dimensional convection–diffusion equation:

$$\frac{\partial N(z, t)}{\partial t} = D \frac{\partial^2 N(z, t)}{\partial z^2} - v \frac{\partial N(z, t)}{\partial z} \tag{4.126}$$

Here, z corresponds to the vertical direction, v is the convective flow velocity in the z direction, and N is, for example, the concentration of a pollutant that is dissolved in water. See [177, 178] for the application of SoPhy to pollutant transport in soils.

4.10.2
Flow in Porous Media

A great number of systems in nature and technology involve porous media. For example, the soil below our feet is a good example of a porous medium: it consists of solid material such as sand and stones as well as of "empty space" (pores) between the solid material. Other examples of porous media are: biological tissues such as bones, fleece materials such as diapers or other hygienic products, all kinds of textile materials, cements, foams, ceramics, and so on. *Panta rhei* (everything flows) is an insight that is usually attributed to the Greek philosopher Heraclitus and applied to porous media it means that the pore space within these materials is not just a useless "empty space". Rather, it serves as the flow domain of one or even several fluids in many cases, and indeed many porous media related questions and problems that are investigated in science and engineering refer to flows through porous media. The first systematic investigation of fluid flow through porous media has been performed by the French scientist and engineer Henry Darcy when he was involved in the construction and optimization of the fountains of the city of Dijon in France [179]. His experiments led him to what is called *Darcy's law* today, which can be written in modern notation as [69, 180, 181]

$$q(x, t) = -\frac{1}{\mu}K(x, t) \cdot \nabla p(x, t) \tag{4.127}$$

where $q \in \mathbb{R}^3$ is the Darcy velocity ($m\,s^{-1}$), μ is the (dynamic) viscosity (Pa·s), $K \in \mathbb{R}^{3 \times 3}$ is the permeability matrix (m^2), and p is the (fluid) pressure (Pa).

Here, the *viscosity* μ basically expresses the internal friction of a fluid or its resistance to deformations as a result of stresses such as shear or extensional stresses [180, 182]. Low-viscosity values mean that a liquid is "thin" and flows easily (such as water for which $\mu = 10^{-3}$ Pa·s), while high viscosity values refer to "thick" fluids such as corn syrup ($\mu > 1$ Pa·s) that exhibit a much larger resistance to flow. Obviously, this is a really important quantity in the entire field of computational fluid dynamics (e.g. it will appear in Section 4.10.3 as a part of the Navier–Stokes equations).

The *Darcy velocity* q is also known as *seepage velocity, filtration velocity, superficial velocity*, or *volumetric flux density* in the literature. To understand it, imagine a one-dimensional flow experiment comprising of a water column on top of a porous material. As the water flows through the porous medium e.g. driven by gravity, the top surface of the water column will gradually come down toward the porous medium, and it is exactly this *superficial velocity* of the water column that is described by the Darcy velocity. This *macroscopic* velocity must be distinguished from the *microscopic*, intrinsic average velocity of the fluid within the pores of the porous medium. If we use v to denote the latter velocity, the relation between v and q can be expressed as

$$q(x) = \phi(x)v(x) \tag{4.128}$$

where ϕ is the *porosity* of the porous medium, which expresses the fraction of the pore space within the porous medium in percent. All these quantities are usually defined throughout the porous medium. For example, the Darcy velocity is defined even outside the pore spaces of the medium, which is achieved using an averaging over so-called *representative elementary volumes* [69, 180]. Darcy formulated Equation 4.127 just as a phenomenological model that fits the data, but it has been shown that this equation can be interpreted as expressing conservation of momentum [183]. As in the case of the diffusion equation (Section 4.10.1), a second equation expressing mass conservation is needed. Again, this can be done in a similar way as in Section 4.2.2, that is, by balancing the mass flows in a small control volume, which leads to

$$\phi \frac{\partial \rho}{\partial t} + \nabla \cdot (\rho \mathbf{q}) = 0 \tag{4.129}$$

where ρ (e.g. in $(\mathrm{g\,m^{-3}})$) is the fluid density. Assuming an incompressible fluid, we have $\partial \rho / \partial t = 0$, and hence Equations 4.127 and 4.129 imply

$$\nabla \left(\mathbf{K}(\mathbf{x}) \cdot \nabla p(\mathbf{x}) \right) = 0 \tag{4.130}$$

if we assume stationary conditions and a constant viscosity. Again, we have a perfect analogy with the stationary heat equation: Darcy's law, Equation 4.127, corresponds to Fourier's law, Equations 4.22, and 4.130 corresponds to the stationary heat equation (Equation 4.23 with $\partial T / \partial t = 0$).

The *permeability matrix* $\mathbf{K}$ in Equation 4.130 expresses the ease of flow through the porous medium. Basically, relatively small pressure gradients will suffice to initiate a flow with some given velocity in the case of high permeability values, while larger pressure gradients will be needed in the case of low permeability values. As a matrix, $\mathbf{K}$ can be interpreted similar to the thermal conductivity matrix that was discussed in Section 4.2.5. Again, Equation 4.130 can be solved, for example, using *Salome-Meca* and the procedure described in Section 4.9.

4.10.2.1 Impregnation Processes

In [184–186], Equation 4.130 has been used to optimize the impregnation of *mica tape-based insulations*. These insulations are used to insulate steel bars inside large turbines that are used for high-voltage electrical power generation. They are manufactured in two steps. In a first step, a mica tape is wound in several layers around the steel bar. Figure 4.22a shows a schematic cross section through such a mica tape winding. As a result of the winding process, the mica tapes – which correspond to the black lines in the figure – form staircase-like structures. In the second step of the manufacturing procedure, the mica tape winding is impregnated with an epoxy resin. During impregnation, the main resin flow is through the pore spaces between the mica tapes (see the arrows in Figure 4.22a). Now the problem is that the permeability of the mica tapes as well as the permeability of the winding as a whole is extremely small, which means that the impregnation is proceeding extremely slow. In the worst case, some regions of the winding may remain

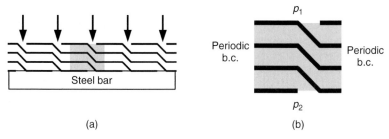

(a)   (b)

Fig. 4.22 (a) Cross section through mica tape insulation
with periodicity cell (gray rectangle). (b) Boundary conditions
in the periodicity cell.

unimpregnated, which can cause expensive electrical failures during the operation
of the turbines. In [184–186], simulations of the impregnation process based on
Equation 4.130 have been used to optimize the impregnation process in a way that
helps to avoid this kind of impregnation failures.

To simulate the fluid flow in the winding using Equation 4.130, boundary
conditions must be applied at the boundaries of the winding. At the top surface
of the winding (where the epoxy resin enters the winding) and at its bottom
surface (where the winding is in contact with the steel bar) the pressures can be
prescribed that are driving the fluid flow, for example, a constant p_1 at the top
of the winding and $p_2 < p_1$ at its bottom (Figure 4.22b). Since the winding has
a periodic structure, it suffices to do the flow computation in a periodicity cell
(Figure 4.22a). This periodicity cell represents the winding in the sense that the
whole winding can be generated by successively attaching copies of the periodicity
cell. When the flow computation is done in the periodicity cell, so-called *periodic
boundary conditions* are applied at its left and right ends, that is, in the direction of
the periodicity of the structure. Mathematically, these periodic boundary conditions
basically identify the corresponding periodic boundaries, in this case by equating
the pressure values at the left end of the periodicity cell with the corresponding
values at its right end. As the example shows, the size of the flow domain and,
hence, the resulting computational effort can be reduced substantially by the
application of periodic boundary conditions. Before setting up a flow computation,
one should thus always analyze the periodicity of the system under investigation,
and apply periodic boundary conditions if possible. A great number of systems in
science and engineering are periodic similar to the example.

Many other industrial processes involve the impregnation of a porous material
with a fluid. An example is the *resin transfer molding (RTM) process*, which is
used to produce fiber-reinforced plastic materials that are used for all kinds of
high-technology applications such as aerospace structures. In this process, a textile
preform is placed into a closed mold which is then impregnated with low-viscosity
(easily flowing) thermosetting polymers. Similar to above, computer simulations
have been used to optimize this process in order to avoid problems such as
incomplete impregnation. Again, Equation 4.130 can be used in many cases [91,
187–189].

4.10.2.2 Two-phase Flow

The above porous media flow model can be used only for one-phase flow, that is, in situations where there is just a single fluid in the pores of the porous medium. Of course, there are many situations which involve the flow of several fluids through a porous medium at the same time. An important application that has substantially driven the development of multiphase porous media flow models in the past is oil exploration [190], which involves the simultaneous flow of oil, water, and gases through porous soil and rock structures. As an example, let us consider the two-phase flow of *water and air* in a porous medium (e.g. in a soil). One of the simplest assumptions that one can make here is that the air pressure is approximately constant, which leads to the *Richard's equation*

$$\frac{\partial \Theta(\psi(\mathbf{x}, t))}{\partial t} - \nabla \cdot \left(\mathbf{K}(\psi(\mathbf{x}, t)) \nabla(\psi(\mathbf{x}, t) + z) \right) = 0 \tag{4.131}$$

where Θ is the volumetric water content (1), $\psi = p/(\rho g)$ is the pressure head (m), p is the pressure in the water phase (Pa), ρ is the water phase density (kg m^{-3}), $g \approx 9.81$ is the gravitational acceleration (m s^{-2}), and $\mathbf{K} \in \mathbb{R}^{3 \times 3}$ is the hydraulic conductivity (m s^{-1}).

Basically, this equation is derived similar to Equation 4.130, using the above assumptions and a separate application of momentum and mass conservation to each of the fluid phases [191, 192]. Note that ψ is just the usual pressure p rescaled in a way that is frequently applied by people who are using Richard's equation (such as soil scientists). This rescaling allows ψ to be interpreted in terms of water columns [191].

4.10.2.3 Water Retention and Relative Permeability

$\Theta(\psi)$ and $\mathbf{K}(\psi)$ are empirical relations that express material properties of the flow domain. $\Theta(\psi)$ is what soil scientists call the *water retention curve*. It expresses the external pressure that needs to be applied to a soil in order to obtain some given value of volumetric moisture content, which is an important soil characteristic particularly with respect to the water supply to plants or with respect to soil aggregate stability.

Figure 4.23 shows example water retention curves for a sand and a sintered clayey material. As the figure shows, increasingly high pressures need to be applied in order to achieve small moisture values. Also, it can be seen that higher pressures are required to achieve small moisture values in the clayey material. Basically, this expresses the simple fact that sand soils are more coarse grained compared to clayey materials, which means that the capillary forces that are retaining the water in a porous material are smaller in a sand soil [194]. Figure 4.23 has been drawn using the code `Mualem.r` in the book software (Appendix A), which is based on the Mualem/van Genuchten model [194]

$$\Theta(\psi) = \Theta_r + \frac{\Theta_s - \Theta_r}{\left(1 + (\alpha|\psi|^n)\right)^{1-1/n}} \tag{4.132}$$

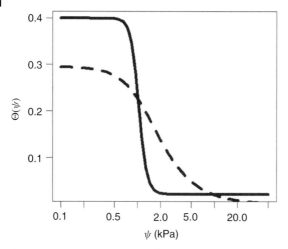

Fig. 4.23 Water retention curves for a sand (solid line, $\Theta_r = 0.022$, $\Theta_s = 0.399$, $\alpha = 0.93$, $n = 8.567$) and a sintered clayey material (dashed line, $\Theta_r = 0$, $\Theta_s = 0.295$, $\alpha = 0.605$, $n = 2.27$). Parameters from [193], figure drawn using Mualem.r.

where Θ_r is the residual water content (1), Θ_s is the saturated water content (1), α is an empirical parameter (m^{-1}), and $n > 1$ is an empirical parameter related to the pore-size distribution (1). The residual water content Θ_r is the volumetric water content that cannot be removed from the soil even if very high external pressure is applied, whereas the saturated water content Θ_s is the volumetric water content in the case where the soil is fully saturated (i.e. no air is present in the pore space). α is related to the inverse of the air entry suction, which is the smallest external pressure that must be applied to remove water from a fully saturated soil.

The empirical relation $\mathbf{K}(\psi)$ in Equation 4.131 is related with the notion of *relative permeability*. In multiphase flow situations, two things must be distinguished:

- K_1: the permeability of a porous medium with respect to a fluid that entirely fills the pores of the porous medium (fully saturated case);
- K_2: the permeability of a porous medium with respect to a fluid in the presence of one or several other fluids (unsaturated case).

Usually, K_2 will be smaller than K_1 since less pore space is available for the fluid under consideration in the unsaturated case. The corresponding reduction of the saturated permeability is expressed as the relative permeability, K_2/K_1. Applied to Equation 4.131, this means that the permeability $\mathbf{K}$ will depend on the water saturation, or, via Equation 4.132, on the pressure head, ψ. This dependence can again be described using appropriate empirical relations such as the Corey equation [183, 190].

Note that Richard's equation 4.131 is a *nonlinear* PDE since the unknown ψ appears in the general nonlinear functions $\Theta(\psi)$ and $\mathbf{K}(\psi)$, see the discussion of nonlinearity in Section 4.3.1.3. Appropriate numerical methods thus need to be applied to linearize Equation 4.131 [139].

4.10.2.4 Asparagus Drip Irrigation

Figure 4.24 shows an example application of Richards equation, which was computed using the "earth science module" of the commercial FE software *Comsol Multiphysics* (since this model is currently unavailable in *CAELinux*). Asparagus is cultivated in ridges having the shape that is indicated in Figure 4.24a. To control the moisture levels within these ridges, drip irrigation through water pipes is used in some cases. Figure 4.24a shows some possible locations of these water pipes below the surface level. The pipes release water into the soil through special drip generating devices that generate a sequence of water drops at the desired rate (e.g. there may be one of these drip generators per 20 cm pipe length). Now the question is *where* the exact location of the pipes should be and *how much* water should be released through the drip generators so as to achieve an optimum distribution of soil moisture within the soil at minimal costs. To solve this optimization problem, a mathematical model based on Richard's equation and the Mualem/van Genuchten model, Equations 4.131 and 4.132, has been developed in [195]. Figure 4.24b shows isosurfaces of the volumetric soil moisture content around one of the drip generators in a water pipe. The same figure is also shown on the title page of this book. Referring to the colors on the title page, it can be seen that the soil moisture is highest (red colors) close to the water pipe, while it decreases gradually

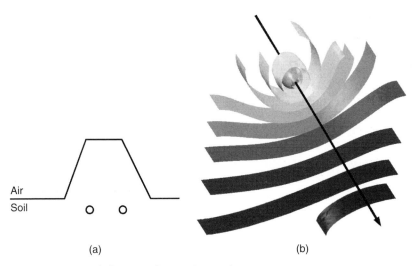

      (a)   (b)

Fig. 4.24 (a) Example location of water pipes under an asparagus ridge. (b) Isosurfaces of the volumetric moisture content in an asparagus ridge (arrow indicates the water pipe).

with increasing distance toward the pipe (the dark blue color indicates the lowest moisture value).

4.10.2.5 Multiphase Flow and Poroelasticity

Richard's equation can be generalized in a number of ways. First, we can drop the assumption that the pressure in one of the fluid phases is constant. If we still consider two-phase flow, this leads to two equations similar to Equation 4.131 for each of the fluid phases or – in the case of more than two fluid phases – to as many equations of this type as there are fluid phases [183, 196]. Using the mechanics of mixtures approach described in [183, 197], this can be further extended to cases where one of the phases is a solid. An approach of this kind has been used, for example, in [5] to describe the wet pressing of paper machine felts, which involves two fluid phases (water and air) and the felt itself as a solid phase. See [198] for a number of other approaches to describe *poroelasticity*, that is, the deformation of a porous material, and in particular the coupling of such a deformation with the fluid flow inside the porous medium, which is required in a great number of applications. Finally, we remark that Richard's equation can also be applied to situations where the flow domain consists of layered materials that are saturated in some parts and unsaturated in other parts of the flow domain, such as technical textiles, diapers, and other hygienic materials [199].

4.10.3
Computational Fluid Dynamics (CFD)

CFD involves all kinds of problems where mathematical models are used to describe fluid flow. As was already mentioned above, CFD is perhaps the "most classical" application of PDEs in science and technology in the sense that it involves many applications that have gained attention in the public, such as meteorological flow simulations on the earth's surface, the simulation of air flow around cars, airplanes or space shuttles. Its special importance is underlined by the fact that an abundant number of systems in science and technology involve fluid flow (we could quote Heraclitus *panta rhei* here again, similar to Section 4.10.2). Of course, the problems involving flow in porous media that were treated in Section 4.10.2 are already a part of CFD in its general sense.

4.10.3.1 Navier–Stokes Equations

In a narrower sense, people often use "CFD" as a synonym for applications of the *Navier–Stokes equations* and its generalizations. In the simplest case (incompressible, Newtonian fluid) these equations can be written as [182, 200]

$$\rho \frac{D\mathbf{v}}{Dt} = -\nabla p + \mu \nabla^2 \mathbf{v} + \mathbf{f} \tag{4.133}$$

$$\nabla \cdot \mathbf{v} = 0 \tag{4.134}$$

where ρ is the density (kg m^{-3}), $\mathbf{v} = (v_x, v_y, v_z)$ is the velocity (m s^{-1}), p is the pressure (Pa), μ is the (dynamic) viscosity (Pa·s), and $\mathbf{f} = (f_x, f_y, f_z)$ is a body force (N m^{-3}).

As before, $\partial\rho/\partial t = 0$ is due to the incompressibility assumption, and this is why ρ does not appear in the time derivative of Equation 4.133. If compressible fluids such as gases are considered, time derivatives of ρ will appear in Equations 4.133 and 4.134 similar to Equation 4.129. Gas flow is often described based on the *Euler equations*, which assume inviscid flow, that is, $\mu = 0$ [182]. The "Newtonian fluid" assumption pertains to the fluids stress–strain behavior. In short, Newtonian fluids flow "like water" in the sense that, like water, they exhibit a linear stress–strain relationship with the dynamic viscosity as constant of proportionality [182]. Non-Newtonian fluids, on the other hand, do not have a well-defined viscosity, that is, the viscosity changes depending on the shear stress that is applied. Examples include blood, toothpaste, mustard, mud, paints, polymers, and so on.

D/Dt in Equation 4.133 is the *material derivative* (which is also called the *convective* or *substantive* derivative). It expresses the time derivative taken with respect to a coordinate system that is moving along the velocity field $\mathbf{v}$ [139]. The *body force* vector $\mathbf{f}$ typically expresses gravitation, but it may also express other forces that are acting on the fluid such as electromagnetic forces. As before, these equations can be interpreted in terms of conservation principles: Equation 4.133 as conservation of momentum and Equation 4.134 as conservation of mass, which can be easily shown based on the control volume approach used in Section 4.2.2.

A major advantage of the material derivative notation is that it can be seen very easily here that Equation 4.133, indeed, expresses conservation of momentum (note that this is much less obvious in the case of Darcy's law, Equation 4.127): The left-hand side of this equation basically is the time derivative of momentum, that is, of $\rho\mathbf{v}$ since everything is expressed on a per-volume basis in the Navier–Stokes equations. You may imagine that all quantities refer to a small fluid volume, that is, to a control volume as was discussed in Section 4.2.2. Now we know from Newton's second law that the rate of change of momentum of a body is proportional to the resultant force acting on that body. Exactly this is expressed by Equation 4.133, since its right-hand side summarizes all the forces that are acting on the control volume, which are forces exerted by pressure gradients and viscous and body forces based on the assumptions made above. Equation 4.133, thus, can be thought of as expressing Newton's second law for a small control volume in a fluid.

In a standard coordinate system, the material derivative can be written as

$$\frac{D\mathbf{v}}{Dt} = \frac{\partial\mathbf{v}}{\partial t} + (\mathbf{v} \cdot \nabla)\mathbf{v} \tag{4.135}$$

which shows that Equation 4.133 again is a *nonlinear* PDE in the sense discussed in Section 4.3.1.3, since derivatives of the unknown $\mathbf{v}$ are multiplied by $\mathbf{v}$ itself. Thus, again, specific numerical methods need to be applied to solve the Navier–Stokes equations based on an appropriate linearization [201–203].

The incompressible Navier–Stokes equations 4.133 and 4.134 can be generalized in a number of ways. As was mentioned above, the incompressibility assumption can be dropped, for example, when one is concerned with gas flow. Other important generalizations include non-Newtonian flow, the consideration of temperature fluctuations and its interactions with the flow (e.g. via density variations in fermentation processes, see the above discussion of fermentation), the modeling of turbulence phenomena, and so on. *Turbulence models* are of particular importance since turbulent flows are rather the rule than the exception in the applications. Equations 4.133 and 4.134 assume *laminar* flow conditions, which correspond to what may be described as a "smooth" flow pattern, in contrast to *turbulent* flow regimes that are characterized by chaotic, stochastic changes of state variables such as fluid velocity and pressure. The open-source CFD software *Code-Saturne* (Section 4.10.3.3 and Appendix A) includes a number of turbulence models such as the *Reynolds-averaged Navier–Stokes equations* which are also known as the *RANS equations* [204].

4.10.3.2 Backward Facing Step Problem

We will use Code-Saturne now to solve a standard problem of fluid mechanics: the *backward facing step problem* [201]. Referring to the geometry shown in Figure 4.25a, the backward facing step problem is characterized by the following boundary conditions:

- 1: inflow boundary, constant inflow velocity $\mathbf{v} = (1, 0, 0)$;
- 2,3,4,6: walls, no-slip condition;
- 5: outflow boundary, pressure condition $p = 0$;
- top and bottom surface: symmetry condition.

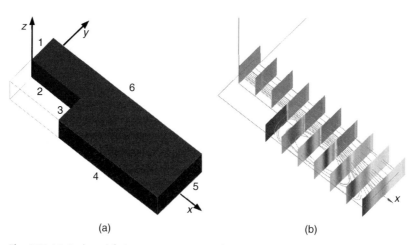

(a)   (b)

Fig. 4.25 (a) Backward facing step geometry with enumerated boundaries. (b) Solution of the backward facing step problem obtained with *Code-Saturne*: combination of cut planes (absolute value of the velocity) with streamlines.

The *no-slip boundary conditions* at the boundaries 2,3,4,6 forbid any flow through these boundaries similar to the "no-flow condition" discussed in Section 4.3.2. Additionally, they impose zero flow velocity relative to the boundary immediately adjacent to the boundary, which expresses the fact that the flow velocity of viscous fluids such as water will always be almost zero close to a wall [182]. The *symmetry boundary conditions* at the top and bottom surfaces basically mean that a geometry is assumed here that extends infinitely into the positive and negative z directions (similar conditions have been considered in Section 4.3.3). The symmetry conditions forbid any flow through these boundaries similar to the no-slip condition, but they do not impose a zero flow velocity close to the boundaries. The fluid can slip freely along symmetry boundaries, and this is why the symmetry boundary condition is also known as the *slip boundary condition*. Owing to the symmetry conditions, everything will be constant along the z direction, that is, there will be no changes, for example, in the fluid flow velocity as we move into the positive or negative z directions. This means that this problem is a 2D problem in the sense explained in Section 4.3.3. To demonstrate *Code-Saturne*'s 3D facilities, it will, nevertheless, be solved in 3D.

4.10.3.3 Solution Using Code-Saturne

Let us now see how the backward facing step problem can be solved using *Code-Saturne*. Code-Saturne is open-source software that is a part of the CAELinux distribution (Appendix A). Like Code_Aster, it has been developed by EDF, a French electricity generation and distribution company. On the basis of the finite volume method, it is able to treat 3D compressible and incompressible flow problems with and without heat transfer and turbulence [204]. It can be run on parallel computer architectures, which is an important benefit since CFD problems can be very demanding in terms of computation time and memory requirements. See [204] and *www.code-saturne.org* for more details.

To solve the backward facing step problem described above, the same steps will be applied that have already been used in Section 4.9: geometry definition, mesh generation, problem definition, solution, and postprocessing. Only those steps of the solution procedure are addressed here that differ substantially from the procedure in Section 4.9. The *geometry definition step* and *mesh generation step* are skipped here since this can be done very similar to the corresponding steps in Sections 4.9.1 and 4.9.2. After mesh generation is finished, we have to leave *Salome-Meca* since the problem definition and solution steps will be done in the separate *Code-Saturne* GUI. As a last step in *Salome-Meca*, the mesh must be exported in a .med file. You will find an appropriate mesh corresponding to Figure 4.25a in the file flowstep.med in the book software (Appendix A).

The *problem definition step* and the *solution step* will now be performed within the *Code-Saturne* GUI. This GUI should be started using the *CFD-Wizard*, which can be accessed if you select "CAE-software/Code-Saturne/CFD-Wizard" under the PC button in *CAELinux*. Figure 4.26a shows appropriate settings within this wizard. After confirming the wizard, the *Code-Saturne* GUI will appear (Figure 4.26b). Within this GUI, select the "open a new case" symbol (directly below the "file"

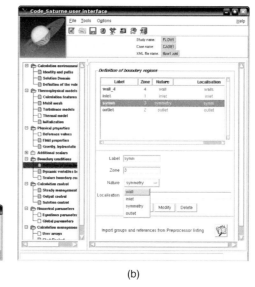

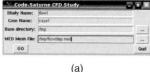

(a)   (b)

Fig. 4.26 (a) CFD-wizard in *CAELinux*. (b) *Code-Saturne GUI*.

menu), which automatically selects the appropriate data that have been generated by the CFD-wizard based on the mesh in `flowstep.med`. After this, you just have to follow the steps listed in the left, vertical subwindow of the *Code-Saturne* GUI. The following must be done:

- *Calculation environment/solution domains*: under "stand alone running", choose Code-Saturne preprocessor batch running.
- *Thermophysical models/calculation features*: choose "steady flow".
- *Physical properties/fluid properties*: set the density to 1000 ($kg\ m^{-3}$), the viscosity to 10^{-3} (Pa s), and the specific heat to 4800 ($J\ kg^{-1}\ K^{-1}$) (which means that we will simulate water flow).
- *Boundary conditions/definition of boundary regions*: select "Import groups and references from preprocessor listing", then choose file `listenv.pre`.

You will then see a list with boundary regions `wall_1`, `wall_2`, `wall_3`, and `wall_4` which correspond to the boundaries in Figure 4.25a as follows:

- `wall_1`: inflow boundary, boundary 1 in Figure 4.25a;
- `wall_2`: outflow boundary, boundary 5 in Figure 4.25a;
- `wall_3`: symmetry boundaries, top and bottom surfaces of the geometry in Figure 4.26a;
- `wall_4`: no-slip boundaries, boundaries 2,3,4,6 in Figure 4.25a.

After selecting the appropriate "Nature" for boundaries wall_1–wall_3, go on as follows:

- *Boundary conditions/dynamic variables b.c.*: select wall_1 and then set $U = 1(\text{m s}^{-1})$ and the hydraulic diameter for the turbulence model to 0.1 m (see [204] for details).
- *Calculation control/steady management*: set the iterations number to 20.
- *Calculation control/output control*: set format to MED_file.
- *Calculation management/prepare batch calculation*: select batch script file lance; then choose /tmp as the prefix of the temporary directory in the advanced options; choose "file/save" in the *Code-Saturne*'s main menu, and save the Salome GUI case file under the name flow1.
- Press the "Run Batch Script" button to start the computation.

You will find the results in the *CAELinux* directory /tmp/FLOW1/CASE1/RESU. Among the results you will, for example, find a text file beginning with listing... which contains details about the solution process. The result will be stored in a .med file.

4.10.3.4 Postprocessing Using Salome-Meca

After this, the postprocessing step is performed within *Salome-Meca* again. First, choose the *Post-Pro* module described in Section 4.9.4. Then you can import the .med file that was generated by *Code-Saturne*. If you do not want to do the above computation yourself, you can also use the file result.med in the book software at this point, which contains the result of a *Code-Saturne* computation as described above. The generation of graphical plots from the data then goes along the same lines as described in Section 4.9.4.

Figure 4.25b shows the solution displayed using a combination of a *cut plane plot* with a *streamline plot*. One can see here, for example, that the darkest color of the cut planes – which corresponds to the smallest velocity – is located immediately behind the step, that is, exactly where physical intuition would tell us that it should be. The cut planes also show that the velocity distribution is indeed two-dimensional (see the above discussion), since it can be seen that the colors remain unchanged as we move into the positive or negative z directions. Using the procedure described above, you may look at this picture in full colors on your computer screen if you want to see the velocity distribution in more detail. The streamline plot within Figure 4.25 (b) shows the recirculation region just downstream of the step, which is caused by the sudden widening of the flow domain at the step.

The *vector plot* in Figure 4.27a shows this recirculation region in some more detail. It shows the exact flow directions at each particular point. Note that the length of the velocity vectors in Figure 4.27a is proportional to the absolute value of the velocity. Finally, Figure 4.27b shows an *isosurface plot* of the absolute value of the velocity, similar to the isosurface plots discussed in Section 4.9.4. The "tape-like"

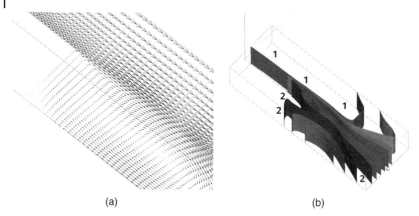

(a)   (b)

Fig. 4.27 Solution of the backward facing step problem displayed using (a) a vector plot of the velocity and (b) an isosurface plot of the absolute value of the velocity.

character of these isosurface (i.e. no bending in the z direction) confirms again that we have solved a two-dimensional problem. In this case, the isosurface plot helps us, for example, to distinguish between regions of high velocity (labeled "1" in the plot) and regions of low velocity (labeled "2" in the plot). One can also see the effect of the no-slip boundary conditions in this plot: for example, the isosurfaces ending on boundary 6 (Figure 4.25a) are bent against the flow direction, which means that the flow velocity decreases close to the boundary, as is required by the no-slip boundary condition.

4.10.3.5 Coupled Problems

Many CFD models are coupled with mathematical models from other fields. A problem of this kind has already been mentioned in Section 4.10.2.5: the wet pressing of paper machine felts, where the porous media flow equations are coupled with equations describing the mechanical deformation of the felt. Problems of this kind are usually classified as *fluid–structure interaction* problems.

There is also a great number of problems where the Navier–Stokes equations are coupled with the porous media flow equations. An example are *filtration processes*, where the flow domain can usually be divided into two parts:

- a *porous flow domain*, where the fluid flows inside the so-called (porous) filter cake that is made up of the solid particles that are deposited at the filtering device and
- a *free flow domain*, where the fluid flows "freely" outside the filter cake.

In [173, 205], such a coupled problem is solved in order the optimize candle filters that are used, for example, for beer filtration.

Another example is the industrial *cleaning of bottles*. In this process, a fluid is injected into a bottle at high velocity (Figure 4.28). The fluid contains cleaning agents that are intended to remove any bacteria from the inside surface of the

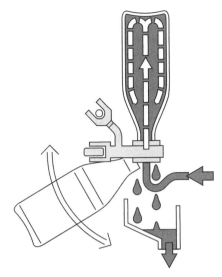

Fig. 4.28 Industrial cleaning of bottles (© 2008 KHS AG, Bad Kreuznach, Germany).

bottle. Owing to the high speed of this process, it is a difficult task for the engineers to ensure an effective cleaning of the entire internal surface of the bottles. Again, computer simulations based on appropriate mathematical models are used to optimize this process [206]. In this case, an appropriate generalization of the Navier–Stokes equations that includes a turbulence model (Section 4.10.3.1) is coupled with a model of the microbial dynamics inside the bottle in the form of ODEs.

4.10.4
Structural Mechanics

The wet pressing of paper machine felts that was mentioned in Section 4.10.2.5 is an example of a problem in the field of *structural mechanics*, since it involves the computation of the mechanical deformation of the felt material. Beyond this, there is again an abundant number of other systems in science and technology which involve mechanical deformations. We have cited Heraclitus *panta rhei* ("everything flows") several times above – in this case, it would be valid to say something like "everything deforms" or *(ta) panta paramorfonontai* (translation by my Greek colleague, A. Kapaklis – thank you). Before looking at more examples, however, let us write down the governing equations in a simple case.

4.10.4.1 **Linear Static Elasticity**
Deformations of elastic solids are caused by forces that are acting on the solid. These forces may act on any point within the body (e.g. body forces such as gravitation) or across its external boundaries. If the solid is in equilibrium, the resultant forces will vanish at any point within the solid. This is expressed by the

following *equilibrium equations* [207]

$$-\nabla \cdot \sigma = \mathbf{F} \tag{4.136}$$

where σ is the stress tensor ($N\,m^{-2}$), and $\mathbf{F}$ is the body force ($N\,m^{-3}$).

The *stress tensor* σ expresses the forces that are acting inside the body. σ is a so-called rank-two tensor quantity, which can be expressed as a 3×3 matrix:

$$\sigma = \begin{pmatrix} \sigma_{11} & \sigma_{12} & \sigma_{13} \\ \sigma_{21} & \sigma_{22} & \sigma_{23} \\ \sigma_{31} & \sigma_{32} & \sigma_{33} \end{pmatrix} \tag{4.137}$$

To understand the meaning of σ, let $\mathbf{n} = (n_1, n_2, n_3)^t$ be the normal vector of a plane through a particular point $\mathbf{x} = (x_1, x_2, x_3)$ within the solid, and let $\sigma(\mathbf{x})$ be the stress at that particular point. Then,

$$\mathbf{T} = \mathbf{n}^t \cdot \sigma(\mathbf{x}) \tag{4.138}$$

is the force that is acting on the plane. Hence, we see that σ basically describes the forces that are acting inside the solid across arbitrarily oriented planes.

Now in order to compute the deformed state of a solid, it is of course not sufficient if we just know the forces expressed by σ. We also need to know how the body reacts on these forces. As we know from our everyday experience, this depends on the material: the same force that substantially deforms a rubber material may not cause the least visible deformation of a solid made of concrete. Thus, we need here what is called a *material law*, that is, an equation that expresses the specific way in which the solid material under investigation "answers" to forces that are applied to the solid. Consider the simple case of a cylinder that is made of a homogeneous material, and that is deformed along its axis of symmetry by a force F (Figure 4.29). Assuming that the force reduces the initial length of the cylinder from L to $L - \Delta L$ as shown in the figure, the *strain*

$$\epsilon = \frac{\Delta L}{L} \tag{4.139}$$

characterizes the deformation of the cylinder. As Equation 4.139 shows, ϵ is a dimensionless quantity. The stress σ (Pa) caused by the force F (N) can be written as

$$\sigma = \frac{F}{A} \tag{4.140}$$

where A (m^2) is the cross-sectional area of the cylinder. Now writing down a material law for the cylinder in this situation amounts to writing down an equation that relates the force that causes the deformation (described by σ) with the "answer" of

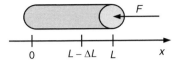

Fig. 4.29 One-dimensional deformation of a cylinder.

the cylinder in terms of its strain ϵ. In the simplest case, this can be written as a linear equation in the form of the well-known *Hooke's law*:

$$\sigma = E \cdot \epsilon \tag{4.141}$$

where E (Pa) is the so-called Young's modulus of elasticity. Young's moduli of elasticity are listed for most relevant materials in books such as [208]. Equation 4.141 can be generalized in a number of ways [207]. In particular, you should note that the material law typically will be nonlinear, and that ϵ will be a tensorial quantity in general, which describes the deformed state in a direction-dependent way similar to σ. In the case of a homogeneous, isotropic material (i.e. a material with direction-independent properties), Hooke's law can be written as [207]

$$\sigma = 2\mu\epsilon + \lambda tr(\epsilon) \cdot \mathbf{I} \tag{4.142}$$

where σ is the stress tensor (Equation 4.137) and ϵ is the strain tensor

$$\epsilon = \begin{pmatrix} \epsilon_{11} & \epsilon_{12} & \epsilon_{13} \\ \epsilon_{21} & \epsilon_{22} & \epsilon_{23} \\ \epsilon_{31} & \epsilon_{32} & \epsilon_{33} \end{pmatrix} \tag{4.143}$$

$tr(\epsilon)$, the *trace* of ϵ, is defined as

$$tr(\epsilon) = \epsilon_{11} + \epsilon_{22} + \epsilon_{33} \tag{4.144}$$

In Equation 4.142, μ is the *shear modulus* or *modulus of rigidity* and λ is *Lame's constant*. μ and λ describe the material properties in Equation 4.142. In many books (such as [208]), these parameters are given in terms of Young's modulus E and *Poisson's ratio* v as follows [207]:

$$\mu = \frac{E}{2 + 2v} \tag{4.145}$$

$$\lambda = \frac{Ev}{(1 + v)(1 - 2v)} \tag{4.146}$$

Now let us assume that some material point is at the coordinate position $\mathbf{x} = (x, y, z)$ in the undeformed state of the material (i.e. with no forces applied), and that this material point then moves to the coordinate position (ξ, η, ζ) in the deformed state of the material. Then, the *displacement vector* $\mathbf{u} = (u, v, w)$ describes the overall displacement of the material point as follows:

$$u = \xi - x \tag{4.147}$$

$$v = \eta - y \tag{4.148}$$

$$w = \zeta - z \tag{4.149}$$

Expressing the strain tensor using the displacements and then inserting the material law, Equation 4.142, into the equilibrium condition, Equation 4.136, the following PDEs are obtained [207]:

$$\mu\nabla^2 u + (\lambda + \mu)\frac{\partial}{\partial x}\left(\frac{\partial u}{\partial x} + \frac{\partial v}{\partial y} + \frac{\partial w}{\partial z}\right) + F_x = 0 \tag{4.150}$$

$$\mu\nabla^2 v + (\lambda + \mu)\frac{\partial}{\partial y}\left(\frac{\partial u}{\partial x} + \frac{\partial v}{\partial y} + \frac{\partial w}{\partial z}\right) + F_y = 0 \tag{4.151}$$

$$\mu\nabla^2 w + (\lambda + \mu)\frac{\partial}{\partial z}\left(\frac{\partial u}{\partial x} + \frac{\partial v}{\partial y} + \frac{\partial w}{\partial z}\right) + F_z = 0 \tag{4.152}$$

These equations are usually called the *Navier's equations* or *Lame's equations*. Along with the appropriate boundary conditions describing the forces and displacements at the external boundaries of the elastic solid, these equations can be used to compute the deformed state of a linearly elastic isotropic solid. Alternatively, these PDEs can also be formulated, for example, using the stresses as unknowns [207]. It depends on the boundary conditions which of these formulations is preferable. Note that Equations 4.150–4.152 describe what is called *linear static elasticity* since a linear material law was used (Equation 4.142), and since stationary or static conditions are assumed in the sense that the elastic body is in equilibrium as expressed by Equation 4.136, that is, all forces on the elastic body sum to zero, and the displacements are not a function of time.

In *CAELinux*, structural mechanical problems such as the linear isotropic Equations 4.150–4.152 can be solved using *Code_Aster* similar to the procedure described in Section 4.9.3. *Code_Aster* can be accessed e.g. as a submodule of *Salome_Meca* (within this submodule, choose the "linear elasticity" wizard to solve a linear isotropic problem).

4.10.4.2 Example: Eye Tonometry

Glaucoma is one of the main reasons of blindness in the western world [209]. To avoid blindness, it is of great importance that the disease is detected at an early stage. Some of its forms are associated with a raised intraocular pressure (IOP), that is, with a too high pressure inside the eye, and, thus, IOP monitoring is an important instrument in glaucoma diagnosis [210]. A traditional measurement method that is still widely used is *Goldmann applanation tonometry*, which measures the IOP based on the force that is required to flatten a circular area of the human cornea with radius $r = 1.53$ mm [211]. Figure 4.30a shows the measurement device, a cylindrical tonometer head, as it is moved against the cornea of the human eye.

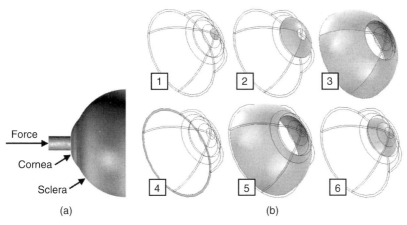

Fig. 4.30 (a) Applanation tonometry measurement procedure. (b) Boundaries of the three-dimensional eye model.

Applanation tonometry is based on Goldmann's assumption that the rigidity of the human cornea does not vary much between individuals. Recently, however, it has been shown that this assumption is wrong. Indeed, the rigidity of the human cornea varies substantially between individuals due to natural variations of the thickness of the cornea [212], and due to variations of the biomechanical properties (such as Young's modulus) of the corneal tissue [213, 214]. As a consequence of the variable rigidity of the human cornea, the measurement values obtained by Goldmann applanation tonometry can deviate substantially from the real IOP [214, 215].

To be able to correct applanation tonometry measurements for the effects of corneal rigidity variations, a mathematical model is needed that is able to predict tonometry measurements depending on a given corneal geometry and given biomechanical properties of the corneal tissue. With this objective, a finite-element model has been developed based on the equations discussed in the last section and on a three-dimensional model of the human eye [215, 216]. An "average" three-dimensional eye geometry based on the data in [213] was used in the simulations (Figures 4.30b and 4.31). Note that as Figure 4.30a shows, the *cornea* sits on top of the *sclera*, and both structures together form a thin, shell-like structure, which is known as the *corneo-scleral shell*. Figure 4.30b shows the boundaries of the eye model where the following boundary conditions are applied:

- Boundary 1: Corneal area that is flattened by the tonometer head. A "flatness condition" is imposed here, which can be realized iteratively as described in [217]
- Boundaries 2,3: Outer surface of the eye. Atmospheric pressure is prescribed here.
- Boundary 4: Here, the outer eye hemisphere is in contact with the inner hemisphere. Since this is "far away" from the cornea, a *no displacement* condition or $\mathbf{u} = 0$ is assumed here.

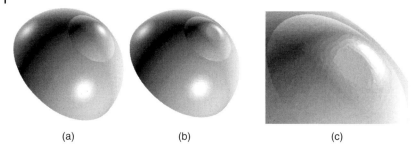

(a)   (b)   (c)

Fig. 4.31 Three-dimensional eye model: undeformed (a) and deformed by the tonometer head (b and c).

- Boundaries 5,6: Internal surface of the corneo-scleral shell.
  The IOP acts on these surfaces, so the IOP is prescribed there.

To simulate the deformation of the corneo-scleral shell based on these boundary conditions and the equations described in the last section, an appropriate material model for the stress–strain relation is needed. Typically, stress–strain relations of biological tissues are nonlinear [218]. Unfortunately, the nonlinear stress–strain behavior of the corneo-scleral shell has not yet been sufficiently characterized by experimental data [214]. Hence, the simulation shown in Figure 4.31 is based on a linear material model that uses Young's moduli within the range that is reported in the literature: a constant Young's modulus of 0.1 MPa in the cornea [213, 214], and 5.5 MPa in the sclera [213, 219]. Following [214], Poisson's ratio was set to a constant value of $\nu = 0.49$ in the entire corneo-scleral shell. Note that this simulation has been performed using the "structural mechanics module" of the commercial *Comsol Multiphysics*. You can do the same simulation using *Code_Aster* based on *CAELinux* as described above (*Comsol Multiphysics* was used here to test some nonlinear material models that are currently unavailable in *Code_Aster*).

The simulations can be used to estimate the measurement error of Goldmann applanation tonometry. Figure 4.32 shows the effect of scleral rigidity variations on the simulated IOP reading. In this figure, a scleral rigidity corresponding to a Young's modulus of 5.5 MPa is assumed as a "reference rigidity". Then, the scleral Young's modulus is varied between 1 and 10 MPa, which is the (approximate) range of scleral Young's moduli reported in the literature [213, 219]. What the figure shows is the percent deviation of the simulated IOP reading from the IOP reading that is obtained in the reference situation. As the figure shows, the IOP reading can deviate by as much as 7% based on the variations of the *scleral* rigidity only. As expected, the effect of the corneal rigidity is even higher and can amount to 25–30% within the range of rigidities that is reported in the literature.

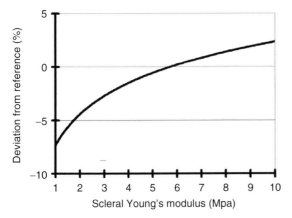

Fig. 4.32 Effect of scleral rigidity variations on the simulated IOP reading.

4.11
Other Mechanistic Modeling Approaches

At this point, similar remarks apply as in the case of phenomenological modeling (Section 2.7): Again, you should be aware of the fact that there is a great number of mechanistic modeling approaches beyond those that are discussed in this chapter and in chapter 3. We will confine ourselves here to a few examples, which are by no means exhaustive, but which may demonstrate how mathematical structures different from the ones discussed above may arise.

4.11.1
Difference Equations

Consider a host–parasite system, where parasites use host plants to deposit their eggs. Let

- N_t: number of host species in the tth breeding season
  $(t = 1, 2, 3, \ldots)$;
- P_t: number of parasite species in the tth breeding season.

Then it can be argued that [220]

$$N_{t+1} = \lambda e^{-\gamma P_t} N_t \tag{4.153}$$

$$P_{t+1} = c N_t \left(1 - e^{-\gamma P_t}\right) \tag{4.154}$$

where $e^{-\gamma P_t}$ is the fraction of hosts not parasitized (the particular form of this term is a result of probabilistic considerations as explained in [220]), c is the average number of eggs laid by surviving parasites, and λ is the host growth rate, given that all adults die before their offspring can breed.

This model is known as the *Nicholson–Bailey model*. Although we will not go into a detailed discussion of these equations here, it is easy to understand the message in qualitative terms: Equation 4.153 says that the hosts grow proportional to the existing number of hosts, that is, in an exponential fashion similar to the description of yeast growth in Section 3.10.2. If there are many parasites, $e^{-\gamma P_t}$ will be close to zero, and hence the host growth rate will also go to zero. In a similar way, Equation 4.154 expresses the fact that the number of parasites will increase with the number of surviving eggs, the number of host species, and the number of parasite species in the previous breeding season.

Mathematically, Equations 4.153 and 4.154 are classified as *difference equations*, *recurrence relations*, or *discrete models* [114, 220]. Models of this kind are characterized by the fact that the model equations can be used to set up an iteration that yields a sequence of states such as $(N_1, P_1), (N_2, P_2), \ldots$. In the above example and many other applications of this kind, the iteration number $t = 1, 2, 3, \ldots$ corresponds to time, that is, time is treated as a discrete variable. Note the difference to the differential equation models above, in which time and other independent variables were treated as continuous quantities. As the above example shows, finite difference models provide a natural setting for problems in the field of population dynamics, but they can also be used to model other inherently discrete phenomena, e.g. in the field of economics, traffic, or transportation flows [220, 221]. Difference equations such as Equations 4.153 and 4.154 can be easily implemented using *Maxima* or *R* as described above. The iterations can be formulated similar to the book software program `HeatClos.r` that was discussed in Section 4.6.3.

4.11.2
Cellular Automata

The concept of cellular automata was developed by John von Neumann and Stanislaw Ulam in the early 1950s, inspired by the analogies between the operation of computers and the human brain [222, 223]. We begin with a definition of cellular automata and then consider an illustrative example [224]:

Definition 4.11.1 (Cellular automaton)   A *cellular automaton* consists of
- a regular, discrete lattice of cells (which are also called *nodes* or *sites*) with boundary conditions;
- a finite – typically small – set of states that characterizes the cells;
- a finite set of cells that defines the interaction neighborhood of each cell; and
- rules that determine the evolution of the states of the cells in discrete time steps $t = 1, 2, 3, \ldots$.

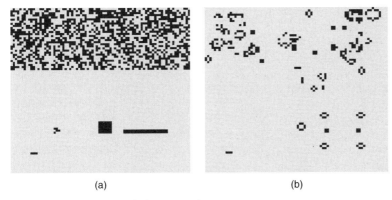

| (a) | (b) |

Fig. 4.33 Conway's game of life computed using Conway.r:
(a) random initial state and (b) state after 100 iterations.

A simple example is the famous *Conway's Game of Life* [225]. In this case, the discrete lattice of cells is a square lattice comprising $n \times n$ cells, which we can think of as representing individuals. These cells are characterized by two states called *live* or *dead*. Figure 4.33 visualizes the states of each cell in such a square lattice using the colors black (for life cells) and gray (for dead cells). The interaction neighborhood of each cell comprises its eight immediate neighbors, and the interaction rules are as follows:

- Live cells with fewer than two live neighbors die (as if by loneliness).
- Live cells with more than three live neighbors die (as if by overcrowding).
- Live cells with two or three live neighbors survive.
- Dead cells with exactly three live neighbors come to live (*almost* as if by...).

Starting with some initial distribution of life and dead cells, these rules determine iteratively the state of the "cell colony" at times $t = 1, 2, 3, \ldots$. This algorithm has been implemented by Petzoldt in [141] using *R*'s simcol package. You may find this example in the file Conway.r in the book software. Figure 4.33a shows the initial state of the cellular automaton generated by Conway.r, and Figure 4.33b shows the state that is attained after 100 iterations. Using Conway.r on your own computer, you will also be able to see the various intermediate states of the cellular automaton on your computer screen. The example shows that a quite complex behavior of a system may arise by the application of very simple rules.

Generally, cellular automata can be used to explain spatiotemporal patterns that are caused by the interaction of cell-like units. Again, there is an abundant number of applications in many fields. Cellular automata, for example, have been used to explain surface patterns on seashells [227]. Seashells are covered with pigment cells that excrete a pigment depending on the activity of the neighboring pigment cells, which corresponds exactly to the way in which an abstract cellular automaton works. In [228], cellular automata have been used to explain the spreading of genes conferring herbicide resistance in plant populations. Recently, it has been shown that the gas exchange of plants can be explained based on cellular automata [229]. See [224] for many more biological applications and [230] for applications in other fields such as the modeling of chemical reactions or of fluids.

4.11.3
Optimal Control Problems

In many practical applications of ODEs, one wants to control the process that is expressed by the ODE in a way such that it behaves optimal in the sense that it maximizes or minimizes some performance criterion. As an example, we may think of the wine fermentation model that was discussed in Section 3.10.2. In that case, it is important to control the temperature $T(t)$ inside the fermenter in an optimal way, for example, in a way such that the amount of residual sugar is minimized.

Beyond a really abundant number of examples of this kind in the field of technology (but also, e.g. in the field of economics, see [231]), there is also a great number of applications pertaining to natural systems. Indeed, optimality is important in nature as well as in technology – just think of Darwin's theories of evolution and the underlying idea of the "survival of the fittest". As an example, let us consider the *maintenance investment problem* in plants stressed by air pollutants. Generally, plants use the carbohydrates that are constantly produced in the process of photosynthesis for two main purposes:

- *for maintenance processes*: that is, to supply energy that is used to repair damaged cell components and to resynthesize degraded enzymes.
- *for growth*: to build up new cells and plant structures.

In the presence of air pollutants, a decrease in the rate of photosynthesis (measured e.g. as g CO_2/g dry matter/day) is often observed. In many cases, this is caused by the fact that the air pollutants destroy important enzymes such as the RuBisCo enzyme [232]. This means that plants stressed by air pollutants need to do more maintenance work, and hence they need more carbohydrates to supply the necessary energy for maintenance. These carbohydrates are no longer available for growth. Hence, the plant has to solve the following problem:

How much carbohydrates should be invested into maintenance and growth, respectively?

In view of Darwin's theory of evolution, this can be formulated as an optimization problem as follows:

How much carbohydrates should be invested into maintenance and growth such that the plant maximizes its reproductive success?

Based on a very simple optimal control problem, it can be shown that real plants indeed determine their carbohydrate investment into maintenance by solving a problem of this kind [134, 135]. Let

- N (g dry matter): "nonactive" overall biomass of the plant; basically, a measure of the size of the plant
- d (g dry matter/g dry matter): concentration of "degradable biomass" within the plant; basically the concentration of enzymes and other "degradable biomass" that is constantly repaired and resynthesized in the plants maintenance operations

The dynamical behavior of these two quantities can be described by the following ODE system:

$$\dot{N}(t) = r(1 - u(t))\phi(d(t))N(t) \tag{4.155}$$

$$\dot{d}(t) = u(t)\rho\phi(d(t)) - \sigma d(t) - r(1 - u(t))\phi(d(t))d(t) \tag{4.156}$$

where t is time and the dot on the left-hand side of the equations denotes the time derivative [134, 135]. In this ODE system, $u(t)$ describes the amount of carbohydrates that is invested into maintenance processes in percent, that is, $u(t) \in [0, 1]$ (see [135] for interpretations of the other parameters appearing in the equations). When the plant solves the above optimization problem, it must "select" an optimal function $u(t)$, and since the plant, thus, virtually controls the performance of the system via the selection of the optimal function $u(t)$, the overall problem is called a *control problem* or *optimal control problem*. Of course, a criterion must be specified as a part of an optimal control problem that characterizes optimality. Above, it was said that the plant decides about its maintenance investment in a way that maximizes its reproductive success. In the framework of the above simple plant model, Equations 4.155 and 4.156, this can be expressed as

$$N(T) \rightarrow \max \tag{4.157}$$

if the problem is solved in the time interval $[0, T] \subset \mathbb{R}$. According to Equation 4.157, the plant is required to choose the maintenance investment $u(t)$ in a way that maximizes its nonactive biomass, that is, in a way that maximizes its size (which is the best approximation of reproductive success maximization within this model). In [135], this problem is solved in closed form using a technique called *Pontryagin's principle*. See [33, 233] for this and other closed form and numerical solution techniques. These techniques apply to general systems of ODEs of the form

$$\mathbf{y}'(t) = \mathbf{F}(t, \mathbf{y}(t), \mathbf{u}(t)) \tag{4.158}$$

$$\mathbf{y}(0) = \mathbf{y}_0 \tag{4.159}$$

where $t \in [0, T]$, $\mathbf{y}(t) = (y_1(t), y_2(t), \ldots, y_n(t))$ is the vector of state variables and $\mathbf{u}(t) = (u_1(t), u_2(t), \ldots, u_m(t))$ is a vector of (time-dependent) control variables. Usually, optimality is expressed in terms of a cost functional that is to be minimized:

$$\phi(\mathbf{y}(T)) + \int_0^T L(\mathbf{y}(t), \mathbf{u}(t), t) \, dt \to \min \tag{4.161}$$

which would give $\phi(N(T), d(T)) = -N(T)$ in the maintenance investment problem discussed above.

4.11.4
Differential-algebraic Problems

In Section 3.5.7, a system of first-order ODEs was defined to be an equation of the form:

$$\mathbf{y}'(t) = \mathbf{F}(t, \mathbf{y}(t)) \tag{4.162}$$

In the applications, this equation may also appear in the more general form

$$\mathbf{F}(t, \mathbf{y}(t), \mathbf{y}'(t)) = 0 \tag{4.163}$$

If Equation 4.163 cannot be brought into the form of Equation 4.162, it is called an *implicit ODE*, while Equation 4.162 is an *explicit ODE*. As can be expected, more sophisticated solution procedures are needed to solve implicit ODEs. Many equations of the form (4.163) can be described as a combination of an ODE with algebraic conditions. Equations of this type are also known as *differential-algebraic equations (DAEs)* [107, 112, 234, 235]. They may appear in a number of different fields, such as mechanical multibody systems (includes robotics applications), electrical circuit simulation, chemical engineering, control theory, and fluid dynamics.

4.11.5
Inverse Problems

Remember *Problem* 2 from Section 4.1.3, which asked for the three-dimensional temperature distribution within a cube. This problem was solved in Section 4.9 based on the stationary heat equation:

$$\nabla \left(\mathbf{K}(\mathbf{x}) \cdot \nabla T(\mathbf{x}, t) \right) = 0 \tag{4.164}$$

As a result, the three-dimensional temperature distribution $T(\mathbf{x})$ within the cube was obtained. As was already explained in Section 1.7.3, a problem of this kind is called a *direct problem*, since a mathematical model (the heat equation) is used here to obtain the output of the system "cube" (temperature in this case) based on a given input (the boundary conditions imposed in *Problem* 2) and based on given system parameters (the thermal conductivity **K**). Now suppose that the cube is a black box to us in the sense that nothing is known about its internal structure except for the fact that it is made up of two materials A and B. Suppose that all we can do to explore its internal structure is to perform experiments of the following kind:

- Fixed temperatures are imposed at some of the cube surfaces.
- The "answer" (output) of the system is determined in terms of the temperatures that are observed at the remaining cube surfaces. In addition to this, we may also have data from a few temperature sensors inside the cube. Let us denote the resulting dataset as D.

In this situation, we may use the heat equation in a different way. Since we know that the inside of the cube is made up of the two materials A and B, the thermal conductivity will be of the form

$$K(x) = \begin{cases} K_A & \text{if A is the material at position } x \\ K_B & \text{otherwise} \end{cases} \tag{4.165}$$

Thus, the material structure inside the cube will be known if we find a way to compute $K(x)$ (and, of course, if $K_A \neq K_B$). In principle, this can be done by solving the following

Problem 7:
Determine $K(x)$ according to Equation (4.165) such that an "optimal" approximation of the data D is obtained if the heat equation (4.164) is solved using $K(x)$.

This type of problem is called an *inverse problem* since we are now looking for parameters of the mathematical model (the thermal conductivity) based on a given output of the system (the temperature measurements in the dataset D). Inverse problems of this kind are abundant in all fields of science and technology. They are used, for example, in geophysics, medical imaging, remote sensing, ocean acoustic tomography, nondestructive testing, and astronomy [32]. Particularly, the medical imaging applications have gained great public attention. These techniques use e.g. X rays to produce a dataset (corresponding to D above), which is then used to determine parameters of appropriate mathematical models (corresponding to the heat equation and its parameter $K(x)$), and from this three-dimensional reconstructions of internal structures of human bodies are obtained. The *Radon integral transform* is one of the mathematical techniques that is used to solve this reconstruction problem [236].

Inverse problems include *parameter estimation problems* such as the regression problems that were treated in Chapter 2. Of course, it must be made precise in the above problem formulation what is meant by an "optimal" approximation before the problem can be solved. In the regression problems treated in Chapter 2, the minimization of the residual sum of squares was used to express an optimal approximation of the data.

Note that it is often much more difficult to solve an inverse problem compared to the direct solution of a mathematical model. Inverse problems often are ill posed, that is, they may violate Hadamard's well-posedness criteria that were discussed in Section 4.3.2.1. Based on the above example, it is easy to understand why inverse problems may suffer from ambiguities, since it obviously may be possible to explain a particular dataset D based on two or more different distributions of the thermal conductivity within the cube. Using a priori knowledge – such as Equation (4.165) in the above example, which expresses the fact that the inside of the cube consists of the materials A and B with thermal conductivities $\mathbf{K}_A$ and $\mathbf{K}_B$, respectively – can help to turn ill-posed inverse problems into well-posed problems [32, 236].

A

CAELinux and the Book Software

If you want to work with the examples and programs described in this book on your own computer, you need *OpenOffice/Calc, Maxima, R, Salome-Meca, Code-Saturne* and a zip file `MMS.zip` which contains the programs and data files used in the book. All this is open-source software that can be used by anybody free of charge. Note, however, that the author makes no warranty, implied or expressed, that the programs and procedures described in this book will work safely on your system, and also no warranty is made for the recency, correctness, completeness or quality of the software-related information provided in this book.

It is recommended to use *Calc, Maxima, R, Salome-Meca* and *Code-Saturne* based on *CAELinux*, a Linux operating system which can be obtained as a Live-DVD under *www.caelinux.com*. Based on the Live-DVD, *CAELinux* can be used without the necessity to install anything on the hard drives of your computer, and without affecting the original operating system installed on your computer. Once you have obtained the *CAELinux* Live-DVD from *www.caelinux.com*, reboot your computer using the Live-DVD. This will turn most machines (including most computers running under Microsoft Windows) into *CAELinux* workstations.

> **Note** In *CAELinux*, the main software instruments used in this book (*Calc, R, Maxima, Salome-Meca, Code-Saturne*) can be used without further installations. All contributed *R* packages used in the book – which are not a part of *R*'s standard distribution – are also included.

If *CAELinux* should not work on your computer, you can of course install *Calc, R, Maxima, Salome-Meca* and *Code-Saturne* separately on your computer, using the installation procedures described at the appropriate internet sites (*Calc*: *http://www.openoffice.org*, *R*: *www.r-project.org*, *Maxima*: *maxima.sourceforge.net*, *Salome-Meca*: *www.caelinux.com*, *Code-Saturne*: *www.code-saturne.org*).

`MMS.zip` is available at the author's homepage under *www.fbg.fh-wiesbaden.de/velten*. It contains the programs and data files listed in Table A.1. The programs are either *R* programs (extension ".r") or *Maxima* programs (extension ".mac"). See the Appendices B and C for details on how you run the programs in *R* or *Maxima*. In these Appendices it will be assumed that you have unzipped `MMS.zip` on some

Mathematical Modeling and Simulation: Introduction for Scientists and Engineers. Kai Velten
Copyright © 2009 WILEY-VCH Verlag GmbH & Co. KGaA, Weinheim
ISBN: 978-3-527-40758-8

Table A.1 Files in MMS.zip.

File	Location	Description
Euler.mac	MechODE/MAC	Euler method example (page 177).
Farm.mac	Solution of the wheat/barley problem (page 31).	
FeverDat.mac	MechODE/MAC	Body temperature example (page 120).
FeverExp.mac	MechODE/MAC	Body temperature example (page 121).
FeverODE.mac	MechODE/MAC	Body temperature example (page 180).
FeverSolve.mac	MechODE/MAC	Body temperature example (page 121).
Label.mac	Principles	Solution of tank labeling problem (page 29).
Mix.mac	Principles	Solution of a mixture problem (page 24).
Mix1.mac	Principles	Solution of a mixture problem (page 26).
ODEEx1.mac-ODEEx16.mac	MechODE/MAC	ODE closed form solution examples (pages 159-173).
RoomDat.mac	MechODE/MAC	Alarm clock example (page 122).
RoomExp.mac	MechODE/MAC	Alarm clock example (page 126).
RoomODE.mac	MechODE/MAC	Alarm clock example (page 183).
RoomODED.mac	MechODE/MAC	Alarm clock example (page 141).
Stiff.mac	MechODE/MAC	Stiff ODE example (page 179).
Tin.mac	Principles	Solution of the tin problem (page 18).
VolPhase.mac	MechODE/MAC	Phase plot example (page 210).
Anova.r	PhenMod/Stat	Analysis of variance. (page 64).
Conway.r	PhenMod	Example cellular automaton. (page 311).
CRD.r	PhenMod/DOE	Completely randomized experimental design (page 102).
FacBlock.r	PhenMod/DOE	Randomized factorial block design (page 107).
Fermentation.r	MechODE/R	Wine fermentation model (pages 203, 217).
HeatClos.r	MechPDE	Solution of heat equation using FD (page 261).
LinRegEx1.r	PhenMod/LinReg	Linear regression using spring.csv (page 67).
LinRegEx2.r	PhenMod/LinReg	Multiple regression using volz.csv (page 76).
LinRegEx3.r	PhenMod/LinReg	Cross validation using volz.csv (page 79).
LinRegEx4.r	PhenMod/LinReg	Linear regression using gag.csv (page 73).
LinRegEx5.r	PhenMod/LinReg	Polynomial regression using gag.csv (page 74).
LSD.r	PhenMod/DOE	Latin square design (page 105).
Mualem.r	MechPDE	Water retention curves (page 293).
NonRegEx1.r	PhenMod/NonReg	Nonlinear regression using klein.csv (page 82).
NonRegEx2.r	PhenMod/NonReg	Nonlinear regression using stormer.csv (page 84).

Table A.1 *(continued)*

NNEx1.r	PhenMod/NN	Neural network using klein.csv (page 91).
NNEx2.r	PhenMod/NN	Neural network using rock.csv (page 98).
ODEEx1.r	MechODE/R	Body temperature ODE model (page 184).
ODEEx2.r	MechODE/R	Alarm clock ODE model (page 191).
ODEEx3.r	MechODE/R	Pharmacokinetic model (page 224).
ODEFitEx1.r	MechODE/R	Parameter estimation in the alarm clock ODE model (page 195).
ODEFitEx2.r	MechODE/R	Parameter estimation in the alarm clock ODE model (page 198).
ODEFitEx3.r	MechODE/R	Parameter estimation in the alarm clock ODE model (page 199).
Plant1.r	MechODE/R	Exponential growth model (page 226).
Plant2.r	MechODE/R	Logistic growth model (page 226).
Plant3.r	MechODE/R	Asparagus growth model (page 228).
RCBD.r	PhenMod/DOE	Randomized complete block design (page 103).
RNumbers.r	PhenMod/Stat	Probability density plots (page 55).
TTest.r	PhenMod/Stat	t-Test (page 62).
Volterra.r	MechODE/R	Predator-prey model (page 206).
VolterraND.r	MechODE/R	Predator-prey model (page 209).
asparagus.csv	MechODE/R	Asparagus growth data (page 228).
crop.csv	Phenmod/Stat	Crop yield data (page 61).
fermentation.csv	MechODE/R	wine fermentation data (page 218).
fever.csv	MechODE/R	Body temperature data (page 120).
fungicide.csv	PhenMod/Stat	Fungicide data (page 64).
gag.csv	PhenMod/LinReg	GAG urine concentration data (page 72).
klein.csv	PhenMod/LinReg	US investment data (pages 81,91).
rock.csv	PhenMod/LinReg	Rock permeability data (page 97).
room.csv	MechODE/R	Room temperature data (page 122).
spring.csv	PhenMod/LinReg	Spring elongation data (pages 32, 67).
stormer.csv	PhenMod/LinReg	Stormer viscometer data (page 84).
volz.csv	PhenMod/LinReg	Rose wilting data (page 75).
flowstep.med	MechPDE/Backward	Backward step problem (page 299).
result.med	MechPDE/Backward	Backward step problem (page 301).
CRD.ods	PhenMod/DOE	Completely randomized design example (page 101).

external universal serial bus (USB) device, and that you are using the programs and data files on that external USB device together with CAELinux as described above (note that everything has been tested by the author using the *CAELinux 2008* Live-DVD). Slight modifications of the procedures described there may be necessary if you have installed *Calc, R, Maxima, Salome-Meca* and *Code-Saturne* on the hard disk of your computer (based on CAELinux or a different operating system).

Many of the programs can be used as *software templates* for the solution of problems that you may have in one of the various problem classes treated in this book. For example, to estimate parameters in an ordinary differential equation

(ODE) system from data, you can take one of the R programs discussed in Section 3.9 (e.g. ODEFitEx1.r) as a template. To solve your problem, you will then just have to replace the ODE and the parameters in the template with your own ODE and parameters.

B

R (Programming Language and Software Environment)

This section gives you some information on *R*, focusing on the procedures used in the book. Readers who need more information on *R* are referred to *R*'s help pages, to the documentation available under *www.r-project.org*, and to a vast literature on *R* (books such as [45, 237–241]).

> **Note** It is assumed here that you have prepared your computer as described in Appendix A.

B.1
Using *R* in a *Konsole* Window

B.1.1
Batch Mode

There are several *R* programs (which are also called *R* scripts) in the book software which can be identified by their file extension ".r" (see Appendix A). Suppose you want to run e.g. the neural network program NNEx2.r. In Table A.1 you see that this program is in the PhenMod/NN directory of MMS.zip. After you have unzipped MMS.zip to the home directory of some external universal serial bus (USB) device which appears under a name such as USBDEVICE on your computer, NNEx2.r will appear in the *CAELinux* directory /media/USBDEVICE/MMS/PhenMod/NN (replace "USBDEVICE" by the appropriate name of your USB device). Open a Linux *Konsole* window e.g. by clicking on the Konsole symbol on the desktop. Within the *Konsole* window, navigate to /media/USBDEVICE/MMS/PhenMod/NN using the cd ("change directory") command. Note that you can use the ls ("list files") command in the Konsole window if you want to see the files within a particular directory. After arriving in /media/USBDEVICE/MMS/PhenMod/NN, enter R to start the *R* software. Wait a moment until you see the ">" command prompt of the *R* software. To execute NNEx2.r, enter the command: source("NNex2.r"). If you prefer a GUI-based procedure to start your programs, you may use the RKWard program which you find in CAELinux if you select CAE Software/Math under the

Mathematical Modeling and Simulation: Introduction for Scientists and Engineers. Kai Velten
Copyright © 2009 WILEY-VCH Verlag GmbH & Co. KGaA, Weinheim
ISBN: 978-3-527-40758-8

PC button of the desktop.

B.1.2
Command Mode

As an alternative to running *R* programs in batch mode as described above, you can also enter *R* commands directly into the *Konsole* window. For example, the program NNEx2.r begins with the command:

```
require(rgl)
```

After you have started the *R* software in the *Konsole* window, you may enter this command directly after the *R* prompt in the *Konsole* window. If you go on and enter the remaining commands of NNEx2.r line by line, you will obtain the same result that is obtained using the batch mode procedure described above. The advantage of this "command mode" operation of *R* is that you can immediately see the results of each of your commands and that you can then use this information to decide how you go on. Beyond this, the command mode operation of *R* is useful if you just want to perform a quick analysis that involves a few commands only. In such cases, the command mode operation can be faster than the batch mode operation where you have to invoke an editor to write the .r program, save that program somewhere, navigate to that directory within the *Konsole* window and so on.

B.2
R Commander

The *R Commander* is a GUI (graphical user interface) that facilitates the use of *R* for basic statistical applications as described in Section 2.1. To start the *R Commander*, open a *Konsole* window within *CAELinux* (in an arbitrary directory), start the *R* software as described in Section B.1 and then enter the command library(Rcmdr).

Several other GUIs for R are available, see *www.r-project.org*.

C

Maxima

This section gives you some information on *Maxima*, focusing on the procedures used in the book. Readers who need more information on *Maxima* may refer to books such as [18] which focus on *Macsyma*, a very similar commercial version of *Maxima*, to the very comprehensive *Maxima* manual by W. Schelter [106], to*Maxima*'s help pages and to the documentation available under *http://maxima.sourceforge.net/*.

> **Note** It is assumed here that you have prepared your computer as described in Appendix A.

C.1
Using *Maxima* in a *Konsole* Window

C.1.1
Batch Mode

There are several *Maxima* programs in the book software which can be identified by their file extension ".mac" (see Appendix A). Suppose we want to run the program `Tin.mac` which you find in the `Principles` directory of `MMS.zip`. The procedure to run this program is very similar to the procedure to run *R* programs that is described in Appendix B.1. As it is described there, open a *Konsole* window within *CAELinux* and navigate into the appropriate directory on your universal serial bus (USB) device. Enter `maxima` in the *Konsole* window to start the *Maxima* software. Then, enter the command `batch("Tin.mac")` to run the program `Tin.mac`. If you prefer a graphical user interface (GUI)-based procedure to start your programs, you may use the *wxMaxima* program described below.

C.1.2
Command Mode

As an alternative to running *Maxima* programs in batch mode, you can also enter *Maxima* commands directly into the *Konsole* window. For example, the program

Mathematical Modeling and Simulation: Introduction for Scientists and Engineers. Kai Velten
Copyright © 2009 WILEY-VCH Verlag GmbH & Co. KGaA, Weinheim
ISBN: 978-3-527-40758-8

`Tin.mac` begins with the command:

```
kill(all)$
```

After you have started the *Maxima* software in the *Konsole* window, you may enter this command directly after the *Maxima* prompt in the *Konsole* window. If you go on and enter the remaining commands of `Tin.mac` line by line, you will obtain the same result that is obtained using the batch mode procedure described above. Again, the advantage of this "command mode" operation of *Maxima* is that you can immediately see the results of each of your commands and that you can then use this information to decide how you go on. Beyond this, the command mode operation of *Maxima* is useful if you just want to perform a quick analysis that involves a few commands only. In such cases, the command mode operation can be faster than the batch mode operation where you have to invoke an editor to write the `.mac` program, save that program somewhere, navigate to that directory within the *Konsole* window and so on.

C.2
wxMaxima

wxMaxima is a GUI that facilitates the use of *Maxima* in its batch mode as well as in its command mode. It provides access to *Maxima*'s most important commands through menus, a 2D formatted display of mathematical formulas, the possibility to create documents made up of text and calculations and so on. For example, to run the program `Tin.mac` using *wxMaxima*, you would use its menu option `File/Batch File`. If you do this, *wxMaxima* will also display the command `batch("Tin.mac")` that can be used in a *Konsole* window as described in Section C.1 above, that is, using *wxMaxima* is also a way for beginners to get acquainted with *Maxima* terminal commands. *wxMaxima* can be accessed in *CAELinux* if you select `CAE Software/Math` under the `PC` button of the desktop.

References

1 Bockhorn, H. (**2005**) *Ullmann's Encyclopedia of Industrial Chemistry*, Chapter mathematical modeling, Ullmann.

2 Minsky, M.L. (**1965**) Matter, minds and models, *Proceedings of the International Federation for Information Processing (IFIP)Congress*, Spartan Books, Washington, DC, pp. 45–49.

3 Cellier, F.E. (**1991**) *Continuous System Modeling*, Springer.

4 Rosenberg, A. (**2000**) *The Philosophy of Science: A Contemporary Introduction*, Routledge.

5 Velten, K. and Best, W. (**2000**) "Rolling of unsaturated porous materials-on the evolution of a fully saturated zone". *Physical Review E*, **62**(3), 3891–99.

6 Banks, J. (**1998**) *Handbook of Simulation*, EMP Books.

7 Fritzson, P. (**2004**) *Principles of Object-Oriented Modeling and Simulation with Modelica 2.1*, John Wiley & Sons, Ltd.

8 Ingham, J., Dunn, I.J., Heinzle, E., Prenosil, J.E. and Snape, J.B. (**2007**) *Chemical Engineering Dynamics: An Introduction to Modelling and Computer Simulation*, Wiley-VCH Verlag GmbH.

9 Macchietto, S. (**2003**) *Dynamic Model Development: Methods, Theory and Applications*, Elsevier Science.

10 Law, A.M. and Kelton, W.D. (**2000**) *Simulation, Modeling and Analysis*, McGraw-Hill.

11 Haefner, J.W. (**2005**) *Modeling Biological Systems*, Springer.

12 Ashby, W.R. (**1956**) *An Introduction to Cybernetics*, John Wiley & Sons, Ltd.

13 Popper, K.R. (**2001**) *All Life is Problem Solving*, Routledge.

14 Preziosi, L. and Bellomo, L.N. (**1995**) *Modelling, Mathematical Methods and Scientific Computation*, CRC Press.

15 Bender, E.A. (**2000**) *An Introduction to Mathematical Modeling*, Dover Publications.

16 Lang, S. (**1998**) *A First Course in Calculus*, Springer.

17 Peterson, J.C. (**2003**) *Technical Mathematics with Calculus*, CENGAGE Delmar Learning.

18 Ben-Israel, A. and Gilbert, R. (**2002**) *Computer-Supported Calculus*, Springer.

19 Triola, M.F. (**2007**) *Elementary Statistics*, Addison-Wesley.

20 Seppelt, R. (**2003**) *Computer-Based Environmental Management*, Wiley-VCH Verlag GmbH.

21 Velten, K., Reinicke, R. and Friedrich, K. (**2000**) "Wear volume prediction with artificial neural networks". *Tribology International*, **33**(10), 731–36.

22 Smith, K.A. and Gupta, J.N.D. (**2002**) *Neural Networks in Business: Techniques and Applications*, IGI Global.

23 Huang, G., Bryden, K.M. and McCorkle, D.S. (**2004**) "Interactive Design Using CFD and Virtual Engineering." *Proceedings of the 10th AIAA/ISSMO Multidisciplinary Analysis and Optimization Conference, AIAA-2004-4364*, Albany, NY.

24 da Silva Bartolo, P.J. (**2007**) *Virtual Modelling and Rapid Manufacturing: Advanced Research in Virtual and Rapid Prototyping*, CRC Press.

Mathematical Modeling and Simulation: Introduction for Scientists and Engineers. Kai Velten
Copyright © 2009 WILEY-VCH Verlag GmbH & Co. KGaA, Weinheim
ISBN: 978-3-527-40758-8

25 Kai, L.K., Fai, C.C. and Chu-Sing, L. (**2003**) *Rapid Prototyping: Principles and Applications*, World Scientific.

26 Breitenecker, F. (**2004**) Mathematische Grundlagen in Modellbildung und Simulation. Presentation at the Wissenschaftstagung im Hauptverband der Sozialversicherungsträger, Wien, Austria, 30.11.2004 (unpublished).

27 Chen, M. (**2004**) *In Silico Systems Analysis of Biopathways*, PhD Thesis, University of Bielefeld.

28 Gertsev, V.I. and Gertseva, W. (**2004**) "Classification of mathematical models in ecology". *Ecological Modelling*, **178**, 329–34.

29 Hritonenko, N. and Yatsenko, Y. (**2006**) *Applied Mathematical Modelling of Engineering Problems*, Kluwer Academic Publishers Group.

30 Matko, D., Zupancic, B. and Karba, R. (**1992**) *Simulation and Modelling of Continuous Systems*, Prentice Hall.

31 Sulis, W.H. (**2001**) *Nonlinear Dynamics in the Life and Social Sciences*, IOS Press.

32 Aster, R., Borchers, B. and Thurber, C. (**2005**) *Parameter Estimation and Inverse Problems*, Academic Press.

33 Kirk, D.E. (**2004**) *Optimal Control Theory: An Introduction*, Dover Publications.

34 Golomb, S.W. (**1970**) "Mathematical models - uses and limitations". *Simulation*, 4(14), 197–98.

35 *Plato, The Republic*, Penguin Classics (**2003**).

36 Greiner, W. (**2000**) *Quantum Mechanics*, Springer.

37 Devore, J.L. (**2007**) *Probability and Statistics for Engineering and the Sciences*, Duxbury Press.

38 Tanner, M. (**1995**) *Practical Queueing Analysis*, McGraw-Hill.

39 Sahai, H. and Ageel, M.I. (**2000**) *Analysis of Variance: Fixed, Random and Mixed Models*, Birkhäuser.

40 Galton, F. (**1886**) "Regression towards mediocrity in hereditary stature". *Journal of the Anthropological Institute*, **15**, 246–63.

41 Richter, O. and Soendgerath, D. (**1990**) *Parameter Estimation in Ecology. The Link between Data and Models*, Wiley-VCH Verlag GmbH.

42 Rabinovich, S. (**2005**) *Measurement Errors and Uncertainties: Theory and Practice*, Springer.

43 Chambers, J.M. and Hastie, T.J. (**1992**) *Statistical Models in S*, Chapman & Hall.

44 de Jong, J.G., Wevers, R.A. and Liebrand-van Sambeek, R. (**1992**) "Measuring urinary glycosaminoglycans in the presence of protein: an improved screening procedure for mucopolysaccharido ses based on dimethylmethylene blue". *Clinical Chemistry*, 38(6), 803–7.

45 Venables, W.N. and Ripley, B.D. (**2002**) *Modern Applied Statistics with S*, Springer, Berlin, Germany.

46 Green, P.J. and Silverman, B.W. (**1994**) *Nonparametric Regression and Generalized Linear Models. A Roughness Penalty Approach*, Chapman & Hall.

47 Volz, P. (**2007**) Bedeutung des Kohlenhydrat-Gehaltes als haltbarkeitsbeeinflussender Faktor bei Schnittrosen. PhD thesis, Humboldt-University, Berlin, Germany.

48 Witten, I.A. and Frank, E. (**2005**) *Data Mining: Practical Machine Learning Tools and Techniques*, Morgan Kaufmann.

49 Greene, W.H. (**2003**) *Econometric Analysis*, Prentice Hall.

50 Klein, L. (**1950**) *Economic Fluctuations in the United States, 1921-1941*, John Wiley & Sons, Ltd.

51 Bates, D.M. and Watts, D.G. (**1988**) *Nonlinear Regression Analysis and Its Applications*, John Wiley & Sons, Ltd.

52 Williams, E.J. (**1959**) *Regression Analysis*, John Wiley & Sons, Ltd.

53 Fausett, L.V. (**1994**) *Fundamentals of Neural Networks*, Prentice Hall.

54 Haykin, S. (**1998**) *Neural Networks: A Comprehensive Foundation*, Prentice Hall.

55 Reinicke, R., Friedrich, K. and Velten, K. (**2001**) "Neuronale netze - ein innovatives Verfahren zur Vorhersage von Verschleißvorgängen". *Tribologie und Schmierungstechnik*, 48(6), 5–7.

56 Reinicke, R., Friedrich, K. and Velten, K. (**2000**) "Modelling Friction and Wear Properties with Artificial Neural Networks". *Proceedings of the 7th International Conference on Tribology*, Vol. 9, Budapest, Hungary, pp. 408–10.

57 Velten, K., Reinicke, R. and Friedrich, K. (**2000**) "Artificial Neural Networks to Predict the Wear Behaviour of Composite Materials." *Proceedings der 9th European Conference on Composite Materials (ECCM 9)* Brighton, UK, erschienen auf CD-ROM, June 2000.

58 Velten, K., Reinicke, R. and Friedrich, K. (**2000**) "Neuronale Netze zur Vorhersage und Analyse des Verschleissverhaltens polymerer Gleitwerkstoffe". *Materialwissenschaft und Werkstofftechnik*, **31**, 715–18.

59 Zhang, Z., Friedrich, K. and Velten, K. (**2002**) "Prediction of tribological properties of short fibre composites using artificial neural networks". *Wear*, **252**(7-8), 668–75.

60 Zhang, Z., Reinicke, R., Klein, P., Friedrich, K. and Velten, K. (**2001**) Wear Prediction of Polymeric Composites Using Artificial Neural Networks. Proceedings of the International Conference on Composites in Material and Structural Engineering, Vol. **6**, Prague, Czech Republic, pp. 203–6.

61 Cybenko, G. (**1989**) "Approximation by superpositions of a sigmoidal function". *Mathematics of Control, Signals, and Systems*, **2**(4), 303–14.

62 Funahashi, K. (**1989**) "On the approximate realization of continuous mappings by neural networks". *Neural Networks*, **2**(3), 183–92.

63 Hornik, K., Stinchcombe, M. and White, H. (**1989**) "Multilayer feedforward networks are universal approximators". *Neural Networks*, **2**(5), 359–66.

64 Ripley, B.D. (**1996**) *Pattern Recognition and Neural Networks*, Cambridge University Press.

65 Binmore, K. and Davies, J. (**2002**) *Calculus: Concepts and Methods*, Cambridge University Press.

66 Mitchell, T.M. (**1997**) *Machine Learning*, Mc Graw-Hill.

67 Ripley, B.D. (**1993**) Statistical aspects of neural networks, *Networks and Chaos - Statistical and Probabilistic Aspects*, Chapman & Hall, pp. 40–123.

68 Ripley, B.D. (**1994**) Neural networks and flexible regression and discrimination, *Advances in Applied Statistics*, Vol. **2**, Carfax, pp. 39–57.

69 Bear, J. (**1990**) *Introduction to Modeling of Transport Phenomena in Porous Media*,

Kluwer Academic Publishers, Dordrecht, Netherlands.

70 Lasaga, A. (**1998**) *Kinetic Theory in the Earth Sciences*, Princeton University Press.

71 Abebe, A., Daniels, J. and McKean, J.W. (**2001**) *Statistics and Data Analysis*. Technical report. Statistical Computation Lab (SCL), Western Michigan University, Kalamazoo, MI.

72 Montgomery, D.C. (**2005**) *Design and Analysis Experiments*, John Wiley & Sons, Ltd.

73 Soravia, S. and Orth, A. (**2007**) Design of Experiments, *Ullmann's Modeling and Simulation*, John Wiley & Sons, Ltd, pp. 363–400.

74 Inuiguchi, M., Tsumoto, S. and Hirano, S. (eds) (**2007**) *Rough Set Theory and Granular Computing*, Springer.

75 Tettamanzi, A. and Tomassini, M. (**2001**) *Soft Computing: Integrating Evolutionary, Neural, and Fuzzy Systems*, Springer.

76 Falkenauer, E. (**1997**) *Genetic Algorithms and Grouping Problems*, John Wiley & Sons, Ltd.

77 Hill, T., Lundgren, A., Fredriksson, R. and Schiöth, H.B. (**2005**) "Genetic algorithm for large-scale maximum parsimony phylogenetic analysis of proteins". *Biochimica et Biophysica Acta*, **1725**, 19–29.

78 Kjellström, G. (**1996**) "Evolution as a statistical optimization algorithm". *Evolutionary Theory*, **11**, 105–17.

79 Eberhart, R.C., Shi, Y. and Kennedy, J. (**2001**) *Swarm Intelligence*, Morgan Kaufmann.

80 Zadeh, L.A. (**1965**) "Fuzzy sets". *Information and Control*, **8**, 338–53.

81 Ibrahim, A. (**2003**) *Fuzzy Logic for Embedded Systems Applications*, Newnes.

82 Marshall, C. and Rossman, G.B. (**2006**) *Designing Qualitative Research*, Sage Publications.

83 Scholz, R.W. and Tietje, O. (**2002**) *Embedded Case Study Methods: Integrating Quantitative and Qualitative Knowledge*, Sage Publications.

84 Yin, R.K. (**2003**) *Case Study Research, Design and Methods*, Sage Publications.

85 Rubinstein, R.Y. and Melamed, B. (**1998**) *Modern Simulation and Modeling*.

86 Bratley, P., Fox, B.L. and Schrage, L.E. (**1987**) *A Guide to Simulation*, Springer.

87 Petzoldt, T. and Rinke, K. (**2007**) "An object-oriented framework for ecological

modeling in R". *Journal of Statistical Software*, **22**(9): *http://www.jstatsoft.org/v22/i09/*.

88 Hamming, R.W. (**1997**) *Digital Filters*, Dover Publications.

89 Savitzky, A. and Golay, M.J.E. (**1964**) "Smoothing and differentiation of data by simplified least squares procedures". *Analytical Chemistry*, **36**(8), 1627–39.

90 Ziegler, H. (**1981**) "Properties of digital smoothing polynomial (dispo) filters". *Applied Spectroscopy*, **35**, 88–92.

91 Velten, K. (**1998**) "Inverse problems in resin transfer molding". *Journal of Composite Materials*, **32**(24), 2178–202.

92 Press, W.H., Teukolsky, S.A., Vetterling, W.T. and Flannery, B.P. (**2002**) *Numerical Recipes in C - The Art of Scientific Computing*, Cambridge University Press.

93 Bose, N.K. and Rao, C.R. (eds) (**1993**) *Handbook of Statistics 10: Signal Processing and its Applications*, Elsevier Science Publishing.

94 Derrida, J. (**1997**) *Archive Fever: A Freudian Impression (Religion & Postmodernism)*, University of Chicago Press.

95 Popper, K. (**2002**) *The Logic of Scientific Discovery*, Routledge.

96 Eymard, P., Lafon, J.P. and Wilson, S.S. (**2004**) *The Number Pi*, American Mathematical Society.

97 Incropera, F. (**2001**) *Fundamentals of Heat and Mass Transfer*, John Wiley & Sons, Ltd.

98 Hartman, P. (**2002**) *Ordinary Differential Equations*, Society for Industrial & Applied Mathematics.

99 Chicone, C. (**2006**) *Ordinary Differential Equations with Applications*, Springer, Berlin, Germany.

100 Swift, R.J. and Wirkus, S.A. (**2006**) *A Course in Ordinary Differential Equations*, Chapman & Hall/CRC.

101 Evans, L.C. (**1998**) *Partial Differential Equations*, Oxford University Press.

102 Ascher, U.M. and Russell, R.D. (**1995**) *Numerical Solution of Boundary Value Problems for Ordinary Differential Equations*, Society for Industrial & Applied Mathematics.

103 Raabe, D., Franz Roters, F., Barlat, F. and Chen, L.Q. (eds) (**2004**) *Continuum Scale Simulation of Engineering Materials: Fundamentals - Microstructures - Process Applications*, Wiley-VCH Verlag GmbH.

104 Temam, R. and Miranville, A. (**2005**) *Mathematical Modeling in Continuum Mechanics*, Cambridge University Press.

105 Monahan, J.F. (**2001**) *Numerical Methods of Statistics*, Cambridge University Press.

106 Schelter, W. (**2000**) *Maxima Manual*, *http://maxima.sourceforge.net*.

107 Hairer, E. and Wanner, G. (**1996**) *Solving Ordinary Differential Equations, Part II: Stiff and Differential-Algebraic Problems*, Springer-Verlag, Berlin, Germany.

108 Hindmarsh, A.C. (**1983**) ODEPACK, a systematized collection of ODE solvers, *Scientific Computing*, North-Holland, Amsterdam, The Netherlands, pp. 55–64.

109 Petzold, L.R. (**1983**) "Automatic selection of methods for solving stiff and nonstiff systems of ordinary differential equations". *SIAM Journal on Scientific and Statistical Computing*, **4**, 136–48.

110 Hairer, E., Norsett, S.P. and Wanner, G. (**1993**) *Solving Ordinary Differential Equations: Nonstiff Problems*, Springer Series in Computational Mechanics, Vol **8**, Springer, New York.

111 Landau, R.H., Paez, M.J. and Bordeianu, C.C. (**2007**) *Computational Physics*, Wiley-VCH Verlag GmbH.

112 Brenan, K.E., Campbell, S.L. and Petzold, L.R. (**1989**) *Numerical Solution of Initial-Value Problems in Differential Algebraic Equations*, Elsevier, New York.

113 Bock, H.G. (**1987**) *Randwertproblemmethoden zur Parameteridentifizierung in Systemen nichtlinearer Differentialgleichungen*. PhD thesis, University of Bonn, Bonner Mathematische Schriften Nr. 183.

114 Murray, J.D. (**1989**) *Mathematical Biology*, Springer.

115 Volterra, V. (**1926**) "Variazione fluttuazioni del numero d'individui in specie animali conviventi". *Memorie Della Academia Nazionale dei Lincei*, **2**, 31–113.

116 Yeargers, E.K., Shonkwiler, R.W. and Herod, J.V. (**1996**) *An Introduction to the Mathematics of Biology*, Birkhäuser.

117 Lotka, A.J. (**1925**) *Elements of Physical Biology*, Williams and Wilkins, Baltimore, MD.

118 Fleet, G.H. (**1993**) *Wine Microbiology and Biotechnology*, CRC Press.

119 Hutkins, R.W. (**2006**) *Microbiology and Technology of Fermented Foods*, Wiley-Blackwell.

120 Bisson, L.F. (**1999**) "Stuck and sluggish fermentations". *American Journal of Enology and Viticulture*, **50**, 107–19.

121 Cramer, A.C., Vlassides, S. and Block, D.E. (**2002**) "Kinetic model for nitrogen-limited wine fermentations". *Biotechnology and Bioengineering*, **1**.

122 Blank, A. (**2007**) *Gärmodellierung und Gärprognose: Der Weg zur besseren Gärsteuerung*. Master's thesis, Fachhochschule Wiesbaden, Fachbereich Geisenheim.

123 Voet, D. and Voet, J.G. (**1995**) *Biochemistry*, John Wiley & Sons, Ltd.

124 Boulton, R.B., Singleton, V.L., Bisson, L.F. and Kunkee, R.E. (**1999**) *Principles and Practices of Winemaking*, Springer.

125 Boulton, R.B. (**1980**) "The prediction of fermentation behaviour by a kinetic model". *American Journal of Enology and Viticulture*, **31**, 40–45.

126 Malherbe, S.S., Fromion, V., Hilgert, N. and Sablayrolles, J.-M. (**2004**) "Modeling the effects of assimilable nitrogen and temperature on fermentation kinetics in enological conditions". *Biotechnology and Bioengineering*, **86**(3), 261–72.

127 Julien, A., Roustan, J.L., Dulau, L. and Sablayrolles, J.M. (**2000**) "Comparison of nitrogen and oxygen demands of enological yeasts: technological consequences". *American Journal of Enology and Viticulture*, **51**(3), 215–22.

128 Jiranek, V., Langridge, P. and Henschke, P.A. (**1995**) "Amino acid and ammonium utilization by saccharomyces cerevisiae wine yeasts from a chemically defined medium". *American Journal of Enology and Viticulture*, **46**(1), 75–83.

129 Spitznagel, E. (**1992**) *Two-Compartment Pharmacokinetic Models*. Technical report, Harvey Mudd College, Claremont, CA.

130 Strauss, S. and Bourne, D.W. (**1995**) *Mathematical Modeling of Pharmacokinetic Data*, CRC Press.

131 Overman, A.R. and Scholtz, R.V. III. (**2002**) *Mathematical Models of Crop Growth and Yield*, CRC Press.

132 Schröder, U., Richter, O. and Velten, K. (**1995**) "Performance of the plant growth models of the special collaborative project 179 with respect to winter wheat". *Ecological Modelling*, **81**, 243–50.

133 Velten, K. and Richter, O. (**1995**) "Optimal root/shoot-partitioning of carbohydrates in plants". *Bulletin of Mathematical Biology*, **57**(1), 99–109.

134 Velten, K. (**1994**) *Optimalität und Pflanzenwachstum (plant growth and optimality)*. PhD thesis, TU Braunschweig.

135 Velten, K. and Richter, O. (**1993**) "Optimal maintenance investment of plants and its dependence on environmental conditions". *Bulletin of Mathematical Biology*, **55**(5), 953–71.

136 Velten, K., Paschold, P.J. and Rieckmann, U. (**2004**) "Analysis of the subsurface growth of white asparagus". *Acta Horticulturae*, **654**, 97–103.

137 Weinheimer, S. (**2008**) *Einfluss eines differenzierten Wasserangebotes auf Wurzelwachstum und Reservekohlenhydrathaushalt von Bleichspargel (Asparagus officinalis L.)*. PhD thesis, Humboldt-University, Berlin, Germany.

138 Grossmann, C. and Roos, H.G. (**2007**) *Numerical Treatment of Partial Differential Equations*, Springer.

139 Knabner, P. and Angermann, L. (**2003**) *Numerical Methods for Elliptic and Parabolic Partial Differential Equations*, Springer.

140 Morton, K.W. and Mayers, D.F. (**2005**) *Numerical Solution of Partial Differential Equations*, Cambridge University Press.

141 Sewell, G. (**2005**) *The Numerical Solution of Ordinary and Partial Differential Equations*, John Wiley & Sons, Ltd.

142 Strauss, W.A. (**1992**) *Partial Differential Equations: An Introduction*, John Wiley & Sons, Ltd.

143 Velten, K., Jaki, J. and Paschold, P.J. (**2002**) "Optimierung temperaturwirksamer Anbaumethoden". *Berichte der Gesellschaft für Informatik in der Land-, Forst- und Ernährungswirtschaft*, **15**, 209–12.

144 Velten, K., Paschold, P.J. and Stahel, A. (**2003**) "Optimization of cultivation measures affecting soil temperature". *Scientia Horticulturae*, **97**(2), 163–84.

145 Serway, R.A. and Jewett, J.W. (**2004**) *Physics for Scientists and Engineers*, Brooks/Cole.

146 Hornung, U. (**1998**) *Homogenization and Porous Media*, Interdisciplinary Applied Mathematics, Vol **6**, Springer.

147 Rief, S. (**2006**) Nonlinear flow in porous media - numerical solution of the navier-stokes system with two pressures and application to paper making. PhD thesis, Technical University of Kaiserslautern, Kaiserslautern, Germany.

148 Epperson, J.F. (**2007**) *An Introduction to Numerical Methods and Analysis*, John Wiley & Sons, Ltd.

149 Chakrabarti, S.K. (**1990**) *Theory of Transonic Astrophysical Flows*, World Scientific.

150 Lapidus, L. and Pinder, G.F. (**1999**) *Numerical Solution of Partial Differential Equations in Science and Engineering*, John Wiley & Sons, Ltd.

151 Rosen, J. (**1995**) *Symmetry in Science: An Introduction to the General Theory*, Springer.

152 Polyanin, A.D. (**2001**) *Handbook of Linear Partial Differential Equations for Engineers and Scientists*, Chapman & Hall.

153 Schiesser, W.E. (**1991**) *The Numerical Method of Lines*, Academic Press.

154 Canuto, C., Hussaini, M.Y., Quarteroni, A. and Zang, T.A. (**2006**) *Spectral Methods: Fundamentals in Single Domains*, Springer-Verlag.

155 LeVeque, R.J. (**2002**) *Finite Volume Methods for Hyperbolic Problems*, Cambridge Texts in Applied Mathematics.

156 Versteeg, H.K. and Malalasekera, W. (**2007**) *An Introduction to Computational Fluid Dynamics: The Finite Volume Method*, Prentice-Hall.

157 Strikwerda, J. (**2004**) *Finite Difference Schemes and Partial Differential Equations*, SIAM: Society for Industrial and Applied Mathematics.

158 Ames, W.F. (**1992**) *Numerical Methods for Partial Differential Equations*, Academic Press.

159 Lay, D.C. (**2005**) *Linear Algebra and its Applications*, Addison-Wesley.

160 Saad, Y. (**2008**) *Iterative Methods for Sparse Linear Systems*, Cambridge University Press.

161 Brenner, S. and Scott, R.L. (**2005**) *The Mathematical Theory of Finite Element Methods*, 2nd edn, Springer.

162 Ern, A. and Guermond, J.L. (**2004**) *Theory and Practice of Finite Elements*, Springer.

163 Frey, P.J. and George, P.L. (**2007**) *Mesh Generation: Application to Finite Elements*, ISTE Publishing Company.

164 Banerjee, P.K. (**1994**) *The Boundary Element Methods in Engineering*, McGraw-Hill.

165 Wrobel, L.C. and Aliabadi, M.H. (**2002**) *The Boundary Element Method*, John Wiley & Sons, Ltd.

166 Zienkiewicz, O.C. and Taylor, R.L. (**2005**) *The Finite Element Method Set*, Butterworth Heinemann.

167 Morand, H.J. (**1995**) *Fluid-Structure Interaction*, John Wiley & Sons, Ltd.

168 Hetland, K.L. (**2005**) *Beginning Python: From Novice to Professional*, Apress.

169 Schöberl, J. (**1997**) "Netgen: an advancing front 2d/3d-mesh generator based on abstract rules". *Computing and Visualization in Science*, **1**(1), 41–52.

170 Chriss, N.A. (**1996**) *Black-Scholes and Beyond: Option Pricing Models*, McGraw-Hill.

171 Korn, R. and Korn, E. (**2001**) *Option Pricing and Portfolio Optimization: Modern Methods of Financial Mathematics*, American Mathematical Society.

172 Thoma, M. and Priesack, E. (**1993**) "Coupled porous and free flow in structured media". *Zeitschrift fur Angewandte Mathematik und Mechanik*, **73**(8), T566–69.

173 Velten, K., Günther, M., Lindemann, B. and Loser, W. (**2004**) "Optimization of candle filters using three-dimensional flow simulations". *Filtration -Transactions of the Filtration Society*, **4**(4), 276–80.

174 Banks, R.B. (**1993**) *Growth and Diffusion Phenomena: Mathematical Frameworks and Applications*, Springer.

175 Briggs, D.E. (**2004**) *Brewing: Science and practice*, Woodhead Publishing.

176 Schlather, M. (**2005**) *Sophy: some soil physics tools for R [www document]*. *http://www.r-project.org/*, contributed extension package (accessed on April 2008).

177 Schlather, M. and Huwe, B. (**2005**) "A risk index for characterising flow pattern in soils using dye tracer distributions". *Journal of Contaminant Hydrology*, **79**(1-2), 25–44.

178 Schlather, M. and Huwe, B. (**2005**) "A stochastic model for 3-dimensional flow patterns in infiltration experiments". *Journal of Hydrology*, **310**(1-4), 17–27.

179 Darcy, H. (**1856**) *Les Fontaines Publiques de la Ville de Dijon*, Victor Dalmont.

180 Bear, J. (**1988**) *Dynamics of Fluids in Porous Media*, Dover Publications.

181 Nield, D.A. and Bejan, A. (**1998**) *Convection in Porous Media*, Springer.

182 Douglas, J.F., Gasoriek, J.M., Swaffield, J. and Lynne Jack, L. (**2006**) *Fluid Mechanics*, Prentice Hall.

183 Allen, M.B., Behie, G.A. and Trangenstein, J.A. (**1988**) *Multiphase Flow in Porous Media*, Springer, Berlin, Germany.

184 Velten, K. and Schattauer, D. (**1998**) "A computer experiment to determine the permeability of mica tape based insulations". *IEEE Transactions on Dielectrics and Electrical Insulation*, **5**(6), 886–91.

185 Velten, K. and Schattauer, D. (**2000**) "Drag Measurement and Evaluation in Impregnation Processes." *Proceedings der 9th European Conference on Composite Materials (ECCM 9)*, Brighton, UK, June 2000. erschienen auf CD-ROM.

186 Velten, K., Schattauer, D., Zemitis, A., Pfreundt, F.-J. and Hennig, E. (**1999**) "A method for the analysis and control of mica tape impregnation processes". *IEEE Transactions on Dielectrics and Electrical Insulation*, **6**(3), 363–69.

187 Lutz, A., Velten, K. and Evstatiev, M. (**1998**) A new tool for fibre bundle impregnation: experiments and process analysis, in *Proceedings of the 8th European Conference on Composite Materials (ECCM-8)* (ed. I. Crivelli-Visconti), Vol. 2, Naples, Italy, pp. 471–77.

188 Velten, K. (**1998**) "Quantitative analysis of the resin transfer molding process". *Journal of Composite Materials*, **32**(20), 1865–92.

189 Velten, K., Lutz, A. and Friedrich, K. (**1999**) "Quantitative characterization of porous materials in polymer processing". *Composites Science and Technology*, **59**(3), 495–504.

190 Fanchi, J.R. (**2005**) *Principles of Applied Reservoir Simulation*, Gulf Professional Publishing.

191 Logan, J.D. (**2001**) *Transport Modeling in Hydrogeochemical Systems*, Springer.

192 Bitterlich, S. and Knabner, P. (**2002**) "Adaptive and formfree identification of nonlinearities in fluid flow from column experiments". *Contemporary Mathematics*, **295**, 63–74.

193 van Cotthem, A., Charlier, R., Thimus, J.F. and Tshibangu, J.P. (eds) (**2006**) *Eurock 2006 Multiphysics Coupling and Long Term Behaviour in Rock Mechanics*. Taylor & Francis.

194 Brady, N.C. (**1999**) *The Nature and Properties of Soils*, Prentice-Hall.

195 Velten, K. and Paschold, P.J. (**2009**) "Mathematical models to optimize asparagus drip irrigation". *Scientia Horticulturae*, (in preparation).

196 Chen, Z., Huan, G. and Ma, Y. (**2006**) *Computational Methods for Multiphase Flows in Porous Media*, Society for Industrial and Applied Mathematics.

197 Rajagopal, K.R. (**1995**) *Mechanics of Mixtures*, World Scientific.

198 Coussy, O. (**2004**) *Poromechanics*, John Wiley & Sons, Ltd.

199 Neunzert, H., Zemitis, A., Velten, K. and Iliev, O. (**2003**) Analysis of transport processes for layered porous materials used in industrial applications, *Mathematics - Key Technology for the Future*, Springer, pp. 243–51.

200 Paschedag, A.R. (**2007**) Computational fluid dynamics, *Ullmann's Modeling and Simulation*, Wiley-VCH Verlag GmbH, pp. 341–61.

201 Chung, T.J. (**2002**) *Computational Fluid Dynamics*, Cambridge University Press.

202 Fletcher, C.A.J. (**1991**) *Computational Techniques for Fluid Dynamics*, Springer.

203 Temam, R. (**2001**) *Navier-Stokes Equations: Theory and Numerical Analysis*, American Mathematical Society.

204 Archambeau, F., Mechitoua, N. and Sakiz, M. (**2004**) "Code saturne: a finite volume code for the computation of turbulent incompressible flows". *International Journal on Finite Volumes*, **1**, 1–62.

205 Velten, K., Lindemann, B. and Günther, M. (**2003**) "Optimization of Candle Filters Using Three-Dimensional Flow Simulations." *Proceedings of the FILTECH EUROPA 2003 Scientific Conference*, CCD Dusseldorf, Germany, October 21-23

206 Velten, K., Spindler, K.H. and Evers, H. (**2009**) Mathematical Models to Optimize the Industrial Cleaning of

Bottles. Technical report, FH Wiesbaden. Report of Project 17080061, financed by the German Federation of Industrial Research Associations (AiF) and the KHS AG, Bad Kreuznach.

207 Sadd, M.H. (**2004**) *Elasticity: Theory, Applications, and Numerics*, Academic Press.

208 Beitz., W. and Karl-Heinz.Kuttner K.H. (eds) (**1994**) *DUBBEL's Handbook of Mechanical Engineering*, Springer.

209 Quigley, H.A. and Broman, A.T. (**2006**) "The number of people with glaucoma worldwide in 2010 and 2020". *British Journal of Ophthalmology*, **90**, 262–67.

210 Netland, P.A. (**2007**) *Glaucoma Medical Therapy: Principles and Management*, Oxford University Press.

211 Goldmann, H. and Schmidt, T. (**1957**) "Über Applanationstonometrie". *Ophthalmologica*, **134**, 221–42.

212 Doughty, M.J. and Zaman, M.L. (**2000**) "Human corneal thickness and its impact on intraocular pressure measures: a review and meta-analysis approach". *Survey of Ophthalmology*, **44**(5), 367–408.

213 Charbonnier, J.B., Charmet, J.C., Vallet, D., Dupoisot, M. and Martinsky, H. (**1995**) "Aphakia correction by synthetic intracorneal implantation: A new tool for quantitative predetermination of the optical quality". *Journal of Biomechanics*, **28**(2), 167–71.

214 Liu, J. and Roberts, C.J. (**2005**) "Influence of corneal biomechanical properties on intraocular pressure measurement quantitative analysis". *Journal of Cataract and Refractive Surgery*, **31**(1), 146–55.

215 Velten, K., Günther, M. and Lorenz, B. (**2006**) "Finite-element simulation of corneal applanation". *Journal of Cataract and Refractive Surgery*, **32**(7), 1073–74.

216 Velten, K., Günther, M., Cursiefen, C. and Beutler, T. (**2008**) "Assessment of biomechanical effects on applanation tonometry using finite elements". *Investigative Ophthalmology*, Submitted to.

217 Vito, R.P. and Carnell, P.H. (**1992**) "Finite element based mechanical models of the cornea for pressure and indenter loading". *Refractive Corneal Surgery*, **8**(2), 146–51.

218 Fung, Y.C. (**1993**) *Biomechanics: Mechanical Properties of Living Tissues*, Springer.

219 Bellezza, A.J., Hart, R.T. and Burgoyne, C.F. (**2000**) "The optic nerve head as a biomechanical structure: Initial finite element modeling". *Investigative Ophthalmology and Visual Science*, **41**, 2991–3000.

220 Fulford, G., Forrester, P. and Jones, A. (**1997**) *Modelling with Differential and Difference Equations*, Cambridge University Press.

221 Yang, Z., Wang, K.C.P and Mao, B. (eds) (**1998**) *Traffic and Transportation Studies: Proceedings of Ictts'98*, July 27-29, 1998, Beijing, People's Republic of China, American Society of Civil Engineers.

222 Neumann, J.V. (**1966**) *Theory of Self-Reproducing Automata*, University of Illinois Press.

223 Sipper, M. (**1999**) "The emergence of cellular computing". *Computer*, **32**(7), 18–26.

224 Deutsch, A. and Dormann, S. (**2004**) *Cellular Automaton Modeling of Biological Pattern Formation*, Birkhäuser.

225 Gardner, M. (**1970**) "The fantastic combinations of john conway's new solitaire game 'life'". *Scientific American*, **223**, 120–23.

226 Petzoldt, T. (**2003**) "R as a simulation platform in ecological modelling". *R News*, **3**(3), 8–16.

227 Fowler, R., Meinhardt, H. and Prusinkiewicz, P. (**1992**) "Modeling seashells". *Computer Graphics (SIGGRAPH 92 Proceedings)*, **26**, 379–87.

228 Richter, O., Zwerger, P. and Böttcher, U. (**2002**) "Modelling spatio-temporal dynamics of herbicide resistance". *Weed Research*, **42**(1), 52–64.

229 Peak, D., West, J.D. and Messinger, K.A. and Mott, S.M. (**2004**) "Evidence for complex, collective dynamics and emergent, distributed computation in plants". *Proceedings of the National Academy of Sciences of the United States of America*, **101**(4), 918–22.

230 Wolfram, S. (**2002**) *Cellular Automata and Complexity*, Westview Press.

231 Sethi, S.P. and Thompson, G.L. (**2000**) *Optimal Control Theory - Applications to Management Science and Economics*, Springer.

232 Darral, N.M. (**1986**) "The sensitivity of net photosynthesis in several plant species to short-term fumigation with sulphur-dioxide". *Journal of Experimental Botany*, **37**(182), 1313–22.

233 Pytlak, R. (**1999**) *Numerical Methods for Optimal Control Problems with State Constraints*, Springer.

234 Kunkel, P. and Mehrmann, V. (**2006**) *Differential-Algebraic Equations: Analysis and Numerical Solution*, European Mathematical Society Publishing House.

235 Reissig, G. (**1998**) Beiträge zur Theorie und Anwendungen impliziter Differentialgleichungen. PhD thesis, TU Dresden.

236 Kak, A.C. and Slaney, M. (**2001**) *Principles of Computerized Tomographic Imaging*, Society for Industrial Mathematics.

237 Crawley, M.J. (**2006**) *Statistics: An Introduction using R*, John Wiley & Sons, Ltd.

238 Dalgaard, P. (**2002**) *Introductory Statistics with R*, Springer.

239 Ligges, U. (**2006**) *Programmieren Mit R*, Springer, Berlin, Germany.

240 Maindonald, J. and Braun, J. (**2006**) *Data Analysis and Graphics Using R: An Example-based Approach*, Cambridge University Press.

241 Venables, W.N. and Smith, D.M. (**2002**) *An Introduction to R*, Network Theory.

242 Hamming, R.W. (**1987**) *Numerical Methods for Scientists and Engineers*, Dover Publications.

Index

Mathematical Modeling and Simulation: Introduction for Scientists and Engineers. Kai Velten
Copyright © 2009 WILEY-VCH Verlag GmbH & Co. KGaA, Weinheim
ISBN: 978-3-527-40758-8